LONDON MATHEMATICAL SOCIETY LECTURE NOTE SERIES

Managing Editor: Professor J.W.S. Cassels, Department of Pure Mathematics and Mathematical Statistics, University of Cambridge, 16 Mill Lane, Cambridge CB2 1SB, England

The books in the series listed below are available from booksellers, or, in case of difficulty, from Cambridge University Press.

34 Representation theory of Lie groups, M.F. ATIYAH *et al*
36 Homological group theory, C.T.C. WALL (ed)
39 Affine sets and affine groups, D.G. NORTHCOTT
40 Introduction to H_p spaces, P.J. KOOSIS
46 p-adic analysis: a short course on recent work, N. KOBLITZ
49 Finite geometries and designs, P. CAMERON, J.W.P. HIRSCHFELD & D.R. HUGHES (eds)
50 Commutator calculus and groups of homotopy classes, H.J. BAUES
57 Techniques of geometric topology, R.A. FENN
59 Applicable differential geometry, M. CRAMPIN & F.A.E. PIRANI
66 Several complex variables and complex manifolds II, M.J. FIELD
69 Representation theory, I.M. GELFAND *et al*
74 Symmetric designs: an algebraic approach, E.S. LANDER
76 Spectral theory of linear differential operators and comparison algebras, H.O. CORDES
77 Isolated singular points on complete intersections, E.J.N. LOOIJENGA
79 Probability, statistics and analysis, J.F.C. KINGMAN & G.E.H. REUTER (eds)
80 Introduction to the representation theory of compact and locally compact groups, A. ROBERT
81 Skew fields, P.K. DRAXL
82 Surveys in combinatorics, E.K. LLOYD (ed)
83 Homogeneous structures on Riemannian manifolds, F. TRICERRI & L. VANHECKE
86 Topological topics, I.M. JAMES (ed)
87 Surveys in set theory, A.R.D. MATHIAS (ed)
88 FPF ring theory, C. FAITH & S. PAGE
89 An F-space sampler, N.J. KALTON, N.T. PECK & J.W. ROBERTS
90 Polytopes and symmetry, S.A. ROBERTSON
91 Classgroups of group rings, M.J. TAYLOR
92 Representation of rings over skew fields, A.H. SCHOFIELD
93 Aspects of topology, I.M. JAMES & E.H. KRONHEIMER (eds)
94 Representations of general linear groups, G.D. JAMES
95 Low-dimensional topology 1982, R.A. FENN (ed)
96 Diophantine equations over function fields, R.C. MASON
97 Varieties of constructive mathematics, D.S. BRIDGES & F. RICHMAN
98 Localization in Noetherian rings, A.V. JATEGAONKAR
99 Methods of differential geometry in algebraic topology, M. KAROUBI & C. LERUSTE
100 Stopping time techniques for analysts and probabilists, L. EGGHE
101 Groups and geometry, ROGER C. LYNDON
103 Surveys in combinatorics 1985, I. ANDERSON (ed)
104 Elliptic structures on 3-manifolds, C.B. THOMAS
105 A local spectral theory for closed operators, I. ERDELYI & WANG SHENGWANG
106 Syzygies, E.G. EVANS & P. GRIFFITH
107 Compactification of Siegel moduli schemes, C-L. CHAI
108 Some topics in graph theory, H.P. YAP
109 Diophantine analysis, J. LOXTON & A. VAN DER POORTEN (eds)
110 An introduction to surreal numbers, H. GONSHOR
111 Analytical and geometric aspects of hyperbolic space, D.B.A.EPSTEIN (ed)
113 Lectures on the asymptotic theory of ideals, D. REES
114 Lectures on Bochner-Riesz means, K.M. DAVIS & Y-C. CHANG
115 An introduction to independence for analysts, H.G. DALES & W.H. WOODIN
116 Representations of algebras, P.J. WEBB (ed)

117 Homotopy theory, E. REES & J.D.S. JONES (eds)
118 Skew linear groups, M. SHIRVANI & B. WEHRFRITZ
119 Triangulated categories in the representation theory of finite-dimensional algebras, D. HAPPEL
121 Proceedings of *Groups - St Andrews 1985*, E. ROBERTSON & C. CAMPBELL (eds)
122 Non-classical continuum mechanics, R.J. KNOPS & A.A. LACEY (eds)
124 Lie groupoids and Lie algebroids in differential geometry, K. MACKENZIE
125 Commutator theory for congruence modular varieties, R. FREESE & R. MCKENZIE
126 Van der Corput's method for exponential sums, S.W. GRAHAM & G. KOLESNIK
127 New directions in dynamical systems, T.J. BEDFORD & J.W. SWIFT (eds)
128 Descriptive set theory and the structure of sets of uniqueness, A.S. KECHRIS & A. LOUVEAU
129 The subgroup structure of the finite classical groups, P.B. KLEIDMAN & M.W.LIEBECK
130 Model theory and modules, M. PREST
131 Algebraic, extremal & metric combinatorics, M-M. DEZA, P. FRANKL & I.G. ROSENBERG (eds)
132 Whitehead groups of finite groups, ROBERT OLIVER
133 Linear algebraic monoids, MOHAN S. PUTCHA
134 Number theory and dynamical systems, M. DODSON & J. VICKERS (eds)
135 Operator algebras and applications, 1, D. EVANS & M. TAKESAKI (eds)
136 Operator algebras and applications, 2, D. EVANS & M. TAKESAKI (eds)
137 Analysis at Urbana, I, E. BERKSON, T. PECK, & J. UHL (eds)
138 Analysis at Urbana, II, E. BERKSON, T. PECK, & J. UHL (eds)
139 Advances in homotopy theory, S. SALAMON, B. STEER & W. SUTHERLAND (eds)
140 Geometric aspects of Banach spaces, E.M. PEINADOR and A. RODES (eds)
141 Surveys in combinatorics 1989, J. SIEMONS (ed)
142 The geometry of jet bundles, D.J. SAUNDERS
143 The ergodic theory of discrete groups, PETER J. NICHOLLS
144 Introduction to uniform spaces, I.M. JAMES
145 Homological questions in local algebra, JAN R. STROOKER
146 Cohen-Macaulay modules over Cohen-Macaulay rings, Y. YOSHINO
147 Continuous and discrete modules, S.H. MOHAMED & B.J. MÜLLER
148 Helices and vector bundles, A.N. RUDAKOV et al
149 Solitons, nonlinear evolution equations and inverse scattering, M.J. ABLOWITZ & P.A. CLARKSON
150 Geometry of low-dimensional manifolds 1, S. DONALDSON & C.B. THOMAS (eds)
151 Geometry of low-dimensional manifolds 2, S. DONALDSON & C.B. THOMAS (eds)
152 Oligomorphic permutation groups, P. CAMERON
153 L-functions in arithmetic, J. COATES & M.J. TAYLOR (eds)
154 Number theory and cryptography, J. LOXTON (ed)
155 Classification theories of polarized varieties, TAKAO FUJITA
156 Twistors in mathematics and physics, T.N. BAILEY & R.J. BASTON (eds)
157 Analytic pro-*p* groups, J.D. DIXON, M.P.F. DU SAUTOY, A. MANN & D. SEGAL
158 Geometry of Banach spaces, P.F.X. MÜLLER & W. SCHACHERMAYER (eds)
159 Groups St Andrews 1989 Volume 1, C.M. CAMPBELL & E.F. ROBERTSON (eds)
160 Groups St Andrews 1989 Volume 2, C.M. CAMPBELL & E.F. ROBERTSON (eds)
161 Lectures on block theory, BURKHARD KÜLSHAMMER
162 Harmonic analysis and representation theory for groups acting on homogeneous trees, A. FIGÁ-TALAMANCA & C. NEBBIA
163 Topics in varieties of group representations, S.M. VOVSI
164 Quasi-symmetric designs, M.S. SHRIKANDE & S.S. SANE
166 Surveys in combinatorics, 1991, A.D. KEEDWELL (ed)
167 Stochastic analysis, M.T. BARLOW & N.H. BINGHAM (eds)
168 Representations of algebras, H. TACHIKAWA & S. BRENNER (eds)
174 Lectures on mechanics, J.E. MARSDEN
175 Adams memorial symposium on algebraic topology 1, N. RAY & G. WALKER (eds)
176 Adams memorial symposium on algebraic topology 2, N. RAY & G. WALKER (eds)
178 Lower K- and L-theory, A. RANICKI

London Mathematical Society Lecture Note Series, 168

Representations of Algebras and Related Topics

Edited by
H. Tachikawa
Institute of Mathematics, University of Tsukuba
S. Brenner
Department of Applied Mathematics and Theoretical Physics, University of Liverpool

CAMBRIDGE
UNIVERSITY PRESS

CAMBRIDGE UNIVERSITY PRESS
Cambridge, New York, Melbourne, Madrid, Cape Town, Singapore, São Paulo

Cambridge University Press
The Edinburgh Building, Cambridge CB2 2RU, UK

Published in the United States of America by Cambridge University Press, New York

www.cambridge.org
Information on this title: www.cambridge.org/9780521424110

First published 1992

A catalogue record for this publication is available from the British Library

ISBN-13 978-0-521-42411-0 paperback
ISBN-10 0-521-42411-9 paperback

Transferred to digital printing 2006

Contents

Contents

INTRODUCTION

The Tsukuba International Conference on Representations of Algebras and Related Topics(1) took place in the week before the International Congress of Mathematicians in Kyoto (1990). The Conference was preceded by a Workshop in which leading workers in the field gave expository lectures on recent developments in areas covered by the Conference.

The aim of this book is to present the Workshop lectures to a wider audience. The participants at the Workshop were not all specialists in the area and so the speakers aimed to make their talks as self-contained as possible. This is reflected in their papers presented here. Several of the authors have taken the opportunity to update their manuscripts, most of which contain results which have not appeared elsewhere. We have included one paper (Dlab's) which was not presented at the Workshop.

The Tsukuba meetings took place at a time of exciting and highly complex interaction between the representation theory of algebras and other branches of mathematics. Several of the powerful technologies developed within algebra representation theory during its rather introspective period from about 1970 to the mid-1980s are now contributing strongly to other areas. In the opposite direction, new problems, ideas and points of view are coming into the subject from previously unrelated areas. Some of these interactions are reflected in the present volume.

The study of functor categories (categories of abelian group valued functors on module categories) is one of the most successful methodologies of algebra representation theory and underlies key concepts like those of almost split sequence and Auslander-Reiten quiver. The homologically finite subcategories discussed in the paper by Auslander and Reiten are subcategories of modules for which the operation of restriction of finitely presented functors is especially well behaved and, for example, can imply the existence of relative almost split sequences. In their paper Auslander and Reiten show that there is an intimate relation between homologically finite subcategories and the generalised tilting modules of Happel and Miyashita. Tilting modules have been generalised still further to tilting complexes in Rickard's Morita theory for derived categories, and generalised tilting theory has opened the way for the importation to algebra representation theory of ideas and methods from, for example, the theory of vector bundles.

Derived categories do not appear explicitly in this volume. However the attractive and widely occurring class of quasi-hereditary algebras, discussed here by Dlab and Ringel, had its origins in the use of derived categories and tilting modules for the study of representations of semi-simple algebraic groups in characteristic $p > 0$.

Another important methodology has emerged from the study of

matrix methods and their formalisation in the theory of bocses, and is the basis of the only proof of the tame-wild dichotomy for algebras over an algebraically closed field (Drozd's Theorem). In his present paper Drozd describes some recent advances in bocs theory and a new application of it to obtain a description of dense sets of irreducible unitary representations of certain Lie groups of mixed type.

Although the tame-wild dichotomy applies only to finite dimensional modules, the essential definitions use infinite dimensional modules. Crawley-Boevey has used indecomposable modules of infinite length and finite endolength (length over the endomorphism ring) to define concepts of generic tameness and wildness which may provide the basis for a tame-wild dichotomy for wider classes of rings. Progress in this direction is amongst the exciting results contained in his paper in this volume. In it he uses both functor and matrix methods in new and wider contexts.

The rings of pure global dimension zero of Azumaya's article include the rings of finite representation type. Such rings R have been characterised by a set of equivalent conditions which must be satisfied by all R-modules. Azumaya shows that, in each case, it is sufficient to require that the condition be satisfied by all countably generated R-modules.

The use of quivers, in several forms, has become characteristic of algebra representation theory and provides a point of contact with many other branches of mathematics. One notable example is Kac's use of invariant theory to describe the representation theory of quivers in terms of Kac-Moody root systems. A direct connection between representations of a Dynkin quiver and the positive part, U_+, of the universal enveloping algebra of the corresponding Lie algebra was established by Ringel and subsequently extended by him to the 'quantum' situation. This has stimulated other work on the relationship between representations of finite dimensional algebras and Lie algebras and the use of quiver representations in the study of quantum groups. In his present paper Ringel shows how the coalgebra structure of U_+ may be understood in terms of his construction from representations of the corresponding quiver.

Dlab's paper also uses representations of quivers and, more generally, species. He discusses a problem originating in the work of Jones on type II_1 factors which led to the Jones polynomial in knot theory. Values of the Jones index which Dlab derives here using quiver representations are also the singular values of a parameter which occurs in knot theory and quantum groups.

Auslander-Reiten quivers impose an extremely useful geometric structure on the categories of finitely generated modules over an artin algebra. In her study of tame blocks for group algebras, Erdmann uses, with consummate skill, the interaction between this geometric structure and the special properties of the blocks. Her paper describes this work and shows how such methods may be used in somewhat more general situations.

Carlson's paper is concerned with the cohomology theory of finite groups. His aim is to find a zero point for indexing the doubly infinite sequence of Tate cohomology. He defines a new invariant of a module, which may serve this end. His paper includes extremely useful reviews of a wide range of topics. The final section of Carlson's paper is independent of the earlier part. In it he discusses two questions concerning contravariantly finite subcategories which were raised at the Workshop.

The Workshop received substantial support from both the Exchange Fund of the University of Tsukuba and the Hara Research Fund. We are most grateful for their assistance. We also wish to express our thanks to the authors and referees of the papers in this volume and to David Tranah of the Cambridge University Press for his help in preparing it for publication.

Sheila Brenner Hiroyuki Tachikawa

(1)The Proceedings of the Conference will be published separately as Canadian Mathematical Society Proceedings series No. 11.

Homologically finite subcategories

Maurice Auslander* and Idun Reiten

Let Λ be an artin algebra and modΛ the category of finitely generated Λ-modules. Unless stated to the contrary, by a subcategory $\mathcal{C}$ of a category we mean a full subcategory of an additive category, for example modΛ, closed under isomorphisms and direct summands. About ten years ago, in connection with proving the existence of preprojective and preinjective partitions for modΛ as well as the existence of almost split sequences in certain subcategories of modΛ, Auslander and Smalø introduced the notions of contravariantly, covariantly and functorially finite subcategories of modΛ and developed some of their basic properties [5] [6]. We refer to the study of these subcategories as the theory of homologically finite subcategories. At the time Auslander and Smalø pointed out that there is an intimate connection between the tilting theory of Happel and Ringel based on tilting modules of projective dimension at most one and the theory of homologically finite subcategories [6]. Utilizing recent developments in the theory, we have now given precise connections between the tilting theory developed by Miayshita and Happel based on tilting modules of arbitrary finite projective dimension and the theory of homologically finite subcategories in [4]. Inspired by this new point of view on tilting theory, Ringel has shown that associated with a quasi-hereditary algebra Λ and a particular type of ordering of its simple modules is a naturally defined tilting module whose endomorphism ring is again quasi-hereditary. It follows from this that quasihereditary algebras occur in pairs [13].

*Partially supported by NSF-Grant DMS 8904594.

In addition to tilting theory, questions in the theory of homologically finite subcategories have come up in the work of Burt and Butler in showing that some categories of representations of bocses have almost split sequences [8]. This problem was investigated by Bautista-Kleiner in [7] using other methods.

More generally, suppose that Λ is a noetherian R-algebra, i.e. R is a commutative noetherian ring and Λ is an R-algebra which is a finitely generated R-module. There are interesting anologues for noetherian R-algebras of much of the theory of homologically finite subcategories of modΛ when Λ is an artin algebra. In particular, the Auslander–Buchweitz theory of Cohen-Macaulay approximations is proving to be of special interest in the study of complete local Cohen-Macaulay rings [2]. It is also worth noting that the theory of Cohen-Macaulay approximations in general has helped clarify some points in the theory of homologically finite subcategories for artin algebras. However, for the most part we will restrict our attention in this paper to the artin algebra situation.

In section 1 we recall the definitions of contravariantly, covariantly and functorially finite subcategories, and investigate the connections with adjoint functors. In section 2 we discuss the origins of the concepts, in connection with preprojective partitions and almost split sequences in subcategories. In section 3 we study ways of constructing covariantly finite subcategories from contravariantly finite ones, and conversely. In section 4 we give the connections with tilting theory.

In general proofs will only be given when they have not already appeared elsewhere. Usually also dual definitions and statements will be left to the reader.

1 Homologically finite subcategories and adjoint functors

Even though most of this paper is concerned with full subcategories of categories of finitely generated modules over artin algebras, we begin by considering homologically finite subcategories of arbitrary categories. In this way we hope to emphasize the generality of the basic notions of the theory of homologically finite subcategories.

This section is devoted to giving the definitions of contravariantly, covariantly and functorially finite subcategories of arbitrary categories and their connections with adjoint functors. We begin by giving some basic definitions and notation [5].

Let $\mathcal{C}$ be an arbitrary category, not necessarily a category of modules. For each pair of objects C_1 and C_2 in $\mathcal{C}$ we denote the set of morphisms from C_1 to C_2 by $\mathcal{C}(C_1, C_2)$. For each C in $\mathcal{C}$ we denote by $\mathcal{C}(\ ,C)$ the contravariant functor from $\mathcal{C}$ to Sets given by $X \mapsto \mathcal{C}(X, C)$ for all objects X in $\mathcal{C}$ and we denote by $\mathcal{C}(C,\ )$ the covariant functor from $\mathcal{C}$ to Sets given by $X \mapsto \mathcal{C}(C, X)$ for all objects X in $\mathcal{C}$. A contravariant (covariant) functor $F : \mathcal{C} \to$ Sets is said to be representable if $F \cong \mathcal{C}(\ ,C)$ ($F \cong \mathcal{C}(C,\ )$) for some object C in $\mathcal{C}$. A morphism $\alpha : F \to G$ of functors from $\mathcal{C}$ to Sets is said to be surjective if $\alpha_X : F(X) \to G(X)$ is surjective for all objects X in $\mathcal{C}$. A contravariant (covariant) functor F from $\mathcal{C}$ to Sets is said to be finitely generated if there is a surjection $\mathcal{C}(\ ,C) \to F$ ($\mathcal{C}(C,\ ) \to F$) for some C in $\mathcal{C}$.

Suppose $\mathcal{X}$ is a subcategory of $\mathcal{C}$. By a right $\mathcal{X}$-approximation of an object C in $\mathcal{C}$ we mean a morphism $f_C : X_C \to C$ with X_C an object in $\mathcal{X}$ such that the induced morphism $\mathcal{C}(X, X_C) \to \mathcal{C}(X, C)$ is surjective for all objects X in $\mathcal{X}$. It is not difficult to see that an object C in $\mathcal{C}$ has a right $\mathcal{X}$-approximation if and only if $\mathcal{C}(\ ,C)\mid_{\mathcal{X}}$, the restriction to the subcategory $\mathcal{X}$ of the functor $\mathcal{C}(\ ,C)$, is finitely generated, i.e. there is a surjection $\mathcal{X}(\ ,X_C) \to \mathcal{C}(\ ,C)\mid_{\mathcal{X}}$ for some object X_C in $\mathcal{X}$. We denote by $\operatorname{Rapp}\mathcal{X}$ the subcategory of $\mathcal{C}$ consisting of all the objects C in $\mathcal{C}$ which have right $\mathcal{X}$-approximations. Clearly, $\mathcal{X}$ is contained in $\operatorname{Rapp}\mathcal{X}$ and it is easily checked that $\operatorname{Rapp}\mathcal{X} = \operatorname{Rapp}(\operatorname{Rapp}\mathcal{X})$. Finally $\mathcal{X}$ is said to be contravariantly finite in $\mathcal{C}$ if $\operatorname{Rapp}\mathcal{X} = \mathcal{C}$, i.e. every object C in $\mathcal{C}$ has a right $\mathcal{X}$-approximation.

Dually, by a left $\mathcal{X}$-approximation of an object C in $\mathcal{C}$ we mean a morphism $g : C \to X^C$ with X^C in $\mathcal{X}$ such that the induced morphism $\mathcal{C}(X^C, X) \to \mathcal{C}(C, X)$ is surjective for all objects X in $\mathcal{C}$. It is not difficult to see that an object C in $\mathcal{C}$ has a left $\mathcal{X}$-approximation if and only if the retriction $\mathcal{C}(C,\ )\mid_{\mathcal{X}}$ is finitely generated. We denote by $\operatorname{Lapp}\mathcal{X}$ the subcategory of $\mathcal{C}$ consisting of all the objects C in $\mathcal{C}$ which have left $\mathcal{X}$-approximations. Clearly $\mathcal{X}$ is contained in $\operatorname{Lapp}\mathcal{X}$ and it is easily checked that $\operatorname{Lapp}\mathcal{X} = \operatorname{Lapp}(\operatorname{Lapp}\mathcal{X})$. We say that $\mathcal{X}$ is covariantly finite in $\mathcal{C}$ if $\operatorname{Lapp}\mathcal{X} = \mathcal{C}$. Finally we say that $\mathcal{X}$ is functorially finite in $\mathcal{C}$ if it is both contravariantly and covariantly finite in $\mathcal{C}$.

The rest of this section is devoted to explaining various connections between adjoint functors and contravariantly and covariantly finite subcategories which give nontrivial interesting examples of homologically finite subcategories.

Suppose $G : \mathcal{C} \to \mathcal{X}$ is a right adjoint for the inclusion functor $\operatorname{inc} : \mathcal{X} \to \mathcal{C}$. Then there are isomorphisms $\mathcal{C}(X, C) \to \mathcal{X}(X, G(C))$ for all X in $\mathcal{X}$ and C in $\mathcal{C}$ which are functorial in X and C. Therefore for each C in $\mathcal{C}$ we have that the restriction $\mathcal{C}(\ ,C)\mid_{\mathcal{X}}$ is representable in X since it is isomorphic to the representable functor $\mathcal{X}(\ ,G(C))$. It is also not difficult to see that $\operatorname{inc} : \mathcal{X} \to \mathcal{C}$ has a right adjoint if $\mathcal{C}(\ ,C)\mid_{\mathcal{X}}$ is representable in $\mathcal{X}$ for all C in $\mathcal{C}$. Since representable functors are finitely generated, it follows that if $\operatorname{inc} : \mathcal{X} \to \mathcal{C}$ has a right adjoint, then $\mathcal{X}$ is contravariantly finite in $\mathcal{C}$. Thus

$\mathcal{X}$ being contravariantly finite in $\mathcal{C}$ can be viewed as a generalization of the inclusion inc : $\mathcal{X} \to \mathcal{C}$ having a right adjoint.

A similar argument shows that inc : $\mathcal{X} \to \mathcal{C}$ having a left adjoint implies that $\mathcal{X}$ is covariantly finite in $\mathcal{C}$. Thus $\mathcal{X}$ being covariantly finite in $\mathcal{C}$ can be viewed as a generalization of inc : $\mathcal{X} \to \mathcal{C}$ having a left adjoint.

Summarizing our discussion we have the following.

Proposition 1.1 *Let $\mathcal{X}$ be a subcategory of $\mathcal{C}$.*

a) *If* inc : $\mathcal{X} \to \mathcal{C}$ *has a right adjoint, then $\mathcal{X}$ is contravariantly finite in $\mathcal{C}$.*

b) *If* inc : $\mathcal{X} \to \mathcal{C}$ *has a left adjoint, then $\mathcal{X}$ is covariantly finite in $\mathcal{C}$.*

We now discuss some more connections between adjoint functors and contravariantly (covariantly) finite subcategories. Amongst other things, this discussion will show that there are contravariantly (covariantly) finite subcategories $\mathcal{X}$ of $\mathcal{C}$ such that inc : $\mathcal{X} \to \mathcal{C}$ does not have a right (left) adjoint.

Suppose now that we are given categories $\mathcal{C}$ and $\mathcal{D}$ and a pair of adjoint functors $F : \mathcal{C} \to \mathcal{D}$ and $G : \mathcal{D} \to \mathcal{C}$. Then we have isomorphisms $\mathcal{D}(F(C), D) \xrightarrow{\sim} \mathcal{C}(C, G(D))$ for all C in $\mathcal{C}$ and D in $\mathcal{D}$ which are functorial in C and D. Associated with the functor F is the subcategory $\operatorname{Im} F$ of $\mathcal{D}$ consisting of all D in $\mathcal{D}$ isomorphic to $F(C)$ for some C in $\mathcal{C}$ and associated with the functor G is the subcategory $\operatorname{Im} G$ of $\mathcal{C}$ consisting of all C in $\mathcal{C}$ isomorphic to $G(D)$ for some D in $\mathcal{D}$. We now show that $\operatorname{Im} F$ is contravariantly finite in $\mathcal{D}$. The dual argument, which we do not give, shows that $\operatorname{Im} G$ is covariantly finite in $\mathcal{C}$.

Let D be in $\mathcal{D}$. Denote by $f : F(G(D)) \to D$ the map corresponding to the identity $G(D) \to G(D)$ under the adjointness isomorphism $\mathcal{D}(FG(D), D) \xrightarrow{\sim} \mathcal{C}(G(D), G(D))$. Suppose X is in $\operatorname{Im} F$ and we are given a morphism $g : X \to D$. We want to show that g factors through f. Since X is in $\operatorname{Im} F$, we have that $X \simeq F(C)$ for some C in $\mathcal{C}$. Let $h : C \to G(D)$ be the morphism corresponding to $g : F(C) \to D$ under the isomorphism $(F(C), D) \xrightarrow{\sim} (C, G(D))$. Then we obtain the map $F(h) : F(C) \to FG(D)$ which is well known to have the property $fF(h) = g$. Hence the morphism $f : FG(D) \to D$ induces a surjection $\mathcal{D}(X, FG(D)) \to \mathcal{D}(X, D)$ for all X in $\operatorname{Im} F$. Thus we have shown that $\operatorname{Im} F$ is contravariantly finite in $\mathcal{D}$.

Summarizing our discussion we have the following.

Proposition 1.2 *Suppose $\mathcal{C}$ and $\mathcal{D}$ are categories and $F : \mathcal{C} \to \mathcal{D}$ and $G : \mathcal{D} \to \mathcal{C}$ an adjoint pair of functors with F a left adjoint and G a right adjoint. Then we have the following.*

a) $\operatorname{Im} F$ *is contravariantly finite in $\mathcal{D}$ and the usual morphism $FG \to I$ gives a right* $\operatorname{Im} F$*-approximation $FG(D) \to D$ for each D in $\mathcal{D}$.*

b) $\operatorname{Im} G$ *is covariantly finite in $\mathcal{C}$ and the usual morphism $I \to GF$ gives a left* $\operatorname{Im} G$*-approximation $C \to GF(C)$ for each C in $\mathcal{C}$.*

We now want to investigate when $\operatorname{inc} : \operatorname{Im} F \to \mathcal{D}$ and $\operatorname{inc} : \operatorname{Im} G \to \mathcal{C}$ have right and left adjoints respectively. Our results along these lines are based on the following.

Proposition 1.3 *Let $\mathcal{C}$ and $\mathcal{D}$ be categories and $F : \mathcal{C} \to \mathcal{D}$ and $G : \mathcal{D} \to \mathcal{C}$ an adjoint pair of functors with F a left adjoint and G a right adjoint. A functor $H : \mathcal{D} \to \operatorname{Im} F$ is a right adjoint of* $\operatorname{inc} : \operatorname{Im} F \to \mathcal{D}$ *if and only if there is a morphism* $\operatorname{inc} H \to I_{\mathcal{D}}$ *such that the induced morphism* $G \operatorname{inc} H \to G$ *is an isomorphism.*

Proof: Suppose $H : \mathcal{D} \to \operatorname{Im} F$ is a right adjoint of $\operatorname{inc} : \operatorname{Im} F \to \mathcal{D}$. Then there is a morphism $H \to I_{\mathcal{D}}$ with the property that the induced morphisms

$\mathcal{D}(U, H(D)) \to \mathcal{D}(U, D)$ are isomorphisms for all U in $\operatorname{Im} F$ and D in $\mathcal{D}$. Let D be in $\mathcal{D}$. Then the morphism $H(D) \to D$ gives rise to the morphism $GH(D) \to G(D)$ which in turn gives rise to the commutative diagram

$$\begin{array}{ccc} \mathcal{C}(C, GH(D)) & \to & \mathcal{C}(C, G(D)) \\ \downarrow \wr & & \downarrow \wr \\ \mathcal{D}(F(C), H(D)) & \xrightarrow{\sim} & \mathcal{D}(F(C), D) \end{array}$$

for all C in $\mathcal{C}$. Hence the morphism $GH(D) \to G(D)$ is an isomorphism since the induced morphisms $\mathcal{C}(C, GH(D)) \to \mathcal{C}(C, G(D))$ are isomorphisms for all C in $\mathcal{C}$. Therefore the morphism $H \to I_{\mathcal{D}}$ has the property that the induced morphism $G \operatorname{inc} H \to G$ is an isomorphism. □

As an immediate consequence of this result we have the following which we leave to the reader to prove.

Corollary 1.4 *If* $\operatorname{inc} : \operatorname{Im} F \to \mathcal{D}$ *has a right adjoint, then* $G(\operatorname{Im} F) = \operatorname{Im} G$.

Remark It would be interesting to know if the converse of Corollary 1.4 is true.

As another immediate consequence of Proposition 1.3 we have the following.

Corollary 1.5 *Let* $\mathcal{C}$ *and* $\mathcal{D}$ *be categories and* $F : \mathcal{C} \to \mathcal{D}$ *and* $G : \mathcal{D} \to \mathcal{C}$ *an adjoint pair of functors with* F *a left adjoint and* G *a right adjoint. Suppose* $G : \mathcal{D} \to \mathcal{C}$ *has the property that a morphism* $f : D_1 \to D_2$ *in* $\mathcal{D}$ *is an isomorphism whenever* $G(f) : G(D_1) \to G(D_2)$ *is an isomorphism. Then* $\operatorname{inc} : \operatorname{Im} F \to \mathcal{D}$ *has a right adjoint if and only if* $\operatorname{Im} F = \mathcal{D}$.

Proof: Suppose inc : $\operatorname{Im} F \to \mathcal{D}$ has a right adjoint. Then there is functor $H : \mathcal{D} \to \operatorname{Im} F$ giving a morphism $\operatorname{inc} H \to I_{\mathcal{D}}$ such that $G \operatorname{inc} H \to G$ is an isomorphism. But then $\operatorname{inc} H \to I_{\mathcal{D}}$ is an isomorphism, which shows that $\operatorname{Im} F = \mathcal{D}$. □

We now apply Corollary 1.5 in some special cases to determine if inc : $\mathcal{X} \to \mathcal{C}$ has a right adjoint when $\mathcal{X}$ is contravariantly finite in $\mathcal{C}$.

Consider the adjoint pair of functors F : Sets $\to$ Groups and G : Groups $\to$ Sets where $F(S)$ is the free group generated by the set S and G is the forgetful functor. Clearly G has the property that a morphism f in Groups is an isomorphism if $G(f)$ is an isomorphism. Therefore by Corollary 1.5, if inc : $\operatorname{Im} F \to$ Groups has a right adjoint, then $\operatorname{Im} F =$ Groups, which is impossible since $\operatorname{Im} F$ is the subcategory of Groups consisting of the free groups. Therefore the subcategory of Groups consisting of the free groups is contravariantly finite in Groups without the inclusion having a right adjoint.

The following is perhaps a more interesting example for ring and module theorists.

Proposition 1.6 *Let $f : \Lambda \to \Gamma$ be a map of rings with the property that Γ is a finitely presented left Λ-module. Let* $\operatorname{mod}\Lambda$ *and* $\operatorname{mod}\Gamma$ *be the categories of finitely presented Λ and Γ-modules respectively. Finally, let F :* $\operatorname{mod}\Lambda \to \operatorname{mod}\Gamma$ *be given by $F(M) = \Gamma \underset{\Lambda}{\otimes} M$ for all Λ-modules M. Then* $\operatorname{Im} F$, *the subcategory of* $\operatorname{mod}\Gamma$ *of induced modules, is contravariantly finite in* $\operatorname{mod}\Lambda$ *and* inc : $\operatorname{Im} F \to \operatorname{mod}\Gamma$ *has a right adjoint if and only if* $\operatorname{Im} F = \operatorname{mod}\Gamma$, *i.e. every finitely presented Γ-module is induced.*

Proof: Suppose M is in $\operatorname{mod}\Gamma$. Then $\operatorname{res} M$, the restriction of M to Λ is in $\operatorname{mod}\Lambda$ since Γ is a finitely presented Λ-module. But then res : $\operatorname{mod}\Gamma \to \operatorname{mod}\Lambda$

is a right adjoint to $\Gamma \underset{\Lambda}{\otimes}$ with the property that $f : M_1 \to M_2$ in $\mathrm{mod}\Gamma$ is an isomorphism if $\mathrm{res}\, f : \mathrm{res}\, M_1 \to \mathrm{res}\, M_2$ is an isomorphism. Therefore we can apply Corollary 1.5 to get our desired result. □

It is obvious that if every Γ-module is induced from Λ, then the relative global dimension of the ring morphism $f : \Lambda \to \Gamma$ is zero. This already shows that the inclusion of the induced modules rarely has a right adjoint. However there are examples where the inclusion of the induced modules has no right adjoint even though the relative global dimension is zero. While there are examples where every Γ-module is induced, for instance if $f : \Lambda \to \Gamma$ is a ring epimorphism, there do not seem to be any good criteria for when this happens.

Let Λ be a twosided noetherian ring. We say that a Λ-module K is a dth syzygy module ($d \geq 1$) if there is an exact sequence $0 \to K \to P_{d-1} \to \cdots \to P_0 \to C \to 0$ with the P_i projective Λ-modules. We end this section by showing that the category $\Omega^d(\mathrm{mod}\Lambda)$ of dth syzygy modules is covariantly finite in $\mathrm{mod}\Lambda$. This was essentially first proved in [1]. But before doing this, it is convenient to introduce some notation.

Suppose Λ is a twosided noetherian ring and $d \geq 1$. Let $\underline{\mathrm{mod}}\Lambda$ denote the category of finitely generated Λ-modules modulo projectives. Then we denote by $\Omega^d : \underline{\mathrm{mod}}\Lambda \to \underline{\mathrm{mod}}\Lambda$ the dth syzygy functor and by $Tr : \underline{\mathrm{mod}}\Lambda \to \underline{\mathrm{mod}}\Lambda^{op}$ the usual transpose. For A and B in $\underline{\mathrm{mod}}\Lambda$, we denote $\underline{\mathrm{mod}}(\Lambda)(A, B)$ by $\underline{\mathrm{Hom}}_\Lambda(A, B)$. We now prove the following.

Proposition 1.7 *For each $d \geq 1$, the functor $Tr\Omega^d Tr : \underline{\mathrm{mod}}\Lambda \to \underline{\mathrm{mod}}\Lambda$ is a left adjoint for $\Omega^d : \underline{\mathrm{mod}}\Lambda \to \underline{\mathrm{mod}}\Lambda$. Then the usual morphism $I_{\underline{\mathrm{mod}}\Lambda} \to \Omega^d(Tr\Omega^d Tr)$ has the property that $M \to \Omega^d(Tr\Omega^d Tr)(M)$ is a left $\mathrm{Im}\,\Omega^d$-approximation of M for all M in $\underline{\mathrm{mod}}\Lambda$. Therefore $\Omega^d(\underline{\mathrm{mod}}\Lambda)$ is covariantly finite in $\underline{\mathrm{mod}}\Lambda$.*

Proof: In view of Proposition 1.2, it suffices to prove that $Tr\Omega^d Tr$ is a left adjoint for Ω^d. Now we have isomorphisms $\underline{\mathrm{Hom}}(A, \Omega^d B) \xrightarrow{\sim} \mathrm{Tor}_1^\Lambda(TrA, \Omega^d B)) \xrightarrow{\sim} \mathrm{Tor}_1^\Lambda(\Omega^d TrA, B) \xrightarrow{\sim} \underline{\mathrm{Hom}}(Tr\Omega^d TrA, B)$ which are functorial in A and B [3], giving our desired result. □

As a fairly straightforward consequence of this result we have the following.

Corollary 1.8 *$\Omega^d(\mathrm{mod}\Lambda)$ is covariantly finite in* $\mathrm{mod}\Lambda$ *for each $d \geq 0$.*

Proof: When Λ is an artin algebra, it was shown in [5] that if $\mathcal{Y}$ is a covariantly finite subcategory of $\underline{\mathrm{mod}}(\Lambda)$ then the subcategory of $\mathrm{mod}\Lambda$ consisting of all M such that $M \cong Y$ in $\underline{\mathrm{mod}}(\Lambda)$ for some Y in $\mathcal{Y}$ is also covariantly finite. The proof works equally well for Λ a twosided noetherian ring using the fact that $\mathrm{Hom}_\Lambda(A, \Lambda)$ is a finitely generated Λ^{op}-module for all A in $\mathrm{mod}\Lambda$. □

We also cite the following result from [1] (see [10]).

Proposition 1.9 *Let Λ be an Auslander ring, that is, in the minimal injective resolution $0 \to \Lambda \to I_0 \to I_1 \to \cdots \to I_i \to \cdots$ of Λ, flat dim. $I_j \leq j$ for all j. Then for each t, $\{C; \mathrm{pd}_\Lambda C \leq t\}$ is covariantly finite in* $\mathrm{mod}\Lambda$.

2 Connections with preprojective partitions

Throughout this section all rings are artin algebras and all categories are subcategories of categories of finitely generated modules over an artin algebra. Our aim in this section is to review how the notion of homologically finite subcategories of $\mathrm{mod}\Lambda$ arose originally in [5] in connection with proving the existence of preprojective partitions of modΛ and the existence of almost split sequences in functorially finite subcategories which are closed under extensions. Along the way we recall other related concepts and results, in particular minimal maps and generators for a family of maps.

We first recall the notion of right minimal morphisms first introduced in [5].

For each module C in a category $\mathcal{C}$ we denote by $\mathcal{C}/C$ the category given by the following data. The objects in $\mathcal{C}/C$ are the morphisms $f : X \to C$ for all X in $\mathcal{C}$. For two objects $f : X \to C$ and $f' : X \to C$ in $\mathcal{C}/C$ we define the morphisms (f, f') from f to f' in $\mathcal{C}/C$ to be the set of all morphisms $h : X \to X'$ in $\mathcal{C}$ such that $f = f'h$. It is easily checked that h in (f, f') is an isomorphism in $\mathcal{C}/C$ if and only if $h : X \to X'$ is an isomorphism in $\mathcal{C}$. A morphism $f : X \to C$ in $\mathcal{C}/C$ is said to be minimal if (f, f) consists only of automorphisms. A morphism $f : X \to C$ in $\mathcal{C}$ is said to be right minimal if f is minimal in $\mathcal{C}/C$.

While the notion of right minimal morphisms and the dual notion of left minimal morphisms may seem bizarre at first glance, the following result shows this is really not the case since it shows that all morphisms in $\mathcal{C}$ are "essentially" right minimal [5].

Proposition 2.1 *Let $f : B \to C$ be a morphism in $\mathcal{C}/C$.*

a) *Then B can be written as a sum $B_0 \amalg B_1$ in such a way that $f \mid B_0$ is minimal and $f \mid B_1 = 0$.*

b) *If $B = B'_0 \amalg B'_1$ is another decomposition of B such that $f \mid B'_0$ is right minimal and $f \mid B'_1 = 0$, then $f \mid B_0$ and $f \mid B'_0$ are isomorphic in $\mathcal{C}/C$.*

Given a morphism $f : B \to C$ in $\mathcal{C}/C$, a minimal morphism $f_0 : B_0 \to C$ in $\mathcal{C}/C$ is called a minimal version of f if there is a decomposition $B = B'_0 \amalg B'_1$ with $f \mid B'_0$ right minimal and $f \mid B'_1 = 0$ such that f_0 and $f \mid B'_0$ are isomorphic in $\mathcal{C}/C$. Expressing Proposition 2.1 in this terminology we have that each $f : B \to C$ in $\mathcal{C}/C$ has a minimal version in $\mathcal{C}/C$ which is unique up to isomorphism, and is called a right minimal version of f.

The following illustrates the use of right minimal morphisms. Suppose C is a fixed module in $\mathcal{C}$. A morphism $f : X \to C$ in $\mathcal{C}/C$ is said to be a generator for a family of morphisms $\{f_i : X_i \to C\}_{i\in I}$ in $\mathcal{C}/C$ if a) $(f_i, f) \neq \emptyset$ for all i in I and b) there is a commutative diagram

$$\begin{array}{ccc} X & \xrightarrow{f} & C \\ \downarrow & & \| \\ \amalg_{i\in J} X_i & \xrightarrow{\amalg f_i} & C \end{array}$$

for some finite subset J of I. A family of morphisms $\{f_i : X_i \to C\}$ is said to be finitely generated if it has a generator. A generator f for a family of morphisms in $\{f_i\}_{i\in I}$ is said to be a minimal generator if f is minimal in $\mathcal{C}/C$.

The following result is an easy consequence of the definitions involved. For a subcategory $\mathcal{X}$ of an additive category $\mathcal{C}$, add $\mathcal{X}$ denotes the subcategory of $\mathcal{X}$ whose objects are the finite direct sums of modules in $\mathcal{X}$.

Proposition 2.2 *Suppose a family of morphisms $\{f_i : X_i \to C\}$ in $\mathcal{C}/C$ is finitely generated with generator $f : X \to C$.*

a) *If $f_0 : X_0 \to C$ is a minimal version of f, then $f_0 : X_0 \to C$ is a minimal generator for $\{f_i\}_{i\in I}$ and X_0 is in* add$\{X_i\}_{i\in I}$.

b) *Any two minimal generators for the family $\{f_i\}_{i\in I}$ are isomorphic in $\mathcal{C}/C$.*

It is worth observing that this result is equivalent to the following statement about functors. Suppose C is an object in an additive category $\mathcal{C}$. Then associated with a family of morphisms $\{f_i : X_i \to C\}$ in $\mathcal{C}/C$ is the subfunctor F of $(\ , C)$ which is $\operatorname{Im}(\amalg(\ , X_i) \xrightarrow{\amalg(\ , f_i)} (\ , C))$. Now $f : X \to C$

is a generator for $\{f_i : X_i \to C\}$ if and only if $\mathrm{Im}((\ ,X) \xrightarrow{(\ ,f)} (\ ,C))$ is F. Thus F is finitely generated if and only if the family $\{f_i : X_i \to C\}$ is finitely generated. It is also easily seen that a generator $f : X \to C$ is minimal if and only if $(\ ,f) : (\ ,X) \to (\ ,C)$ is a projective cover for F. Thus Proposition 2.2 amounts to saying that every finitely generated subfunctor G of $(\ ,C)$ has a projective cover which is unique up to isomorphism.

As an illustration of how Proposition 2.2 can be used, we point out the following connection with homologically finite subcategories.

Corollary 2.3 *Let $\mathcal{X}$ be an additive subcategory of* $\mathrm{mod}\Lambda$. *Suppose $f : X \to C$ is a right $\mathcal{X}$-approximation of C in* $\mathrm{mod}\Lambda$. *Then we have the following.*

a) *If $f_0 : X_0 \to C$ is a minimal version of f, then $f_0 : X_0 \to C$ is also a right $\mathcal{X}$-approximation of C.*

b) *If $f_0 : X_0 \to C$ and $f_1 : X_1 \to C$ are both right minimal morphisms which are right $\mathcal{X}$-approximations of C, then f_0 and f_1 are isomorphic in* $\mathrm{mod}\Lambda/C$.

Proof: Let $\mathcal{F}$ be the family of all morphisms $g : Z \to C$ with Z in $\mathcal{X}$. Then it is not hard to see that a right $\mathcal{X}$-approximation is nothing more than a generator for $\mathcal{F}$. Our desired result is now a trivial consequence of Proposition 2.2. □

With these preliminary remarks about morphisms in mind, we now turn our attention to discussing preprojective partitions for additive categories, which was one of the original motivations for introducing covariantly finite subcategories of a category [5].

We say that a category $\mathcal{Y}$ is finite if Ind $\mathcal{Y}$, the subcategory of $\mathcal{Y}$ consisting of all indecomposable modules which are isomorphic to summands of modules in $\mathcal{Y}$ has only a finite number of nonisomorphic modules. Suppose $\mathcal{X}$ is an additive category. We here say that a cover of $\mathcal{X}$ is a finite subcategory $\mathcal{C}$ of $\operatorname{Ind}\mathcal{X}$ such that for each X in $\mathcal{X}$ there is a epimorphism $C \to X$ with C in $\operatorname{add}\mathcal{C}$. A minimal cover of $\mathcal{X}$ is a cover $\mathcal{C}$ of $\mathcal{X}$ such that no proper subcategory of $\mathcal{C}$ is a cover of $\mathcal{X}$. The following description of covers in terms of left approximations is basic to our entire treatment of preprojective partitions [5].

Proposition 2.4 *Let $\mathcal{X}$ be an additive category.*

a) *A finite subcategory $\mathcal{C}$ of* $\operatorname{Ind}\mathcal{X}$ *is a cover for $\mathcal{X}$ if and only if there is a left $\mathcal{X}$-approximation $\Lambda \to C$ of Λ with C in* $\operatorname{add}\mathcal{C}$.

b) *If $\Lambda \to C_0$ is a minimal left $\mathcal{X}$-approximation of Λ, then* $\operatorname{Ind} C_0 = \operatorname{Ind}(\operatorname{add} C_0)$ *is a cover for $\mathcal{X}$ which is contained in all covers of $\mathcal{X}$.*

c) *If $\mathcal{X}$ has a cover, then any two minimal covers are the same.*

d) *If $\mathcal{X}$ is covariantly finite in* $\operatorname{mod}\Lambda$, *then $\mathcal{X}$ has a cover.*

If $\mathcal{X}$ has a cover we denote its unique minimal cover by $\mathcal{P}_0(\mathcal{X})$. Whenever we write $\mathcal{P}_0(\mathcal{X})$ we are implicitly assuming that $\mathcal{X}$ has a cover.

Let $\mathcal{X}$ be an additive category. A preprojective partition of $\mathcal{X}$ is an ordered collection of subcategories

$$\mathcal{P}_0(\mathcal{X}), \mathcal{P}_1(X), \ldots, \mathcal{P}_\infty(\mathcal{X})$$

of $\operatorname{Ind}\mathcal{X}$ having the following properties.

a) $\operatorname{Ind}\mathcal{X} = \bigcup_{i=0}^{\infty} P_i(\mathcal{X})$.

b) $\mathcal{P}_i(\mathcal{X}) = \mathcal{P}_0(\operatorname{add}(\bigcup_{j\geq i} P_j(\mathcal{X}))$ for all $i < \infty$.

c) $\mathcal{P}_\infty(\mathcal{X}) = \operatorname{Ind}\mathcal{X} - \bigcup_{i<\infty} P_i(\mathcal{X})$.

The following criterion for the existence of preprojective partitions for categories serves as the basis for all known existence theorems for preprojective partitions [5].

Proposition 2.5 *The following are equivalent for an additive category $\mathcal{X}$.*

a) *$\mathcal{X}$ has a preprojective partition.*

b) *If $\mathcal{U}$ is a finite subcategory of* $\mathrm{Ind}\,\mathcal{X}$*, then Λ has a left* $\mathrm{add}(\mathrm{Ind}\,\mathcal{X} - \mathcal{U})$*-approximation.*

As an application of this result we obtain the following existence theorem for preprojective partitions [5].

Theorem 2.6 *If the additive category $\mathcal{X}$ is covariantly finite in* $\mathrm{mod}\Lambda$*, then $\mathcal{X}$ has a preprojective partition. In particular* $\mathrm{mod}\Lambda$ *has a preprojective partition.*

This theorem is an easy consequence of our previous discussion combined with the following property of contravariantly finite subcategories of $\mathrm{mod}\Lambda$ [5].

Proposition 2.7 *Let $\mathcal{X}$ be an additive category which is covariantly finite in* $\mathrm{mod}\Lambda$*. Then* $\mathrm{add}(\mathrm{Ind}\,\mathcal{X} - \mathcal{U})$ *is covariantly finite in* $\mathrm{mod}\Lambda$ *for all finite subcategories $\mathcal{U}$ of* $\mathrm{Ind}\,\mathcal{X}$.

The proof of Proposition 2.7 depends on the interplay between left almost split morphisms and the existence of covariantly finite subcategories. We recall that a morphism $f : B \to C$ in an additive category $\mathcal{X}$ is a radical morphism if no composition $A \to B \xrightarrow{f} C \to E$, with A and E indecomposable, is an isomorphism. For C indecomposable in $\mathcal{X}$, C then has a left almost split morphism if the family of all radical morphisms $C \to X$ with X in $\mathcal{X}$ is finitely generated. And $\mathcal{X}$ has left almost split morphisms if each indecomposable C in $\mathcal{X}$ does. The first connection between covariantly finite subcategories and left almost split morphisms is the following [5].

Proposition 2.8 *Let $\mathcal{X}$ be an additive category with left almost split morphisms. Then every additive covariantly finite subcategory of $\mathcal{X}$ has left almost split morphisms. In particular, every covariantly finite additive subcategory of* modΛ *has left almost split morphisms.*

Before giving our next connection between covariantly finite subcategories and left almost split morphisms, it is convenient to have the following definitions.

Let $\mathcal{C}$ be a category. We say that the collection of radical morphisms in $\mathcal{C}$, rad $\mathcal{C}$, is right nilpotent if given any chain of radical morphisms $\cdots \to C_i \xrightarrow{f_i} \cdots \to C_1 \xrightarrow{f_1} C_0$ there is an integer n such that the composition $f_1 \cdots f_{n-1} f_n$ is 0.

If $\mathcal{C}$ is a finite category, then rad $\mathcal{C}$ is right nilpotent, and also left nilpotent, which is the dual notion. More generally if the lengths of the modules in Ind $\mathcal{C}$ are bounded, then rad $\mathcal{C}$ is both left and right nilpotent by the Harada-Sai lemma [11]. The next result generalizes a result from [5], with a similar proof.

Proposition 2.9 *Let $\mathcal{X}$ be an additive category which has left almost split morphisms. If $\mathcal{C}$ is an additive subcategory of $\mathcal{X}$ whose radical is right nilpotent, then* add(Ind $\mathcal{X}$ − Ind $\mathcal{C}$) *is covariantly finite in $\mathcal{X}$. In particular if $\mathcal{C}$ is an additive category whose radical is right nilpotent, then* add(Ind Λ − Ind $\mathcal{C}$) *is covariantly finite in* modΛ.

As a consequence of these remarks we have the following generalization of Theorem 2.6 and Proposition 2.7.

Corollary 2.10 *Suppose $\mathcal{X}$ is an additive category which is covariantly finite in* modΛ. *Suppose $\mathcal{C}$ is an additive subcategory of $\mathcal{X}$ whose radical is right nilpotent. Then we have the following.*

a) add(Ind $\mathcal{X}$ − Ind $\mathcal{C}$) *is covariantly finite.*

b) add(Ind $\mathcal{X}$ − Ind $\mathcal{C}$) *has a preprojective partition.*

The same arguments as those used above also give the following.

Proposition 2.11 *Let $\mathcal{X}$ be an additive category with left almost split morphisms. Then $\mathcal{X}$ has a preprojective partition if it has a cover.*

We now give illustrations of some of the things we have been discussing (see [6] for particulars).

Suppose $\mathcal{X}$ is an additive category closed under submodules, i.e. for each monomorphism $0 \to X' \to X$ with X in $\mathcal{X}$ we have that X' is also in $\mathcal{X}$. Then $\mathcal{X}$ is covariantly finite in modΛ because inc : $\mathcal{X} \to$ modΛ has a left adjoint F : mod$\Lambda \to \mathcal{X}$ given as follows. Let M be an arbitrary Λ-module and let $\mathcal{F}$ be the set of all submodules M' of M such that M/M' is in $\mathcal{X}$. Then $\mathcal{F}$ is closed under finite intersections since $\mathcal{X}$ is closed under submodules. Therefore $\mathcal{F}$ has a unique submodule $M_{\mathcal{X}}$ of M such that $M_{\mathcal{X}} \subset M'$ for all M' in $\mathcal{F}$. Then the epimorphism $f_{\mathcal{X}} : M \to M/M_{\mathcal{X}}$ has the property that if $g : M \to X$ is a morphism with X in $\mathcal{X}$, then there is a unique morphism $h : M/M_{\mathcal{X}} \to X$ such that $hf_{\mathcal{X}} = g$. Therefore the functor F : mod$\Lambda \to \mathcal{X}$ given by $F(M) = M/M_{\mathcal{X}}$ has the property that there are isomorphisms

$(M, X) \xrightarrow{\sim} (F(M), X)$ functorial in M in modΛ and X in $\mathcal{X}$. This shows that $\mathcal{X}$ has left almost split morphisms and a preprojective partition.

It is of interest to know when $\mathcal{X}$ is also contravariantly finite. If $\mathcal{X}$ is contravariantly finite then there would be a cocover for $\mathcal{X}$, i.e. there would be a module Z in $\mathcal{X}$ such that $\mathcal{X} = \operatorname{Sub} Z$ where $\operatorname{Sub} Z$ is the additive subcategory of all modules isomorphic to submodules of finite sums of Z. In fact, one can show that $\mathcal{X}$ is contravariantly finite if and only if $\mathcal{X} = \operatorname{Sub} Z$ for some Z in $\mathcal{X}$. Of course it then follows that the categories $\operatorname{Sub} Z$ for Z in modΛ are functorially finite in modΛ and therefore have left and right almost split morphisms as well preprojective and preinjective partitions, where preinjective partitions are the duals of preprojective partitions.

Dually, an additive subcategory $\mathcal{Y}$ of modΛ closed under factor modules is contravariantly finite in modΛ, and hence has right almost split morphisms and a preinjective partition. Also $\mathcal{Y}$ is covariantly finite if and only if $\mathcal{Y} = \operatorname{Fac} Y$ for some Y in modΛ, where $\operatorname{Fac} Y$ is the category consisting of all modules isomorphic to factor modules of finite sums of Y. Hence categories of the form $\operatorname{Fac} Y$ for some Y in modΛ are functorially finite in modΛ, have right and left almost split morphisms and preprojective and preinjective partitions. The additive category generated by the indecomposable modules which are not preprojective over an hereditary algebra of infinite type is an example of a category closed under factors which is not covariantly finite.

We also note that all subcategories of modΛ are covariantly finite if and only if Λ is of finite type. This follows from the fact that $\mathcal{P}_\infty$ is not covariantly finite if it is not empty [5].

We also give the following more recent examples of constructing contravariantly and covariantly finite subcategories.

Proposition 2.12 *[14] Let Λ and Γ be artin R-algebras and ${}_\Gamma M_\Lambda$ a $\Gamma - \Lambda$-bimodule with R acting centrally. Let $\sum$ be the triangular matrix ring $\left(\begin{smallmatrix}\Lambda & 0\\ M & \Gamma\end{smallmatrix}\right)$,*

$\mathcal{X}$ a subcategory of modΛ *and $\mathcal{Y}$ a subcategory of* modΓ. *Let $\mathcal{Z}_{\mathcal{Y}}^{\mathcal{X}}$ be the subcategory of* mod$\sum$ *whose objects are the triples $({}_\Lambda A, {}_\Gamma B, f)$, where $f : M \otimes_\Lambda A \to B$, with A in $\mathcal{X}$ and B in $\mathcal{Y}$.*

Then $\mathcal{X}$ and $\mathcal{Y}$ are contravariantly (covariantly) finite in modΛ *and* modΓ *respectively, if and only if $\mathcal{Z}_{\mathcal{Y}}^{\mathcal{X}}$ is contravariantly (covariantly) finite in* mod$\sum$.

In particular if $\mathcal{X}$ and $\mathcal{Y}$ are finite subcategories, then $\mathcal{Z}_{\mathcal{Y}}^{\mathcal{X}}$ becomes functorially finite.

Proposition 2.13 *[13] Let $A_1, \ldots, A_n$ be Λ-modules with* $\mathrm{Ext}^1(A_i, A_j) = 0$ *for $i \geq j$. Let $\mathcal{F}$ be the category whose objects are summands of objects having filtrations with factors amongst the A_i. Then $\mathcal{F}$ is functorially finite in* modΛ.

We recall that an almost split sequence in an additive category $\mathcal{X}$ is an exact sequence $0 \to A \xrightarrow{f} B \xrightarrow{g} C \to 0$ all of whose modules are in $\mathcal{X}$ such that A and C are indecomposable modules, f is left almost split in $\mathcal{X}$ and g is right almost split in $\mathcal{X}$ [6]. Furthermore, we say that a module C is Ext-projective in $\mathcal{X}$ if $\mathrm{Ext}_\Lambda^1(C, X) = 0$ for all X in $\mathcal{X}$ and a module A is Ext-injective in $\mathcal{X}$ if $\mathrm{Ext}_\Lambda^1(X, A) = 0$ for all X in $\mathcal{X}$. Finally, we say that $\mathcal{X}$ has almost split sequences if there is an almost split sequence $0 \to A \to B \to C \to 0$ for each indecomposable C which is not Ext-projective and for each indecomposable A which is not Ext-injective in $\mathcal{X}$. The following existence theorem for an additive category to have almost split sequences was given in [6].

Theorem 2.14 *Let $\mathcal{X}$ be an additive functorially finite subcategory of* modΛ *which is closed under extensions, i.e. if $0 \to A \to B \to C \to 0$ is an exact sequence with A and C in $\mathcal{X}$, then B is in $\mathcal{X}$. Then $\mathcal{X}$ has almost split sequences.*

We end this section with the following recent result by Carlson-Happel about contravariantly finite subcategories [9], and we give an application of it.

Proposition 2.15 a) *Let $\mathcal{X}$ be a contravariantly finite subcategory of* $\text{mod}\Lambda$ *closed under minimal left almost split maps, and let $\mathcal{C}$ be the additive subcategory of* $\text{mod}\Lambda$ *generated by the indecomposable modules not in $\mathcal{X}$. Then we have* $\text{Hom}(\mathcal{X},\mathcal{C}) = 0$, *that is,* $\text{Hom}(X, C) = 0$ *for all X in $\mathcal{X}$ and C in $\mathcal{C}$.*

b) *If Λ is indecomposable and $\mathcal{X}$ is functorially finite in* $\text{mod}\Lambda$ *and generated as an additive category by a nonempty union of components in the AR-quiver of Λ, the* $\mathcal{X} = \text{mod}\Lambda$.

Proof: We only point out here how b) follows from a). We get $\text{Hom}(\mathcal{X},\mathcal{C}) = 0$ by using that $\mathcal{X}$ is contravariantly finite and $\text{Hom}(\mathcal{C},\mathcal{X}) = 0$ by using that $\mathcal{X}$ is covariantly finite. □

We note that b) does hot hold if functorially finite is replaced by covariantly finite (or dually contravariantly finite). For if Λ is hereditary of infinite type consider the additive category generated by the nonpreprojective modules.

We shall now apply this result to prove the following, where a proof for group algebras was pointed out to us by Külshammer and other participants of this conference.

Proposition 2.16 *Let $i : \Lambda \to \Gamma$ be a ring homomorphism, where Γ is an indecomposable symmetric algebra which is a finitely generated projective right Λ-module and Λ is selfinjective. If the relatively projective Γ-modules $\mathcal{X}$ are closed under extensions, then $\mathcal{X}$ is* $\text{add}\,\Gamma$ *or* $\text{mod}\Gamma$.

This is based upon the following.

Lemma 2.17 *Let* Γ *be a symmetric algebra and* $\mathcal{X}$ *contravariantly finite a subcategory of* $\operatorname{mod}\Gamma$ *which is closed under extensions, syzygies and cosyzygies. Let* $S_1, \ldots, S_t$ *be the simple modules not in* $\mathcal{X}$*, and* $P_1, \ldots, P_t$ *their projective covers. Then* $\mathcal{Z} = \operatorname{ind}\mathcal{X} - \{P_1, \ldots, P_t\}$ *is empty or a union of components in the AR-quiver of* Γ.

Proof: Let P be indecomposable projective in $\mathcal{Z}$. Then $P/\underline{r}P$ is in $\mathcal{X}$, and so $\underline{r}P$ is also in $\mathcal{X}$ since $\mathcal{X}$ is closed under syzygies. Similarly $P/\operatorname{soc} P$ is in $\mathcal{X}$.

Let then C be indecomposable nonprojective in $\mathcal{X}$, and let $0 \to \Omega^2 C \to E \to C \to 0$ be almost split in $\operatorname{mod}\Gamma$. Since by assumption $\Omega^2 C$ is in $\mathcal{X}$ and so in $\mathcal{Z}$, we also have that E is in $\mathcal{X}$. If P_i is a summand of E for some i, then $C \simeq P_i/\operatorname{soc} P_i$, where $\operatorname{soc} P_i = S_i$ is not in $\mathcal{X}$. So this contradicts the fact that C is in $\mathcal{X}$. Hence E is in $\mathcal{Z}$. The almost split sequence $0 \to C \to F \to \Omega^{-2} C \to 0$ is treated similarly, and this finishes the proof. □

Proof of Proposition 2.16 Since Γ_Λ is finitely generated projective, $\Gamma_\Lambda \otimes_\Lambda$ is exact, so that $\mathcal{X}$ is closed under syzygies. Since Λ is selfinjective, the projective and injective Λ-modules coincide, so $\mathcal{X}$ is also closed under cosyzygies. Then $\mathcal{Z} = \operatorname{ind}\mathcal{X} - \{P_1, \ldots, P_t\}$, in the notation of Lemma 2.17, is also contravariantly finite by Proposition 2.7. By Lemma 2.17, $\mathcal{Z}$ is empty or a union of components. We then have the following four cases. (a) If $\mathcal{Z}$ is empty, then clearly $\mathcal{X} = \operatorname{add}\Gamma$. (b) Assume that all indecomposable projective modules are in $\mathcal{Z}$. Then all simple Γ-modules are in $\mathcal{X}$, and hence $\mathcal{X} = \operatorname{mod}\Gamma$ since $\mathcal{X}$ is closed under extensions. (c) Assume that $\mathcal{Z}$ is not empty and has no indecomposable projective modules. Then there must be a nonzero map $Z \to P$ with Z in $\mathcal{Z}$ and P projective. This contradicts Proposition 2.15. (d) Assume that there are indecomposable projective modules P and Q with P in $\mathcal{Z}$ and Q not in $\mathcal{Z}$. Since Γ is symmetric and indecomposable, there is a chain of nonzero maps between indecomposable projective modules from P to Q, again contradicting Proposition 2.15.

3 Extension closed categories

Throughout this section we assume that Λ is an artin algebra and modΛ is the category of finitely generated Λ-modules. Associated with a subcategory $\mathcal{C}$ of modΛ are the categories $\mathcal{X}(\mathcal{C})$ and $\mathcal{Y}(\mathcal{C})$ where $\mathcal{X}(\mathcal{C})$ consists of all X in modΛ such that $\mathrm{Ext}^1_\Lambda(X,\mathcal{C}) = 0$ (where $\mathrm{Ext}^1_\Lambda(X,\mathcal{C}) = 0$ means $\mathrm{Ext}^1_\Lambda(X,C) = 0$ for all C in $\mathcal{C}$), and $\mathcal{Y}(\mathcal{C})$ is the class of all Y in modΛ such that $\mathrm{Ext}^1_\Lambda(\mathcal{C},Y) = 0$. Our aim in this section is to explore some basic results concerning how the homological finiteness properties of $\mathcal{C}$, $\mathcal{X}(\mathcal{C})$ and $\mathcal{Y}(\mathcal{C})$ are related when $\mathcal{C}$ is closed under extensions. These results give useful ways of constructing homologically finite subcategories from others.

The importance of assuming that $\mathcal{C}$ is closed under extensions was first pointed out by Wakamatsu in the following result, which we refer to as Wakamatsu's Lemma. For the convenience of the reader we include a proof.

Lemma 3.1 *Let $\mathcal{C}$ be an additive category closed under extensions.*

a) *Suppose $0 \to Y_M \to C_M \xrightarrow{f} M$ is exact with f a minimal right $\mathcal{C}$-approximation of M. Then Y_M is in $\mathcal{Y}(\mathcal{C})$.*

b) *Suppose $M \xrightarrow{g} C^M \to X^M \to 0$ is exact with g a minimal left $\mathcal{C}$-approximation of M. Then X^M is in $\mathcal{X}(\mathcal{C})$.*

Proof: a). Without loss of generality we can assume that $f : C_M \to M$

is an epimorphism. Suppose we are given an exact sequence $0 \to Y_M \to U \to C \to 0$ with C in $\mathcal{C}$. Then we have the following exact pushout diagram

$$\begin{array}{ccccccccc} & & 0 & & 0 & & & & \\ & & \downarrow & & \downarrow & & & & \\ 0 & \to & Y_M & \to & C_M & \to & M & \to & 0 \\ & & \downarrow & & \downarrow & & \| & & \\ 0 & \to & U & \to & V & \xrightarrow{h} & M & \to & 0 \\ & & \downarrow & & \downarrow & & & & \\ & & C & = & C & & & & \\ & & \downarrow & & \downarrow & & & & \\ & & 0 & & 0. & & & & \end{array}$$

Since $\mathcal{C}$ is closed under extensions, we have that V is in $\mathcal{C}$. Therefore, because $f : C_M \to M$ is a right $\mathcal{C}$-approximation of M, we obtain the following commutative diagram with exact rows

$$\begin{array}{ccccccccc} 0 & \to & Y_M & \to & C_M & \xrightarrow{f} & M & \to & 0 \\ & & \downarrow & & \downarrow & & \| & & \\ 0 & \to & U & \to & V & \to & M & \to & 0 \\ & & \downarrow & & \downarrow & & \| & & \\ 0 & \to & Y_M & \to & C_M & \xrightarrow{f} & M & \to & 0. \end{array}$$

Since f is right minimal, we have that the composition $C_M \to V \to C_M$ is an isomorphism. This implies that the composition $Y_M \to U \to Y_M$ is an isomorphism. Therefore the exact sequence $0 \to Y_M \to U \to C \to 0$ splits, which is our desired result.

b) This is dual of a). □

This result suggests making the following definitions. Let $\mathcal{C}$ be an additive category, not necessarily closed under extensions. Define $\mathcal{K}\mathrm{er}\,\mathcal{C}$ to be the intersection of all the extension closed subcategories of modΛ containing all the modules of the form $\operatorname{Ker} f$ where $f : C \to M$ is a minimal right $\mathcal{C}$-approximation. It is not difficult to see that $\mathcal{K}\mathrm{er}\,\mathcal{C}$ is an extension closed subcategory of modΛ consisting of the modules isomorphic to summands of modules A having finite filtrations $A = A_0 \supset A_1 \supset \cdots \supset A_n = 0$ where each A_i/A_{i+1} is isomorphic to $\operatorname{Ker} f$ for some minimal right $\mathcal{C}$-approximation $f : C \to M$. The extension closed subcategory $\mathcal{C}\mathrm{oker}\,\mathcal{C}$ is defined dually. Since $\mathcal{X}(\mathcal{C})$ and $\mathcal{Y}(\mathcal{C})$ are both extension closed, Wakamatsu's Lemma becomes in this terminology the following. If $\mathcal{C}$ is closed under extensions then $\mathcal{K}\mathrm{er}\,\mathcal{C} \subset \mathcal{Y}(\mathcal{C})$ and $\mathcal{C}\mathrm{oker}\,\mathcal{C} \subset \mathcal{X}(\mathcal{C})$.

We now apply Wakamatsu's Lemma to obtain the following characterization of when contravariantly finite subcategories containing the projective modules are extension closed.

Proposition 3.2 *Suppose $\mathcal{C}$ is a contravariantly finite subcategory of* modΛ *containing the projective modules. Then $\mathcal{C}$ is extension closed if and only if $\mathcal{K}\mathrm{er}\,\mathcal{C}$ is contained in $\mathcal{Y}(\mathcal{C})$.*

Proof: If $\mathcal{C}$ is extension closed, then we know that $\mathcal{K}\mathrm{er}\,\mathcal{C} \subset \mathcal{Y}(\mathcal{C})$ by Wakamatsu's Lemma.

Suppose $\mathcal{K}\mathrm{er}\,\mathcal{C} \subset \mathcal{Y}(\mathcal{C})$ and let $0 \to M \to L \to N \to 0$ be an exact sequence with M and N in $\mathcal{C}$. Then $\operatorname{Ext}^1_\Lambda(M, \mathcal{Y}(\mathcal{C})) = 0 = \operatorname{Ext}^1_\Lambda(N, \mathcal{Y}(\mathcal{C}))$, so that $\operatorname{Ext}^1_\Lambda(L, \mathcal{Y}(\mathcal{C})) = 0$. Consider the exact sequence $0 \to Y_L \to C_L \xrightarrow{f} L \to 0$ where $f : C_L \to L$ is a minimal right $\mathcal{C}$-approximation which is surjective since $\mathcal{C}$ contains the projectives. The sequence splits since $\operatorname{Ext}^1_\Lambda(L, Y_L) = 0$, and hence L is in $\mathcal{C}$. □

It is convenient to make the following definitions before stating our next result. A subcategory of modΛ is said to be preresolving if it is closed under extensions and contains all projective modules in $\mathrm{mod}\Lambda$. A subcategory of $\mathrm{mod}\Lambda$ is said to be precoresolving if it is closed under extensions and contains all injective modules. Clearly $\mathcal{X}(\mathcal{C})$ and $\mathcal{Y}(\mathcal{C})$ are preresolving and precoresolving respectively for all categories $\mathcal{C}$.

Historically, the next result along the lines of interest in this section is the following.

Lemma 3.3 *Suppose $\mathcal{C}$ is an extension closed subcategory of* $\mathrm{mod}\Lambda$.

a) *If $\mathcal{C}$ is covariantly finite, then $\mathcal{X}(\mathcal{C})$ is a contravariantly finite category with $\mathcal{K}\mathrm{er}\,\mathcal{X}(\mathcal{C}) \subset \mathcal{C}$. Further, if $\mathcal{C}$ is precoresolving, then $\mathcal{Y}(\mathcal{X}(\mathcal{C})) = \mathcal{C}$.*

b) *If $\mathcal{C}$ is contravariantly finite, then $\mathcal{Y}(\mathcal{C})$ is covariantly finite with* $\mathrm{Coker}\,\mathcal{Y}(\mathcal{C}) \subset \mathcal{C}$. *Further, if $\mathcal{C}$ is preresolving, then $\mathcal{X}(\mathcal{Y}(\mathcal{C})) = \mathcal{C}$.*

Proof: See [4].

We give an example showing that the hypothesis of being closed under extensions can not be left out in Lemma 3.3.

Example Let Λ be the path algebra of the quiver $\cdot \rightrightarrows \cdot$ over an algebraically closed field k. Let S be a simple regular module and $\mathcal{C} = \mathrm{add}\, S$, which is clearly contravariantly finite. Then $\mathcal{Y}(\mathcal{C}) = \{C; \mathrm{Ext}^1(S, C) = 0\} = \{C; \mathrm{Hom}_\Lambda(C, S) = 0\}$ since $DTrS \simeq S$. Hence $\mathcal{Y}(\mathcal{C})$ is the additive category generated by the preinjective modules and the regular modules which are not in the same component as S. Then it is not hard to see that $\mathcal{Y}(\mathcal{C})$ is not covariantly finite.

As an immediate consequence of Lemma 3.3 we have the following.

Proposition 3.4 *Let $\mathcal{S}$ be the set of contravariantly finite preresolving subcategories of* $\mathrm{mod}\Lambda$ *and let $\mathcal{T}$ be the set of covariantly finite coresolving subcategories of* $\mathrm{mod}\Lambda$. *Then the maps $\mathcal{S} \to \mathcal{T}$ and $\mathcal{T} \to \mathcal{S}$ given by $\mathcal{C} \mapsto \mathcal{Y}(\mathcal{C})$ and $\mathcal{C} \mapsto \mathcal{X}(\mathcal{C})$ respectively are inverse bijections.*

We now point out that the hypothesis that $\mathcal{C}$ is closed under extensions is really necessary for Wakamatsu's Lemma to hold. It is well known and not difficult to see that the finitistic injective dimension of Λ is zero if and only if Λ contains all simple Λ-modules, e.g. Λ is a local ring. Suppose Λ is not selfinjective with finitistic injective dimension of Λ zero. Then we have seen in section 2 that $\operatorname{Sub}\Lambda$ is contravariantly finite in $\operatorname{mod}\Lambda$, and further $\operatorname{Sub}\Lambda$ contains Λ and $\operatorname{Sub}\Lambda \neq \operatorname{mod}\Lambda$. Now obviously $\mathcal{Y}(\operatorname{Sub}\Lambda)$ consists of the modules of injective dimension at most one and hence $\mathcal{Y}(\operatorname{Sub}\Lambda)$ consists of the injective modules in this case. Hence $\mathcal{K}\mathrm{er}\operatorname{Sub}\Lambda$ is not contained in $\mathcal{Y}(\operatorname{Sub}\Lambda)$.

In connection with this discussion, it is of interest to know when $\operatorname{Sub}\Lambda$ is extension closed.

Proposition 3.5 *The following are equivalent for an artin algebra.*

a) $\operatorname{Sub}\Lambda$ *is extension closed.*

b) $\operatorname{Hom}(TrDP, \Lambda) = 0$ *where P is a projective module such that* $\operatorname{add}(P)$ *is a minimal cocover for the category of projective Λ-modules.*

c) $\operatorname{pd}_{\Lambda^{op}} I(\Lambda^{op}) \leq 1$ *where* pd *denotes projective dimension and $I(\Lambda^{op})$ is the Λ^{op}-injective envelope of Λ^{op}.*

Proof: That a) is equivalent to b) follows from [6]. Further $\operatorname{Hom}(TrDP, \Lambda) = 0$ if and only if $\operatorname{pd}_{\Lambda^{op}} DP \leq 1$, that is if and only if $\operatorname{id}_\Lambda P \leq 1$. Now we claim that $\operatorname{add} P = \operatorname{add} Q$ where Q is the projective cover of the injective envelope $I(\Lambda)$ of Λ. Let Q_1 be indecomposable in $\operatorname{add} Q$, and let $i : Q_1 \to P_1$ be a monomorphism with P_1 projective and in $\operatorname{add} P$. Let $f : Q_1 \to I(\Lambda)$ be part of the projective cover. Since $I(\Lambda)$ is injective, there is a map $g : P_1 \to I(\Lambda)$ such that $gi = f$. This shows that i is a split monomorphism, and from this it follows that Q_1 is in $\operatorname{add} P$.

Since we have a commutative diagram

$$\begin{array}{ccc} & Q & \\ \nearrow^{j} & & \searrow \\ \Lambda & \longrightarrow & I(\Lambda), \end{array}$$

where j is a monomorphism, it follows that $\operatorname{add} Q$ is a cocover for the category of projective modules. Hence $\operatorname{add} P \subset \operatorname{add} Q$ because $\operatorname{add} P$ is a minimal cocover.

We now have $\operatorname{id}_\Lambda P = \operatorname{id}_\Lambda Q = \operatorname{pd}_{\Lambda^{op}} I(\Lambda^{op})$, and this finishes the proof. □

We have the following application to a result from [12].

Corollary 3.6 *If* $\operatorname{pd}_\Lambda I(\Lambda) \leq 1$, *then* $\{C; \operatorname{pd} C \leq 1\}$ *is a contravariantly finite subcategory of* $\operatorname{mod}\Lambda$.

Proof: If $\operatorname{pd}_\Lambda I(\Lambda) \leq 1$, the category $\mathcal{X}$ of torsionless Λ^{op}-modules is closed under extensions by Proposition 3.5. Then by Lemma 3.3 $\mathcal{Y}(\mathcal{X}) = \{C; \operatorname{id}_{\Lambda^{op}} C \leq 1\}$ is covariantly finite in $\operatorname{mod}\Lambda^{op}$, and hence $\{C; \operatorname{pd}_\Lambda C \leq 1\}$ is contravariantly finite in $\operatorname{mod}\Lambda$. □

Before giving our next application of these lemmas, it is convenient to make the following observation. For an additive subcategory $\mathcal{C}$ of $\operatorname{mod}\Lambda$, $\operatorname{Rapp}_0 \mathcal{C}$ denotes the category consisting of the modules M having a right $\mathcal{C}$-approximation $f : C_M \to M$ which is an epimorphism and $\operatorname{Lapp}_0 \mathcal{C}$ the category consisting of the modules M having a left $\mathcal{C}$-approximation $g : M \to C^M$ which is a monomorphism.

Lemma 3.7 *Let* $\mathcal{C}$ *be an arbitrary additive category in* $\operatorname{mod}\Lambda$.

a) *$\mathcal{C}$ is contravariantly finite in* modΛ *if and only if $\mathcal{C}$ is contravariantly finite in* Fac$\mathcal{C}$. *Further, if $\mathcal{C}$ is contravariantly finite in* modΛ, *then* Fac$\mathcal{C}$ = Rapp$_0\mathcal{C}$.

b) *$\mathcal{C}$ is covariantly finite in* modΛ *if and only if $\mathcal{C}$ is covariantly finite in* Sub$\mathcal{C}$. *Further, if $\mathcal{C}$ is covariantly finite in* modΛ, *then* Sub$\mathcal{C}$ = Lapp$_0\mathcal{C}$.

Proof: a) We have already remarked in section 2 that Fac$\mathcal{C}$ is always contravariantly finite in modΛ. Therefore it follows that $\mathcal{C}$ is contravariantly finite in modΛ if it is contravariantly finite in Fac$\mathcal{C}$ by the transitivity of contravariantly finite subcategories. Obviously, if $\mathcal{C}$ is contravariantly finite in modΛ it is contravariantly finite in Fac$\mathcal{C}$.

The rest of the proof of a) is a direct consequence of the definitions involved. □

We now give our first application of these lemmas, which gives a way of constructing from an extension closed contravariantly finite subcategory a closely related preresolving, contravariantly finite subcategory.

Proposition 3.8 *Let $\mathcal{C}$ be an extension closed, contravariantly finite subcategory of* modΛ. *Then we have the following.*

a) *$\mathcal{X}(\mathcal{Y}(\mathcal{C}))$ is a contravariantly finite preresolving subcategory of* modΛ *containing $\mathcal{C}$.*

b) *If the exact sequence $0 \to Y_M \to C_M \xrightarrow{f} M \to 0$ is a minimal right $\mathcal{C}$-approximation of M in* Fac$\mathcal{C}$, *then $0 \to Y_M \to C_M \xrightarrow{f} M \to 0$ is also a minimal right $\mathcal{X}(\mathcal{Y}(\mathcal{C}))$-approximation of M.*

c) $\mathcal{K}\mathrm{er}\,\mathcal{X}(\mathcal{Y}(\mathcal{C})) \supset \mathcal{K}\mathrm{er}\,\mathcal{C}$.

Proof: a) Since $\mathcal{C}$ is extension closed and contravariantly finite we have by Lemma 3.3 that $\mathcal{Y}(\mathcal{C})$ is extension closed and covariantly finite in modΛ.

Then $\mathcal{X} = \mathcal{X}(\mathcal{Y}(\mathcal{C}))$ is a contravariantly finite preresolving subcategory of modΛ by Lemma 3.3. It is obvious that $\mathcal{C} \subset \mathcal{X}(\mathcal{Y}(\mathcal{C}))$.

b) By Lemma 3.3, we have that Y_M is in $\mathcal{Y}(\mathcal{C})$. Therefore $\mathrm{Ext}^1_\Lambda(\mathcal{X}, Y_M) = 0$ since $\mathcal{X} = \mathcal{X}(\mathcal{Y}(\mathcal{C}))$, which trivially implies that $0 \to Y_M \to C_M \xrightarrow{f} M \to 0$ is a minimal right $\mathcal{X}$-approximation of M.

c) This is a consequence of b). □

We further have the following result which we refer to as Ringel's Lemma.

Lemma 3.9 *Let $\mathcal{C}$ be a contravariantly finite, preresolving subcategory of* modΛ. *For each extension closed subcategory $\mathcal{Z}$ of* modΛ *satisfying* $\mathcal{K}\mathrm{er}\,\mathcal{C} \subset \mathcal{Z} \subset \mathcal{Y}(\mathcal{C})$ *we have the following:*

a) *$\mathcal{Z}$ is covariantly finite in* modΛ.

b) $\mathcal{X}(\mathcal{Z}) = \mathcal{C}$.

Proof: a) See [4].

b) Since $\mathcal{K}\mathrm{er}\,\mathcal{C} \subset \mathcal{Z} \subset \mathcal{Y}(\mathcal{C})$, we have that $\mathcal{C} \subset \mathcal{X}\mathcal{Y}(\mathcal{C}) \subset \mathcal{X}(\mathcal{Z}) \subset \mathcal{X}(\mathcal{K}\mathrm{er}\,\mathcal{C})$. So we are done if we show that $\mathcal{X}(\mathcal{K}\mathrm{er}\,\mathcal{C}) \subseteq \mathcal{C}$. Let M be in $\mathcal{X}(\mathcal{K}\mathrm{er}\,\mathcal{C})$. Since $\mathcal{C}$ contains the projective modules, $\mathrm{Fac}\,\mathcal{C} = \mathrm{mod}\Lambda$ so that the minimal right approximation sequence $0 \to Y_M \to C_M \to M \to 0$ is exact. But this sequence splits since M is in $\mathcal{X}(\mathcal{K}\mathrm{er}\,\mathcal{C})$. □

Remark The argument Ringel uses in proving part a) of Lemma 3.9 actually proves the following somewhat stronger result, as we now show.

Lemma 3.10 *Let $\mathcal{C}$ be an extension closed subcategory of* modΛ. *Let $\mathcal{Z}$ be an extension closed subcategory of* modΛ *satisfying: a)* $\mathcal{K}\mathrm{er}\,\mathcal{C} \subset \mathcal{Z} \subset \mathcal{Y}(\mathcal{C})$ *and b)* $\mathrm{Fac}\,\mathcal{Z} \subset \mathrm{Rapp}_0\,\mathcal{C}$. *Then $\mathcal{Z}$ is covariantly finite in* modΛ.

Proof: It suffices to show that $\mathcal{Z}$ is covariantly finite in Sub $\mathcal{Z}$. Suppose we have the exact sequence $0 \to Z' \to Z \to Z'' \to 0$ with Z in $\mathcal{Z}$. Since Z'' is in Fac $\mathcal{Z}$ which is contained in $\mathrm{Rapp}_0\,\mathcal{C}$, we have the following exact commutative diagram

$$\begin{array}{ccccccccc}
 & & & & 0 & & 0 & & \\
 & & & & \downarrow & & \downarrow & & \\
 & & & & Y_{Z''} & = & Y_{Z''} & & \\
 & & & & \downarrow & & \downarrow & & \\
0 & \to & Z' & \to & U & \to & C_{Z''} & \to & 0 \\
 & & \| & & \downarrow & & \downarrow & & \\
0 & \to & Z' & \to & Z & \to & Z'' & \to & 0 \\
 & & & & \downarrow & & \downarrow & & \\
 & & & & 0 & & 0 & &
\end{array}$$

with the right hand column a minimal right $\mathcal{C}$-approximation of Z''. Since $\mathcal{Z}$ is extension closed and $\mathcal{Z} \supset \mathcal{K}\mathrm{er}\,\mathcal{C}$, we have that U is in $\mathcal{Z}$. The fact that $\mathcal{Z} \subset \mathcal{Y}(\mathcal{C})$ implies that $0 \to Z' \to U \to C_{Z''} \to 0$ is a left $\mathcal{Z}$-approximation of Z' in $\mathcal{Z}$. This shows that $\mathcal{Z}$ is covariantly finite in Sub $\mathcal{Z}$ and therefore covariantly finite in $\mathrm{mod}\Lambda$. □

As an easy consequence of Wakamatsu's Lemma, Lemma 3.3, and Ringels Lemma we have the following.

Proposition 3.11 *Let $\mathcal{C}$ be a preresolving contravariantly finite subcategory of* $\mathrm{mod}\Lambda$. *An extension closed covariantly finite subcategory $\mathcal{Z}$ of* $\mathrm{mod}\Lambda$ *has the property that $\mathcal{X}(\mathcal{Z}) = \mathcal{C}$ if and only if $\mathcal{K}\mathrm{er}\,\mathcal{C} \subset \mathcal{Z} \subset \mathcal{Y}(\mathcal{C})$.*

The following result is easily deduced from our discussion until now.

Corollary 3.12 *Let $\mathcal{C}$ be an extension closed subcategory of* modΛ.

a) *$\mathcal{C}$ is covariantly finite in* modΛ *if and only if $\mathcal{X}(\mathcal{C})$ is contravariantly finite in* modΛ *with* $\mathcal{K}\mathrm{er}\,\mathcal{X}(\mathcal{C}) \subset \mathcal{C}$.

b) *$\mathcal{C}$ is contravariantly finite in* modΛ *if and only if $\mathcal{Y}(\mathcal{C})$ is covariantly finite and* $\mathrm{Coker}\,\mathcal{Y}(\mathcal{C}) \subset \mathcal{C}$.

4 Contravariantly finite subcategories and tilting theory

In this section we discuss a recently established connection between contravariantly finite subcategories and tilting theory. We also give some illustrative examples after reviewing the general theory [4].

We recall that a subcategory $\mathcal{X}$ of modΛ is resolving if it is preresolving and is closed under kernels of epimorphisms. A subcategory $\mathcal{Y}$ of modΛ is coresolving if it is precoresolving and closed under cokernels of monomorphisms. For a subcategory $\mathcal{Z}$ of modΛ, consider the subcategories $\mathcal{Z}^{\perp} = \{C; \mathrm{Ext}^i(\mathcal{Z}, C) = 0; i > 0\}$ and ${}^{\perp}\mathcal{Z} = \{C; \mathrm{Ext}^i(C, \mathcal{Z}) = 0; i > 0\}$. An important feature of resolving subcategories is that if $\mathcal{Z}$ is resolving, then $\mathcal{Z}^{\perp} = \mathcal{Y}(\mathcal{Z})$, which is coresolving.

An interesting property of resolving subcategories is the following [4].

Theorem 4.1 *Let $\mathcal{X}$ be a contravariantly finite resolving subcategory of* modΛ *and $g_i : X_{S_i} \to S_i$ minimal right $\mathcal{X}$-approximations, where $S_1, \ldots, S_n$ are the simple Λ-modules. Then $\mathcal{X}$ is the category whose objects are summands of modules having filtrations with factors of the form X_{S_j}.*

This result can be used to get some information on the finitistic dimension conjecture [4]. Here we denote $\mathrm{fin.dim.}\,\Lambda = \sup\{\mathrm{pd}\, C; \mathrm{pd}\, C < \infty\}$.

Corollary 4.2 *Assume that* $\mathcal{X} = \{C; \operatorname{pd} C < \infty\}$ *is contravariantly finite in* modΛ, *and let* $h_i : X_{S_i} \to S_i$ *be a minimal right* $\mathcal{X}$*-approximation, where* $S_1, \ldots, S_n$ *are the simple* Λ*-modules. Then* fin.dim. $\Lambda = \max_i \operatorname{pd}(X_{S_i}) < \infty$.

It is however not true in general that $\mathcal{X} = \{C; \operatorname{pd} C < \infty\}$ is contravariantly finite [12], and it would be interesting to know when it is the case.

We now go on to give connections between resolving contravariantly finite subcategories and tilting modules. For an additive subcategory $\mathcal{X}$ of modΛ, let $\hat{\mathcal{X}} = \{C; \exists$ exact sequence $0 \to X_n \to \cdots \to X_1 \to X_0 \to C \to 0$ with X_i in $\mathcal{X}\}$ and $\check{\mathcal{X}} = \{C; \exists$ an exact sequence $0 \to C \to X_0 \to \cdots \to X_n \to 0$ with X_i in $\mathcal{X}\}$. We say that T is a tilting module (in generalized sense) if $\operatorname{Ext}^i_\Lambda(T,T) = 0$ for $i > 0$, $\operatorname{pd}_\Lambda T < \infty$ and Λ is in $\check{\operatorname{add}}\, T$. We say that T is basic if the summands in a direct sum decomposition of T into indecomposable modules are nonisomorphic. Dually T is a cotilting module if $\operatorname{Ext}^i_\Lambda(T,T) = 0$ for $i > 0$, $\operatorname{id} T < \infty$ and $D\Lambda$ is in $\hat{\operatorname{add}}\, T$. The classical case for a tilting module is when $\operatorname{pd}_\Lambda T \leq 1$ and for a cotilting module when $\operatorname{id}_\Lambda T \leq 1$. Note that Λ is always a tilting module and $D\Lambda$ is always a cotilting module. If Λ is basic selfinjective, then $\Lambda \simeq D\Lambda$ is (up to multiplicity of summands) the only tilting module and the only cotilting module. If $\operatorname{id}_\Lambda \Lambda = \infty$, then Λ is a tilting module which is not a cotilting module.

We now give our results concerning cotilting modules and contravariantly finite resolving subcategories [4].

Theorem 4.3 *Let* Λ *be an artin algebra.*

a) *If* T *is a cotilting module, then* ${}^\perp T$ *is a functorially finite resolving subcategory of* modΛ *and* $\hat{\operatorname{add}}\, T = ({}^\perp T)^\perp$ *which is a covariantly finite coresolving subcategory of* modΛ. *Further* $\operatorname{add} T$ *is the Ext-injective modules in* ${}^\perp T$ *and the Ext-projective modules in* $\hat{\operatorname{add}}\, T$.

b) *$T \mapsto {}^{\perp}T$ gives a one-one correspondence between basic cotilting modules and contravariantly finite resolving subcategories $\mathcal{X}$ with $\hat{\mathcal{X}} = \mathrm{mod}\Lambda$.*

c) *$T \mapsto \widehat{\mathrm{add}}\, T$ gives a one-one correspondence between basic cotilting modules and covariantly finite coresolving subcategories $\mathcal{Y}$ with $\mathcal{Y} \subset \{C; \mathrm{id}\, C < \infty\}$.*

Theorem 4.4 *Let Λ be an artin algebra.*

a) *If T is a tilting module, then $T^{\perp}$ is a functorially finite coresolving subcategory of* $\mathrm{mod}\Lambda$ *and $\check{\mathrm{add}}\, T = {}^{\perp}(T^{\perp})$ is contravariantly finite resolving. Further* $\mathrm{add}\, T$ *is the Ext-projective modules in $T^{\perp}$ and the Ext-injective modules in $\check{\mathrm{add}}\, T$.*

b) *$T \mapsto T^{\perp}$ gives a one-one correspondence between basic tilting modules and covariantly finite coresolving subcategories $\mathcal{Y}$ with $\check{\mathcal{Y}} = \mathrm{mod}\Lambda$.*

c) *$T \mapsto \check{\mathrm{add}}\, T$ gives a one-one correspondence between basic tilting modules and contravariantly finite resolving subcategories $\mathcal{X}$ with $\mathcal{X} \subset \{C; \mathrm{pd}\, C < \infty\}$.*

If T is both a tilting and a cotilting module the above theorems associate with T the two contravariantly finite resolving subcategories ${}^{\perp}T$ and $\check{\mathrm{add}}\, T$, which may be different in general. For example if Λ is a selfinjective algebra which is not semisimple, then ${}^{\perp}\Lambda = \mathrm{mod}\Lambda$ and $\check{\mathrm{add}}\, \Lambda = \mathrm{add}\, \Lambda$. Similarly the covariantly finite coresolving subcategories $\Lambda^{\perp} = \mathrm{mod}\Lambda$ and $\widehat{\mathrm{add}}\, \Lambda = \mathrm{add}\, \Lambda$ are different. At the same time this example illustrates that the additional requirements on $\hat{\mathcal{X}}$ and $\check{\mathcal{Y}}$ can not be left out. For $\mathrm{add}\, \Lambda$ is the Ext-injectives both in $\mathrm{mod}\Lambda$ and $\mathrm{add}\, \Lambda$, but $\widehat{\mathrm{mod}}\, \Lambda = \mathrm{mod}\Lambda$, whereas $\widehat{\mathrm{add}}\, \Lambda = \mathrm{add}\, \Lambda$ is different from $\mathrm{mod}\Lambda$.

Even though a selfinjective algebra does not have other tilting and cotilting modules than Λ, the theorem still gives some information in this case: If $\mathcal{X}$

is a contravariantly finite resolving subcategory of modΛ for a selfinjective algebra Λ, with $\hat{\mathcal{X}}=$ modΛ, then $\mathcal{X}$ = modΛ.

Without the assumption $\hat{\mathcal{X}}=$ modΛ we have seen that two different contravariantly finite subcategories have the same associated cotilting module. It can also happen that the Ext-injective modules in a contravariantly finite resolving subcategory do not form a cotilting module.

Example Let Λ be an artin algebra of finite representation type and infinite global dimension. Then clearly $\mathcal{X} = \{C; \operatorname{pd} C < \infty\}$ is contravariantly finite resolving, with $\hat{\mathcal{X}}$ different from modΛ.

If $\operatorname{pd}_\Lambda D\Lambda < \infty$, then add $D\Lambda$ is the Ext-injectives in $\mathcal{X}$, and so $D\Lambda$ is the cotilting module associated both with $\mathcal{X}$ and modΛ.

If $\operatorname{pd}_\Lambda D\Lambda = \infty$, then the Ext-injectives in $\mathcal{X}$ do not form a cotilting module. For then we would have a cotilting module T with $\operatorname{pd}_\Lambda T < \infty$, which is impossible in view of the exact sequence $0 \to T_n \to \cdots \to T_1 \to T_0 \to D\Lambda \to 0$ with the T_i in add T.

In case Λ is of finite global dimension, the extra conditions on $\mathcal{X}$ and $\mathcal{Y}$ can be dropped. In particular we see that the tilting and cotilting modules coincide in this case.

Corollary 4.5 *Assume* gl.dim. $\Lambda < \infty$.

a) $T \mapsto {}^{\perp}T = \check{\text{add}}\, T$ *gives a one-one correspondence between basic cotilting (tilting) modules and contravariantly finite resolving subcategories of* modΛ.

b) $T \mapsto T^{\perp} = \hat{\text{add}}\, T$ *gives a one-one correspondence between basic tilting (cotilting) modules and covariantly finite coresolving subcategories of* modΛ.

If T is a classical cotilting module it is known that ${}^{\perp}T = \operatorname{Sub} T$, and if T is a classical tilting module, that $T^{\perp} = \operatorname{Fac} T$. We have seen in section 2 that for any module C, $\operatorname{Sub} C$ and $\operatorname{Fac} C$ are functorially finite. For hereditary algebras we have only classical tilting or cotilting modules. Here we get the following [4].

Corollary 4.6 *Let Λ be a hereditary algebra.*

a) *$T \mapsto \operatorname{Sub} T$ gives a one-one correspondence between basic cotilting (tilting) modules and contravariantly finite resolving subcategories of* $\operatorname{mod}\Lambda$.

b) *$T \mapsto \operatorname{Fac} T$ gives a one-one correspondence between basic tilting (cotilting) modules and covariantly finite coresolving subcategories of* $\operatorname{mod}\Lambda$.

In view of Theorem 4.3 it is of interest to be able to construct the Ext-projective modules in a covariantly finite coresolving subcategory $\mathcal{Y}$ of $\operatorname{mod}\Lambda$, or the Ext-injective modules in a contravariantly finite resolving subcategory $\mathcal{X}$ of $\operatorname{mod}\Lambda$. For this the following observation is useful.

Proposition 4.7 *Let T be a tilting module and $0 \to \Lambda \xrightarrow{f_0} T_0 \xrightarrow{f_1} T_1 \to \cdots \xrightarrow{f_n} T_n \to 0$ an exact sequence with T_i in* $\operatorname{add} T$, *such that all maps f_i are minimal. Then* $\operatorname{Im} f_i \to T_i$ *is a minimal left $T^{\perp}$-approximation for $i = 0, \ldots, n-1$.*

Proof: Since $\mathcal{X} = \check{\operatorname{add}}\, T$ is resolving, we see that $K_i = \operatorname{Im} f_i$ is in $\mathcal{X}$. Considering $0 \to K_i \xrightarrow{f_i} T_i \to K_{i+1} \to 0$, we see that f_i is a left $T^{\perp}$-approximation since $\operatorname{Ext}^1(K_{i+1}, T^{\perp}) = 0$, using that K_{i+1} is in $\mathcal{X}$. □

We then get the following direct consequence, which is especially useful when $\operatorname{gl.dim.} \Lambda < \infty$.

Corollary 4.8 *Let $\mathcal{Y}$ be a covariantly finite coresolving subcategory of* $\operatorname{mod}\Lambda$ *where the Ext-projectives form a tilting module, and consider the exact sequence $0 \to \Lambda \to T_0 \xrightarrow{f_1} T_1 \to \cdots \xrightarrow{f_n} T_n \to \cdots$ where* $\operatorname{Im} f_i \to T_i$ *is a minimal* *left $\mathcal{Y}$-approximation. Then there is some n such that T_i is zero for $i > n$, and $T = T_0 \amalg \cdots \amalg T_n$ is the tilting module belonging to $\mathcal{Y}$.*

Also in the case of an arbitrary covariantly finite coresolving subcategory $\mathcal{Y}$ we get a similar procedure for constructing Extprojective modules in $\mathcal{Y}$, but now we do not know if we get all.

Proposition 4.9 *Let $\mathcal{Y}$ be covariantly finite coresolving subcategory of* $\operatorname{mod}\Lambda$, *and consider the exact sequence $0 \to \Lambda \xrightarrow{f_0} T_0 \xrightarrow{f_1} T_1 \to \cdots \xrightarrow{f_n} T_n \to \cdots$, where* $\operatorname{Im} f_i \to T_i$ *is a minimal left $\mathcal{Y}$-approximation. Then all T_i are* Ext*projective modules in $\mathcal{Y}$.*

Proof: Write $K_i = \operatorname{Im} f_i$, and consider first the exact sequence $0 \to \Lambda \xrightarrow{f_0} T_0 \to K_1 \to 0$. For Y in $\mathcal{Y}$ we have the exact sequence $0 \to (K_1, Y) \to (T_0, Y) \to (\Lambda, Y) \to \operatorname{Ext}^1(K_1, Y) \to \operatorname{Ext}^1(T_0, Y) \to \operatorname{Ext}^1(\Lambda, Y) = 0$. Since $f_0 : \Lambda \to T_0$ is a minimal left $\mathcal{Y}$-approximation, it follows by Wakamatsu's Lemma that $\operatorname{Ext}^1(K_1, Y) = 0$, so that $\operatorname{Ext}^1(T_0, Y) = 0$ and consequently T_0 is Ext-projective in $\mathcal{Y}$. Continuing the same way we get our desired result. □

In view of Theorem 4.3 it is interesting to have criteria for deciding when a category is resolving or coresolving. We have the following results along these lines.

Proposition 4.10 *Let $\mathcal{X}$ be a contravariantly finite subcategory of* $\operatorname{mod}\Lambda$ *containing Λ. Then $\mathcal{X}$ is resolving if and only if* $\operatorname{Ext}^1(\mathcal{X}, \operatorname{Ker}\mathcal{X}) = 0$ *and $\mathcal{X}$ is closed under syzygies.*

Proof: If $\mathcal{X}$ is resolving, it is clear that $\mathcal{X}$ is closed under syzygies and $\operatorname{Ext}^1(\mathcal{X}, \mathcal{K}\mathrm{er}\,\mathcal{X}) = 0$ by Wakamatsu's Lemma. Assume that $\operatorname{Ext}^1(\mathcal{X}, \mathcal{K}\mathrm{er}\,\mathcal{X}) =$

0 and $\mathcal{X}$ is closed under syzygies. Let $0 \to A \to B \to C \to 0$ be an exact sequence in $\mathrm{mod}\Lambda$. Consider the induced exact sequence $\mathrm{Ext}^1(C, \mathcal{K}\mathrm{er}\,\mathcal{X}) \to \mathrm{Ext}^1(B, \mathcal{K}\mathrm{er}\,\mathcal{X}) \to \mathrm{Ext}^1(A, \mathcal{K}\mathrm{er}\,\mathcal{X}) \to \mathrm{Ext}^2(C, \mathcal{K}\mathrm{er}\,\mathcal{X})$. If A and C are in $\mathcal{X}$, we get $\mathrm{Ext}^1(B, \mathcal{K}\mathrm{er}\,\mathcal{X}) = 0$. Then we consider the exact sequence $0 \to Y_B \to X_B \xrightarrow{f} B \to 0$, where f is a minimal right $\mathcal{X}$-approximation, which is onto since Λ is in $\mathcal{X}$. Since Y_B is by definition in $\mathcal{K}\mathrm{er}\,\mathcal{X}$, the sequence splits, showing that B is in $\mathcal{X}$.

If B and C is in $\mathcal{X}$, we have $\mathrm{Ext}^1(B, \mathcal{K}\mathrm{er}\,\mathcal{X}) = 0$ and also $\mathrm{Ext}^2(C, \mathcal{K}\mathrm{er}\,\mathcal{X}) = 0$ since $\mathcal{X}$ is closed under syzygies. Hence we get $\mathrm{Ext}^1(A, \mathcal{K}\mathrm{er}\,\mathcal{X}) = 0$, so that we can show that A is in $\mathcal{X}$. This finishes the proof that $\mathcal{X}$ is resolving. □

As a consequence we get the following result due to Kleiner.

Corollary 4.11 *Let $i : \Lambda \to \Gamma$ be a ring homomorphism between artin algebras Λ and Γ, and assume that Γ is a finitely generated projective right Λ-module. If the subcategory $\mathcal{X}$ of* $\mathrm{mod}\Gamma$ *of relatively projective modules is closed under extensions, then it is resolving.*

Proof: If $0 \to \Omega C \to P \to C \to 0$ is exact in $\mathrm{mod}\Lambda$, with P projective, then $0 \to \Gamma \otimes_\Lambda \Omega C \to \Gamma \otimes_\Lambda P \to \Gamma \otimes_\Lambda C \to 0$ is exact in $\mathrm{mod}\Gamma$. This shows that $\mathcal{X}$ is closed under syzygies. Clearly Γ is in $\mathcal{X}$, and if $\mathcal{X}$ is closed under extensions, then $\mathrm{Ext}^1(\mathcal{X}, \mathcal{K}\mathrm{er}\,\mathcal{X}) = 0$ by Wakamatsu's Lemma. □

Actually, Kleiners original argument for this result amounts to proving the following more general result. We here say that a contravariantly finite subcategory $\mathcal{X}$ of $\mathrm{mod}\Lambda$ has exact liftings if for each exact sequence $0 \to A \to$

$B \to C \to 0$ in modΛ there is an exact commutative diagram

$$\begin{array}{ccccccccc} 0 & \to & \widetilde{X_A} & \to & \widetilde{X_B} & \to & \widetilde{X_C} & \to & 0 \\ & & \downarrow f & & \downarrow g & & \downarrow h & & \\ 0 & \to & A & \to & B & \to & C & \to & 0 \end{array}$$

where f, g and h are right $\mathcal{X}$-approximations.

Proposition 4.12 *Let $\mathcal{X}$ be a preresolving contravariantly finite subcategory of* modΛ. *Then $\mathcal{X}$ has exact liftings if and only if $\mathcal{X}$ is resolving.*

Proof: Let $\mathcal{X}$ be a preresolving contravariantly finite subcategory of modΛ. If $\mathcal{X}$ is resolving, we know from [4] that $\mathcal{X}$ has exact liftings.

Assume conversely that $\mathcal{X}$ has exact liftings. We want to show that $\mathcal{X}$ is closed under kernels of epimorphisms. So let $0 \to A \to B \to C \to 0$ be an exact sequence in modΛ with B and C in $\mathcal{X}$. Since $\mathcal{X}$ has exact liftings, we have an exact commutative diagram

$$\begin{array}{ccccccccc} & & 0 & & 0 & & 0 & & \\ & & \downarrow & & \downarrow & & \downarrow & & \\ 0 & \to & \widetilde{Y_A} & \xrightarrow{w} & \widetilde{Y_B} & \to & \widetilde{Y_C} & \to & 0 \\ & & \downarrow u & & \downarrow v & & \downarrow & & \\ 0 & \to & \widetilde{X_A} & \xrightarrow{j} & \widetilde{X_B} & \to & \widetilde{X_C} & \to & 0 \\ & & \downarrow f & & \downarrow g & & \downarrow h & & \\ 0 & \to & A & \to & B & \to & C & \to & 0 \\ & & \downarrow & & \downarrow & & \downarrow & & \\ & & 0 & & 0 & & 0 & & \end{array}$$

where f, g and h are right $\mathcal{X}$-approximations. Since B and C are in $\mathcal{X}$, we have that g and h are split epimorphisms. Hence $\widetilde{Y_C}$ is in $\mathcal{X}$ and there is a morphism $s : \widetilde{X_B} \to \widetilde{Y_B}$ such that $sv = \mathrm{id}_{\widetilde{Y_B}}$. Now the exact sequence

$0 \to \widetilde{Y_A} \to \widetilde{X_A} \to A \to 0$ can be written as the direct sum of exact sequences $0 \to Y_A \to X_a \xrightarrow{t} A \to 0$ and $0 \to Z = Z \to 0 \to 0$ where $t : X_A \to A$ is a minimal right $\mathcal{X}$-approximation and Z is in $\mathcal{X}$. Then $\text{Ext}^1(\mathcal{X}, Y_A) = 0$ by Wakamatsu's lemma, and hence $\text{Ext}^1_\Lambda(\widetilde{Y_C}, Y_A) = 0$. Then $w \mid_{Y_A}: Y_A \to \widetilde{Y_B}$ is a split monomorphism, so there is some $t : \widetilde{Y_B} \to Y_A$ with $tw \mid_{Y_A} = \text{id}_{Y_A}$. Then we have $\text{id}_{Y_A} = tsvw \mid_{Y_A} = tsj(u \mid_A)$, so that $u \mid_{Y_A}: Y_A \to X_A$ is a split monomorphism. Then A is a summand of X, so that A is in $\mathcal{X}$. This finishes the proof that $\mathcal{X}$ is resolving. □

We also state the following consequence.

Corollary 4.13 *Let Λ and Γ be artin algebras and (F, G) an adjoint pair of functors $F : \text{mod}\Lambda \to \text{mod}\Gamma$ and $G : \text{mod}\Gamma \to \text{mod}\Lambda$. Assume that F and G are exact functors and that $\mathcal{X} = \text{add}(\text{Im}\, F)$ is preresolving. Then $\mathcal{X}$ is resolving.*

Proof: For C in $\text{mod}\Gamma$, we have seen in section 1 that the natural map $FG(C) \to C$ is a right $\mathcal{X}$-approximation. Since F and G are exact, it then follows that $\mathcal{X}$ has exact liftings. Hence we can apply Proposition 4.12. □

To get situations where the relatively projective modules are resolving, it is in view of Corollary 4.11 important to decide when they are closed under extensions, especially in the case when Γ is a finitely generated projective right Λ-module. In section 2 we have seen that if Γ is symmetric and Λ is selfinjective, it happens only in the trivial cases. However, there is the following interesting sufficient condition by Kleiner, generalizing earlier sufficient conditions by Bautista-Kleiner [7] and Burt-Butler [8] for situations occuring in connection with representations of BOCS's.

Theorem 4.14 *Let $i : \Lambda \to A$ be a morphism of artin algebras over a commutative artin ring R, such that A is projective as a right Λ-module. Let $\mathcal{X}$ be the subcategory of* mod$A$ *consisting of relatively projective A-modules.*

If for each C in $\mathcal{X}$, the kernel $\sum(C)$ of the natural map $A \otimes_\Lambda C \to C$ is an injective Λ-module, then $\mathcal{X}$ is closed under extensions.

In the cases when the relatively projective modules $\mathcal{X}$ is a contravariantly finite resolving subcategory, it is of interest to investigate the connection with tilting theory, in view of Theorems 4.3 and 4.4. The first natural question is when the Ext-injective modules form a cotilting module T. Sufficient conditions for this are given in [8], including that Λ is a Gorenstein ring. It would be interesting to know when $\mathcal{X}$ is one of the two contravariantly finite resolving subcategories associated with the cotilting module T in Theorems 4.3 and 4.4. This is obviously the case when Λ and Γ have finite global dimension.

Another situation where contravariantly finite resolving subcategories have come up is in connection with good modules over quasihereditary algebras, in the work of Ringel in [13], where he gives an application of our correspondence theorems:

Let $P_1, \ldots, P_n$ be an ordered sequence of the indecomposable projective Λ-modules, and let θ_i be the largest factor module of P_i with all composition factors of the form $P_j/\operatorname{rad} P_j$ for $j \leq i$. Consider the category $\mathcal{X}$ whose objects are the Λ-modules having filtrations by the θ_i. If Λ is in $\mathcal{X}$, and the $\operatorname{End}(\theta_i)$ are division rings, Λ is said to be quasihereditary (with the given ordering on the projectives). It is shown in [13] that in this case $\mathcal{X}$ is a resolving subcategory of modΛ, and hence we have an associated (co)tilting module T. In addition the following is shown [13].

Theorem 4.15 *Let Λ be quasihereditary with given ordering on the projectives, and let T be the associated cotilting module. Then $\Gamma = \operatorname{End}(T)^{op}$ is* *again quasihereditary with a natural induced ordering. Denoting by T' the associated cotilting module, we have* $\mathrm{End}_\Gamma(T')^{op} \simeq \Lambda$.

References

[1] M. Auslander and M. Bridger, *Stable module theory*, Mem. of the AMS 94, Providence 1969.

[2] M. Auslander and R.O. Buchweitz, *The homological theory of maximal Cohen-Macaulay approximations*, Soc. Math. de France, Mem. no. 38 (1989) 5-37.

[3] M. Auslander and I. Reiten, *Representation theory of artin algebras III. Almost split sequences*, Comm. Algebra 3 (1975) 239-294.

[4] M. Auslander and I. Reiten, *Applications of contravariantly subcategories*, Adv. in Math. Vol. 86, No. 1 (1991) 111-152.

[5] M. Auslander and S. O. Smalø, *Preprojective modules over artin algebras*, J. Algebra 66 (1980) 61-122.

[6] M. Auslander and S. Smalø, *Almost split sequences in subcategories*, J. Algebra 69 (1981) 426-454; Addendum, J. Algebra 71 (1981) 592-594.

[7] R. Bautista and M. Kleiner, *Almost split sequences for relatively projective modules*, J. Algebra.

[8] W. L. Burt and M. C. R. Butler, *Almost split sequences for Bocses*, preprint 1990.

[9] J. Carlson and D. Happel, *Contravariantly finite subcategories and irreducible maps*, Proc. Amer. Math. Soc.

[10] R. Fossum, Ph. Griffith and I. Reiten, *Trivial extensions of abelian categories*, SLN 456 (1975).

[11] M. Harada and Y. Sai, *On categories of indecomposable modules* I, Osaka J. Math. 7 (1970) 323-344.

[12] K. Igusa, G. Todorov and S. Smalø, *Finite projectivity and contravariant finiteness.* Proc. of AMS, v. 109 no. 4 (1990) 937-941.

[13] C. M. Ringel, *The category of modules with good filtrations over a quasi-hereditary algebra has almost split sequences*, Math. Zeitschrift (to appear).

[14] S. O. Smalø, *Functorial finite subcategories over triangular matrix rings*, Proc. AMS, Vol. 111, No. 3 (1991) 651-656.

[15] T. Wakamatsu, *On constructing stably equivalent functors*, preprint.

Countable Generatedness Version of Rings of Pure Global Dimension Zero

GORO AZUMAYA

Department of Mathematics, Indiana University,
Dedicated to the memory of Professor Robert B. Warfield

INTRODUCTION

In connection with Auslander's theory [2] on rings of finite representation type, the equivalence of the following three conditions for a ring R was established by Zimmermann [15] and Gruson-Jensen [9]: (1) every left R-module is a direct sum of finitely generated submodules, (2) every indecomposable left R-module is finitely presented, (3) for any left R-module M, every pure submodule of M is a direct summand of M. Now, as is easily seen, in terms of pure-projectivity and pure-injectivity the condition (3) can be described either as (4) every left R-module is pure-projective or as (5) every left R-module is pure-injective. Because of this R is called a ring of left pure global dimension zero if it satisfies any of these equivalent conditions. The purpose of the present paper is to show that even if we restrict ourselves to countably generated modules in each of the conditions between (2) and (5) we have equivalent conditions, that is, each of the following conditions characterizes R to be a ring of left pure global dimension zero: (2′) every countably generated indecomposable left R-module is finitely presented, (3′) for any countable generated left R-module M, every countably generated pure submodule of M is a direct summand of M, (4′) every countably generated left R-module is pure-projective, (5′) every countably generated left R-module is pure-injective.

In order to prove the equivalence of these conditions, we need to make use of the proposition that a module M is pure-injective if the countable direct sum $M^{(\mathbf{N})}$ of M is a direct summand of a module L whenever $M^{(\mathbf{N})}$ is a pure submodule of L and $L/M^{(\mathbf{N})}$ is countably generated. This proposition is however obtained by refining a theorem of Zimmermann [16] which characterizes Σ-pure-injective modules. To clarify these situations in full, we make this paper expository. Indeed, the first three sections are devoted to the details of pure-projective, pure-injective and Σ-pure-injective modules, whose theory has chiefly been developed by Warfield [14] and Zimmermann [16], and based on their results our final theorem is proved in the last section. We try to make our paper self-contained, but we have to refer to several results with proof omitted; in particular, a theorem of Fuller [8], which is crucial in the proof of the final theorem, is used without proof. It is to be mentioned that during the preparation of this paper it has been pointed out to the author that Simson proved in [12, Theorem 6.3] that the condition that every countably presented left R-module is pure-projective implies the condition (4), which is certainly stronger than our implication (4′)⇒(4) above.

Throughout this paper, R means a ring with an identity element 1 and R-modules are always unital. Given a family $\{M_i \mid i \in I\}$ of R-modules, where I is an index set, we denote by $\oplus M_i$ the (exterior) direct sum, i.e., the R-module consisting of those vectors $[x_i]$ in the direct product $\prod M_i$ such that the i-th entry $x_i = 0$ for almost all $i \in I$. In case each M_i is a submodule of an R-module M and the sum $\sum M_i$ is direct, i.e., every element of $\sum M_i$ is uniquely expressed as $\sum x_i$ with $x_i \in M$ such that $x_i = 0$ for almost all $i \in I$, we denote it by $\sum \oplus M_i$ and call it the interior direct sum or simply the direct sum of M_i ($i \in I$). The exterior direct sum $\oplus M_i$ and the interior direct sum $\sum \oplus M_i$ are isomorphic canonically. In case modules M_i are all equal to a single module M, we denote by $M^{(I)}$ the corresponding exterior direct sum and by M^I the corresponding direct product. If M is a left R-module, for every vector $[r_i] \in R^{(I)}$ and $[x_i] \in M^I$ we can define $[r_i].[x_i]$ to be $\sum r_i x_i (\in M)$, which is indeed a finite sum. We denote by $\mathbb{N}$ the set $\{1, 2, 3, \ldots\}$ of all natural numbers. Thus $M^{(\mathbb{N})}$ is a countable direct sum of M and $M^{(\mathbb{N})}$ is a countable direct product of M. If n is a natural number, we define M^n to be $M^I (= M^{(I)})$, where $I = \{1, 2, \ldots, n\}$.

1. PURITY AND FINITELY PRESENTED MODULES

Let I and J be sets and let $\Phi = [r_{ij}]$ be a row-finite $I \times J$-matrix over R. Let A and A' be left R-modules and $f : A \to A'$ an epimorphism. We call f a *Φ-pure* epimorphism if for any vector $[x'_j]$ in $(A')^J$ such that $\Phi[x'_j] = 0$ there exists a vector $[x_j]$ in A^J such that $\Phi[x_j] = 0$ and $f(x_j) = x'_j$ for all $j \in J$. Let next B be a submodule of the module A. B is called a *Φ-pure* submodule of A if for any vectors $[x_j]$ in A^J and $[b_i]$ in B^I such that $\Phi[x_j] = [b_i]$ there exists a vector $[y_j]$ in B^J such that $\Phi[y_j] = [b_i]$.

Proposition 1.1. *Let $f : A \to A'$ be an epimorphism with kernel B. Then f is Φ-pure if and only if B is a Φ-pure submodule of A.*

Proof Suppose that f is Φ-pure. Let $[x_j] \in A^J$ and $[b_i] \in B^I$ be such that $\Phi[x_j] = [b_i]$. Then we have $\Phi[f(x_j)] = [f(b_i)] = 0$. Therefore there exists a vector $[z_j] \in A^J$ such that $\Phi[z_j] = 0$ and $f(z_j) = f(x_j)$ for all $j \in J$. If we put $y_j = x_j - z_j$ for each $j \in J$ then we have $f(y_j) = f(x_j) - f(z_j) = 0$ whence $[y_j] \in B$ and $\Phi[y_j] = \Phi[x_j] - \Phi[z_j] = [b_i]$. Thus B is Φ-pure in A.

Suppose conversely that B is a Φ-pure submodule of A. Let $[x'_j] \in (A')^J$ be such that $\Phi[x'_j] = 0$. Since f is an epimorphism, there exists for each

$j \in J$ an $x_j \in A$ such that $f(x_j) = x'_j$. Let $[b_i] = \Phi[x_j]$. Then it follows that $[f(b_j)] = \Phi[f(x_j)] = \Phi[x'_j] = 0$, which implies that $b_i \in B$ for all $i \in I$. Since B is Φ-pure in A, there exists a vector $[y_j] \in B^J$ such that $[b_i] = \Phi[y_j]$. If we put $z_j = x_j - y_j$ for each $j \in J$, we have $f(z_j) = f(x_j) - f(y_j) = f(x_j) = x'_j$. This shows that f is Φ-pure.

Let Φ be a row-finite $I \times J$-matrix over R. For each $i \in I$, denote by g_i the i-th row of Φ. Then each g_i is an element of the free left R-module $F = R^{(J)}$. Let G be the submodule of F generated by all g_i's, i.e., $G = \sum Rg_i$. We denote by $\mathrm{Cok}(\Phi)$ the factor module F/G. Let now M be a left R-module. A row-finite matrix Φ over R is called a *defining matrix* of M if $\mathrm{Cok}(\Phi) \cong M$. Defining matrices of M are not unique but always exist. For, M is a homomorphic image of a free left R-module $F = R^{(J)}$ for some set J. Let G be the kernel of the epimorphism $F \to M$. Then $F/G \cong M$. Let g_i $(i \in I)$ be generators of G, i.e., $G = \sum Rg_i$, and let Φ be the row-finite $I \times J$-matrix whose i-th row is g_i for each $i \in I$. Then $\mathrm{Cok}(\Phi) = F/G$ and so Φ is a defining matrix of M.

Let M, A and A' be left R-modules and $f : A \to A'$ an epimorphism. M is called *f-projective* if $\mathrm{Hom}(M, f) : \mathrm{Hom}_R(M, A) \to \mathrm{Hom}_R(M, A')$ is an epimorphism.

Proposition 1.2. *Let M be the direct sum of left R-modules M_i. Then M is f-projective if and only if M_i is f-projective for all i.*

Proof. Suppose that M_i is f-projective for every i. Let $h : M \to A'$ be a homomorphism. If we denote by $h_i : M_i \to A'$ the restriction of h to M_i, there exists a homomorphism $q_i : M_i \to A$ such that $f \circ q_i = h_i$. The system of homomorphisms $[q_i]$ then defines a homomorphism $q : M \to A$ such that each q_i is the restriction of q to M_i. It is now clear that $f \circ q = h$, which shows that M is f-projective. Suppose conversely that M is f-projective. Consider a direct summand M_i and let $p_i : M \to M_i$ be the projection. Let there be given a homomorphism $h_i : M_i \to A'$. If we put $h = h_i \circ p_i : M \to A'$ then h_i is the restriction of h to M_i and there exists a homomorphism $q : M \to A$ such that $f \circ q = h$. It follows then that if we denote by $q_i : M_i \to A$ the restriction of q to M_i it satisfies $f \circ q_i = h_i$. Thus M_i is f-projective.

Proposition 1.3. *Let M be a left R-module with a defining matrix Φ. Let A and A' be left R-modules and $f : A \to A'$ an epimorphism. Then M is f-projective if and only if f is Φ-pure.*

Proof. We may assume that $M = \mathrm{Cok}(\Phi) = F/G$, where Φ is a row-finite $I \times J$-matrix over R, $F = R^{(J)}$ and G is the submodule of F generated by row vectors g_i of Φ for $i \in I$. Let, for each $j \in J$, e_j be the vector in F such that its j-th entry is 1 and other entries are 0. Then $[e_j]$ forms a free basis of F, and indeed we have $[r_j] = \sum r_j e_j$ for every $[r_j] \in F$. Let $u_j \in M$ be the coset $e_j + G$. Then for any $g = [r_j] \in F$ its coset $g + G$ is $\sum r_j u_j = g.[u_j]$. This shows that u_j's ($j \in J$) generate M and that g is in G if and only if $g.[u_j] = 0$. Now let $q : M \to A$ be a homomorphism and let $x_j = q(u_j)$. Then clearly $g.[x_j] = 0$ for all $g \in G$. So in particular $g_i.[x_j] = 0$ for all $i \in I$, or equivalently, $\Phi[x_j] = 0$. Conversely, suppose that there is given a vector $[x_j] \in A^J$ such that $\Phi[x_j] = 0$, i.e., $g_i.[x_j] = 0$ for all $i \in I$. Since G is generated by g_i's, this implies that $g.[x_j] = 0$ for all $g \in G$. Thus we know that by associating $g + G = g.[u_j]$ ($\in F/G$) with $g.[x_j]$ we have a homomorphism $q : M = F/G \to A$ which therefore satisfies $q(u_j) = x_j$ for all $j \in J$. The same, of course, is true for A'.

Let M be f-projective and let $[x'_j]$ be in $(A')^J$ such that $\Phi[x'_j] = 0$. Then there exists a homomorphism $h : M \to A'$ such that $h(u_j) = x'_j$ for all $j \in J$, as seen above. Since M is f-projective, there is a homomorphism $q : M \to A$ such that $f \circ q = h$. Let $x_j = q(u_j)$ for $j \in J$. Then it follows that $\Phi[x_j] = 0$, as shown above, and moreover $f(x_j) = f(q(u_j)) = h(u_j) = x'_j$ for all $j \in J$. Thus f is Φ-pure. Conversely, let f be Φ-pure and $h : M \to A'$ be a homomorphism. Let $x'_j = h(u_j)$ for $j \in J$. Then $\Phi[x'_j] = 0$, as seen above. Therefore there exists a vector $[x_j] \in A^J$ such that $\Phi[x_j] = 0$ and $f(x_j) = x'_j$ for all $j \in J$. But the first condition implies the existence of a homomorphism $q : M \to A$ such that $q(u_j) = x_j$ for all $j \in J$, as seen above again. It follows that $f(q(u_j)) = h(u_j)$ for all $j \in J$. Since u_j's generate M, this implies that $f \circ q = h$, which shows that M is f-projective.

Corollary 1.4. *Let M be a left R-module with a defining matrix Φ. Let A and A' be left R-modules and $f : A \to A'$ an epimorphism with kernel B. Then the following conditions are equivalent:*

(1) *M is f-projective.*
(2) *f is Φ-pure.*
(3) *B is Φ-pure in A.*

Let M be a left R-module. M is called *finitely presented* if there is a finitely generated free left R-module F and a finitely generated submodule G of F such that $F/G \cong M$. In this case, we may assume that $F = R^n$ for some positive integer n and G is generated by a finite number of vectors $g_1, g_2, \dots, g_m$ in F; let Φ be the $m \times n$-matrix whose i-th row is g_i then $\mathrm{Cok}(\Phi) = F/G \cong M$, and so Φ is a defining matrix of M. Let conversely Φ be a finite, say $m \times n$-, matrix over R and let G be the submodule of R^n generated by the m row vectors of Φ. Then $\mathrm{Cok}(\Phi) = R^n/G$ is finitely presented. Thus we know that a module is finitely presented if and only if it has a finite defining matrix. (Note however that every defining matrix of a finitely presented module is not necessarily a finite matrix.)

Proposition 1.5. *Let M be a finitely generated R-module. Then the following conditions are equivalent:*

(1) *M is finitely presented.*
(2) *M is isomorphic to P/K for a finitely generated projective left R-module P and a finitely generated submodule K of P.*
(3) *If Q is a finitely generated left R-module and $q : Q \to M$ an epimorphism then* $\mathrm{Ker}(Q)$ *is finitely generated.*

Proof. Since every free module is projective, (1)⇒(2) is clear. Assume (2). Then there exists an epimorphism $p : P \to M$ whose kernel is K. Let Q be a finitely generated left R-module and $q : Q \to M$ an epimorphism, and let L be the kernel of q. Since P is projective, there exists a homomorphism $h : P \to Q$ such that $q \circ h = p$, i.e., $q(h(x)) = p(x)$ for all $x \in P$. It follows then that, for an element x of P, $x \in K$ if and only if $h(x) \in L$, and this shows that $h(P) \cap L = h(K)$. Moreover, let y be any element of Q. Since p is an epimorphism, there is an $x \in P$ such that $q(y) = p(x)$. But since $p(x) = q(h(x))$, it follows that $q(y - h(x)) = 0$, i.e., $y - h(x) \in L$, which implies that $Q = h(P) + L$. From these we know that $L/h(K) \cong Q/h(P)$. However both $h(K)$ and $Q/h(P)$ are finitely generated, because they are homomorphic images of the finitely generated modules K and Q respectively. Therefore it follows that L is finitely generated. Thus (2)⇒(3) is proved. (3)⇒(1) is trivial, because every finitely generated module is a homomorphic image of a finitely generated free module.

Proposition 1.6. *Let M be a left R-module and N a sub-module of M.*

(i) *If M is finitely generated and M/N is finitely presented then N is finitely generated.*
(ii) *If M is finitely presented and N is finitely generated then M/N is finitely presented.*
(iii) *If M/N and N are finitely presented then so is M.*

Proof. (i) follows immediately from Proposition 1.5 since N is the kernel of the natural epimorphism $M \to M/N$. Now, since every finitely presented module is finitely generated, we know anyway that M is finitely generated in either case (ii) or (iii). Let F be a finitely generated free left R-module which has an epimorphism $F \to M$. Let G be the kernel and H the inverse image of N. Then we have the canonical isomorphisms $F/G \cong M$, $H/G \cong N$ and $F/H \cong M/N$. Thus, if M is finitely presented then G is finitely generated by (i) and hence if further N whence H/G is finitely generated then H is finitely generated too, which proves (ii). On the other hand, if

M/N is finitely presented then H is finitely generated by (i) and therefore if further N is finitely presented then G is finitely generated again by (i), which proves (iii) this time. We have so far considered left R-modules, but every proposition for left R-modules given above holds for right R-modules in the obvious manner. We should however point out that for a right R-module M its defining matrix Φ is a column-finite matrix, i.e., Φ is a matrix, say of size $I \times J$, over R such that, for each $j \in J$, the j-th column g_j of Φ is in the right free R-module $F = R^{(I)}$ and $\mathrm{Cok}_R(\Phi) = (F/G)_R \cong M_R$. It is to be mentioned that for a finite matrix Φ there correspond both the finitely presented left R-module ${}_R\mathrm{Cok}(\Phi)$ and the finitely presented right R-module $\mathrm{Cok}_R(\Phi)$ so that Φ is a defining matrix of each of them.

Proposition 1.7. *Let M be a finitely presented right R-module and Φ a finite defining matrix of M. Let A be a left R-module and B a submodule of A. Then B is Φ-pure in A if and only if*

$$M \otimes \mathcal{E} : M \otimes_R B \to M \otimes_R A$$

is a monomorphism, where $\mathcal{E}$ is the inclusion map $B \to A$.

Proof. Let $\Phi = [r_{ij}]$ be an $m \times n$-matrix and let, for each $j = 1,2,\ldots,n$, $g_j = [r_{ij} \mid 1 \le i \le m]$ be the j-th column of Φ. Then these vectors are in the finitely generated free right R-module $F = R^m$. Let $G = \sum g_j R$. Then $F/G \cong M$, or what is the same thing, there is an epimorphism $\varphi : F \to M$ whose kernel is G. Let, for each $i = 1,2,\ldots,m$, e_i denote the vector in F whose i-th entry is 1 and other entries are all 0. Then these vectors form a free basis of F, and indeed for any vector $[r_i]$ in F, we have $[r_i] = \sum e_i r_i$. Thus in particular, $g_j = \sum_i e_i r_{ij}$ for $j = 1,2,\ldots,n$. Let next $u_i = \varphi(e_i)$ for $i = 1,2,\ldots,n$. Then clearly $u_i \in M$ and $M = \sum u_i R$. Consider now the tensor product $M \otimes_R A$ with the left R-module A. Then we have $M \otimes_R A = \sum u_i \otimes A$, i.e., every element of $M \otimes_R A$ is expressed as $\sum u_i \otimes a_i$ with $a_i \in A$. Similarly we have $F \otimes_R A = \sum e_i \otimes A$. But since e_i's form a free basis of F, every element of $F \otimes_R A$ is uniquely expressed in the form $\sum e_i \otimes a_i$ with $a_i \in A$, as is well known. Since g_j's generate G, we also know that every element of $G \otimes_R A$ is expressed as $\sum g_j \otimes' a_j$ with $a_j \in A$, where $\otimes'$ denotes the tensor multiplication in $G \otimes_R A$ for convenience. Now the sequence $G \xrightarrow{\eta} F \xrightarrow{\varphi} M \longrightarrow 0$ is exact, where η is the inclusion map. So, by tensoring with A, we have the following exact sequence:

$$G \otimes_R A \xrightarrow{\eta \otimes A} F \otimes_R A \xrightarrow{\varphi \otimes A} M \otimes_R A \longrightarrow 0$$

Let a_i $(i = 1,2,\ldots,n)$ be in A. Then the element $\sum u_i \otimes a_i$ of $M \otimes_R A$ is the image of the element $\sum e_i \otimes a_i$ of $F \otimes_R A$ by $\varphi \otimes A$. Therefore, by the exactness of the above sequence, it follows that $\sum u_i \otimes a_i = 0$ if and only if

$\sum e_i \otimes a_i$ is the image of an element of $G \otimes_R A$. Since however every element of $G \otimes_R A$ is of the form $\sum g_j \otimes' x_j$ with x_j in A and its image by $\eta \otimes A$ is $g_j \otimes x_j = \sum_j(\sum_i e_i r_{ij}) \otimes x_j = \sum_i e_i \otimes (\sum_j r_{ij} x_j)$, we know that $\sum u_i \otimes a_i = 0$ if and only if $a_i = \sum r_{ij} x_j$, i.e., $[a_i] = \Phi[x_j]$ for some x_j $(j = 1,2,\dots,n)$ in A. The same is true for B instead of A, that is, for elements b_i $(i = 1,2,\dots,m)$ of B, $\sum u_i * b_i = 0$ if and only if $[b_i] = \Phi[y_j]$ for some y_j $(j = 1,2,\dots,n)$ in B, where $*$ denotes the tensor multiplication in $M \otimes B$.

Suppose that B is Φ-pure in A. Let $\sum u_i * b_i$ $(b_i \in B)$ be an element of $M \otimes_R B$. Then its image by $M \otimes \mathcal{E}$ is $\sum u_i \otimes b_i \in M \otimes_R A$. Suppose that $\sum u_i * b_i$ is in the kernel of $M \otimes \mathcal{E}$, or equivalently, $\sum u_i \otimes b_i = 0$. Then there exists a vector $[x_j] \in A^n$ such that $\Phi[x_j] = [b_i]$, as seen above. Since B is Φ-pure in A, we can find a vector $[y_j]$ in B^n such that $\Phi[y_j] = [b_i]$. But this implies that $\sum u_i * b_i = 0$, again as seen above. Thus it is shown that $M \otimes \mathcal{E}$ is a monomorphism. Conversely suppose that $M \otimes \mathcal{E}$ is a monomorphism. Let $[x_j] \in A^n$ and $[b_i] \in B^m$ satisfy $\Phi[x_j] = [b_i]$. Then this implies that $\sum u_i \otimes b_i = 0$ in $M \otimes_R A$. Since however $\sum u_i \otimes b_i$ is the image of $\sum u_i * b_i (\in M \otimes_R B)$ by $M \otimes \mathcal{E}$, which is a monomorphism, it follows that $\sum u_i * b_i = 0$. But this is equivalent to the existence of $[y_j]$ in B^n such that $\Phi[y_j] = [b_i]$. Thus we know that B is Φ-pure in A.

Let A be a left R-module and B a submodule of A. B is called a *pure* submodule of A (or *pure* in A) if $M \otimes \mathcal{E} : M \otimes_R B \to M \otimes_R A$ is a monomorphism for every right R-module M, where $\mathcal{E}$ is the inclusion map $B \to A$.

Theorem 1.8. (Cohn [7]) *Let A be a left R-module and B a submodule of A and let $\mathcal{E}$ be the inclusion map $B \to A$. Then the following conditions are equivalent:*

(1) *B is pure in A.*

(2) *$M \otimes \mathcal{E} : M \otimes_R B \to M \otimes_R A$ is a monomorphism for every finitely presented right R-module M.*

(3) *B is Φ-pure in A for every finite matrix Φ over R.*

Proof. Clearly (1) implies (2). The equivalence of (2) and (3) follows immediately from Proposition 1.7. So we need only to prove (3)$\Rightarrow$(1). Let M be any right R-module. Let $\Phi = [r_{ij}]$ be a defining, say $I \times J$-, matrix of M. Then Φ is column-finite, i.e., for each $j \in J$, $r_{ij} = 0$ for almost all $i \in I$. If we denote by g_j the j-th column of Φ for each $j \in J$ then g_j is in the free right R-module $F = R^{(I)}$ and expressed as $g_j = \sum_i e_i r_{ij}$, where e_i is the vector in F whose i-th entry is 1 and other entries are all 0. Let $G = \sum g_j R$. Then G is a submodule of F such that $F/G \cong M$, or equivalently, we have an exact sequence $0 \to G \xrightarrow{\eta} F \xrightarrow{\varphi} M \to 0$, where η is the inclusion map. This sequence yields now the following commutative diagram with exact rows:

$$\begin{array}{ccccccc} G\otimes_R A & \xrightarrow{\eta\otimes A} & F\otimes_R A & \xrightarrow{\varphi\otimes A} & M\otimes_R A & \longrightarrow & 0 \\ \downarrow{\scriptstyle G\otimes\mathcal{E}} & & \downarrow{\scriptstyle F\otimes\mathcal{E}} & & \downarrow{\scriptstyle M\otimes\mathcal{E}} & & \\ G\otimes_R B & \xrightarrow{\eta\otimes B} & F\otimes_R B & \xrightarrow{\varphi\otimes B} & M\otimes_R B & \longrightarrow & 0. \end{array}$$

Let $u_i = \varphi(e_i)$ for each $i \in I$. Then we have $M = \sum u_i R$. Therefore every element of $M \otimes_R A$ is of the form $\sum u_i \otimes a_i$ with $a_i \in A$ such that $a_i = 0$ for almost all $i \in I$. The same is true for $F \otimes_R A$ and $G \otimes_R A$; since, however, e_i's form a free basis of F, every element of $F \otimes_R A$ is uniquely expresed as $\sum e_i \otimes a_i$ with $a_i \in A$. This is also true for the lower exact sequence, and in particular every element of $M \otimes_R B$ is expressed as $\sum u_i * b_i$ with $b_i \in B$ such that $b_i = 0$ for almost all $i \in I$, where $*$ denotes the tensor multiplication in $M \otimes_R B$ or in $F \otimes_R B$. Suppose now that the element $\sum u_i * b_i$ is in the kernel of $M \otimes \mathcal{E}$. Since the image of $\sum u_i * b_i$ by $M * \mathcal{E}$ is $\sum u_i \otimes b_i$, it follows that $\sum u_i \otimes b_i = 0$. Then in exactly the same way in the proof of Proposition 1.7 we can show that there exists an $x_j \in A$ for each $j \in J$ such that $x_j = 0$ for almost all j and $b_i = \sum_j r_{ij} x_j$ for all $i \in I$. Let J_0 be a finite subset of J such that $x_j = 0$ if $j \in J - J_0$, and then choose a suitable finite subset I_0 of I such that $b_i = 0$ and $r_{ij} = 0$ whenever $i \in I - I_0$ and $j \in J_0$. We have then $b_i = \sum_{j \in J_0} r_{ij} x_j$ for all $i \in I_0$. If we denote by Φ_0 the $I_0 \times J_0$-minor matrix of Φ then we know that $[b_i] = \Phi_0[x_j]$, where $[b_i]$ and $[x_j]$ are vectors in B^{I_0} and A^{J_0} respectively obtained by restricting i in I_0 and j in J_0. Since we assume the condition (3), there exists then a vector $[y_j] \in B^J$ such that $[b_i] = \Phi_0[y_j]$, or equivalently, $b_i = \sum_{j \in J_0} r_{ij} y_j$ for all $i \in I_0$. For simplicity, we denote by $\sum_i^0$ and $\sum_j^0$ the summations extended over all i in I_0 and j in J_0. We have then $\sum_i^0 e_i * b_i = \sum_i^0 e_i * \sum_j^0 r_{ij} y_j = \sum_i^0 (\sum_j^0 e_i * r_{ij} y_j) = \sum_j^0 (\sum_i^0 e_i r_{ij}) * y_j$. But since $g_j = \sum_{i \in I} e_i r_{ij} = \sum_i^0 e_i r_{ij}$, it follows that $\sum_i^0 e_i * b_i = \sum_j^0 g_j * y_j$. Applying then the epimorphism $\varphi \otimes B : F \otimes_R B \to M \otimes_R B$, we have $\sum_i^0 u_i * b_i = \sum_j^0 \varphi(g_j) * y_j = 0$, because each g_j is in the kernel G of φ. Since clearly $\sum u_i * b_i = \sum_i^0 u_i * b_i$, we have thus $\sum u_i * b_i = 0$. This shows that the kernel of $M \otimes \mathcal{E}$ is 0, i.e., $M \otimes \mathcal{E}$ is a monomorphism. Thus (3)$\Rightarrow$(1) is proved. If we combine Theorem 1.8 with Corollary 1.4 and the fact that a left R-module is finitely presented if and only if it has a finite defining matrix, we have the following theorem:

Theorem 1.9. **(Warfield [14])** *Let A and A' be left R-modules and $f : A \to A'$ an epimorphism with kernel B. Then the following conditions are equivalent:*

(1) *B is pure in A.*

(2) *Every finitely presented left R-module is f-projective.*

We call an epimorphism $f : A \to A'$ *pure* if either of the equivalent conditions in Theorem 1.9 is satisfied.

Corollary 1.10. *Let A be a left R-module, B a submodule of A and C a submodule of B. Then*

(i) *If B is pure in A then B/C is pure in A/C;*
(ii) *If C is pure in A and B/C is pure in A/C then B is pure in A.*

Proof. Let $f : A \to A/B$ and $g : A \to A/C$ be the natural epimorphisms. Since C is contained in the kernel B of f, f induces an epimorphism $f' : A/C \to A/B$ in the natural manner, whose kernel is B/C. Clearly we have then that $f = f' \circ g$. Suppose that B is pure in A, or equivalently, f is pure. Let M be any finitely presented left R-module and $h : M \to A/B$ a homomorphism. Then, by the preceding theorem, there exists a homomorphism $p : M \to A$ such that $f \circ p = h$. Let $q = g \circ p$. Then q is a homomorphism $M \to A/C$ and satisfies $f' \circ q = f \circ g \circ p = f \circ p = h$, so that M is f'-projective. Thus B/C is pure in A/C again by the preceding theorem, which proves (i). Suppose next that C is pure in A and B/C is pure in A/C, or equivalently, both g and f' are pure. Let M be any finitely presented left R-module, and $h : M \to A/B$ a homomorphism. According to the preceding theorem, the purity of f' implies the existence of a homomorphism $q : M \to A/C$ such that $f' \circ q = h$ and then the purity of g implies the existence of a homomorphism $p : M \to A$ such that $g \circ p = q$. It follows then that $f \circ p = f' \circ g \circ p = f' \circ q = h$, which shows that M is f-projective. Since this is true for all finitely presented left R-modules M, B is pure in A by Theorem 1.9. This proves (ii).

Now let $\mathbf{C}$ be the collection of all left R-modules. Then the relation of being isomorphic is an equivalence relation defined in $\mathbf{C}$, and so $\mathbf{C}$ is divided into equivalent classes with respect to this equivalence relation. Each equivalent class is called a (left R-)*module type*; thus two modules belong to the same module type if and only if they are isomorphic. Let J be a set. A left R-module M is called J-*generated* if there is a vector $[u_j]$ in M^J which generates M, i.e., $M = \sum Ru_j$. This condition is equivalent to the condition that M is a homomorphic image of $R^{(J)}$. In fact, by associating each vector $[r_j] \in R^{(J)}$ with $\sum r_j u_j$ we have an epimorphism $R^{(J)} \to M$. If G is the kernel of this epimorphism M is isomorphic to the factor module $R^{(J)}/G$, i.e., M and $R^{(J)}/G$ belong to the same module type. Thus, if we denote by $\mathbf{C}(J)$ the collection of all module types of J-generated left R-modules, we have a mapping from the set of all submodules G of $R^{(J)}$ onto $\mathbf{C}(J)$ by associating G with the module type of $R^{(J)}/G$ and this shows that $\mathbf{C}(J)$ is a set, as a matter of fact. Let $\mathbb{N} = \{1, 2, \ldots\}$ be the set of all positive integers. Then $\mathbb{N}$-generated modules are usually called *countably generated* modules. Clearly every finitely generated module and hence every finitely presented module is countably generated. Thus all module types of finitely presented modules form a subset of $\mathbf{C}(\mathsf{N})$, which we denote by **FP**.

Proposition 1.11. (Warfield [14]) *Let A be a left R-module. Then there exists a left R-module P which is a direct sum of finitely presented left R-modules and has a pure epimorphism $P \to A$.*

Proof. For each $t \in \mathbf{FP}$, we choose a finitely presented left R-module M_t whose module type is t. Let I be the set of those pairs (t, h), where $t \in \mathbf{FP}$ and $h \in \mathrm{Hom}_R(M_t, A)$. For each element $i = (t, h)$ of I, we define $M_i = M_t$ and $h_i = h$ ($\in \mathrm{Hom}_R(M_i, A)$). Let $P = \sum \oplus M_i$ be the direct sum of M_i for all $i \in I$. Then the family of homomorphisms $h_i : M_i \to A$ for all $i \in I$ defines the homomorphism $p : P \to A$ so that h_i is the restriction of p to M_i. p is however an epimorphism, because if t is the module type of the finitely presented module R then $M_t \cong R$, while for any $a \in A$, a is in the cyclic submodule Ra of A and clearly Ra is a homomorphic image of R whence of M_t. Let t be a module type in **FP** and $h : M_t \to A$ a homomorphism. Then the pair $i = (t, h)$ is in I and so $M_t = M_i$. Thus the identity map of M_t can be regarded as the inclusion map $i_t : M_t \to P$ which clearly satisfies $p \circ i_t = h$. This shows that M_t is p-projective. Since every finitely generated module is isomorphic to M_t for some $t \in \mathbf{FP}$, we know that p is a pure epimorphism.

Let A and A' be left R-modules and $f : A \to A'$ an epimorphism. We call f to *split* if there is a homomorphism $g : A' \to A$ such that $f \circ g$ is the identity map of A'. As is well known, f splits if and only if $\mathrm{Ker}(f)$ is a direct summand of A, or equivalently, there is a submodule of A which is mapped isomorphically onto A' by f. It is also to be mentioned that *the splitness of f is equivalent to the condition that every left R-module is f-projective.* In fact, if A' is f-projective then there exists a homomorphism $g : A' \to A$ such that $f \circ g = 1$, the identity map $A' \to A'$, and conversely if we assume the existence of such a $g : A' \to A$ then for every module M and a homomorphism $h : M \to A'$ the homomorphism $q = g \circ h : M \to A$ satisfies $f \circ q = f \circ g \circ h = 1 \circ h = h$. If we compare these facts with Theorem 1.9, we know that *every direct summand of a module is a pure submodule.*

Now a left R-module M is called *pure-projective* if, for all left R-modules A, A' and all pure epimorphisms $f : A \to A'$, M is f-projective.

Proposition 1.12. *For a left R-module M, the following conditions are equivalent:*

(1) *M is pure-projective.*
(2) *Every pure epimorphism from any left R-module onto M splits.*
(3) *M is isomorphic to a direct summand of a direct sum of finitely presented left R-modules.*

Proof. Suppose that M is pure-projective. Let P be a left R-module and $p : R \to M$ a pure epimorphism. Then it follows that M is p-projective, which implies that p splits. This proves (1)⇒(2). Next assume (2). By Proposition 1.11, there exist a left R-module P which is a direct sum of finitely presented modules and a pure epimorphism $p : P \to M$. Therefore p must split, i.e., there exists a monomorphism $q : M \to P$ such that $p \circ q$ is the identity map of M. As is well known, this implies that P is the direct sum of $\mathrm{Ker}(p)$ and $q(M)$, an isomorphic image of M. Thus we prove (2)⇒(3). By the definition of pure epimorphisms, every finitely presented module is pure-projective. Therefore it follows from Proposition 1.2 that every direct sum of finitely presented modules and more generally every direct summand of such a direct sum is pure-projective. Thus (3)⇒(1) is proved. Finally, we give the following proposition on those modules in which every pure module is a direct summand:

Proposition 1.13.(Stenström [13]) *Let M be a left R-module. Suppose every pure submodule M_0 of M for which M/M_0 is indecomposable is a direct summand of M. Then in fact every pure submodule of M is a direct summand of M and M is a direct sum of indecomposable submodules.*

Proof. First we point out the following fact: Let S be a set. A family of subsets of S is called *directed* if for any members U and V of the family there exists a member W of the family such that $U \subset W$ and $V \subset W$. Let $\mathbf{F}$ be a (non-empty) family of subsets of S, and suppose that for each directed subfamily of $\mathbf{F}$ the union of all members of the subfamily is in $\mathbf{F}$. Then it follows by Zorn's Lemma that $\mathbf{F}$ has a maximal member.

Let N be a proper pure submodule of the left R-module M. Let $x \in M$ be such that $x \notin N$, and consider the family of those submodules of M which (a) contain N, (b) do not contain x and (c) are pure in M. Then, for any subfamily of this family, the union of all its members clearly satisfies (a) and (b); but if the subfamily is directed, the union also satisfies (c) as can be seen by using the equivalence of (1) and (3) in Theorem 1.8. Thus our family has a maximal member, say M_0. M_0 is a pure submodule of M, but we can show that M/M_0 is indecomposable. Suppose that M/M_0 were decomposable, i.e., $M/M_0 = M_1/M_0 \oplus M_2/M_0$ with submodules M_1 and M_2 of M such that $M_0 \subsetneqq M_1$,$M_0 \subsetneqq M_2$ and $M_0 = M_1 \cap M_2$. As direct summands, both M_1/M_0 and M_2/M_0 are then pure in M/M_0. Therefore, by Corollary 1.10, both M_1 and M_2 are pure in M. By the maximality of M_0, it would follow that $x \in M_1$ and $x \in M_2$ and so $x \in M_1 \cap M_2 = M_0$. But this is a contradiction, and thus M/M_0 is indecomposable. Therefore M_0 is a direct summand of M by assumption, i.e., there exists a submodule N' of M such that $M = M_0 \oplus N'$. This implies that $M/M_0 \cong N'$ and so N' is indecomposable. Since $N \subset M_0$

it follows that $N \cap N' = 0$ whence $N + N' = N \oplus N'$. Moreover, since N is pure in M whence in M_0 we know that $N \oplus N'$ is pure in $M_0 \oplus N' = M$.

Now let S be the set of all indecomposable submodules of M. Let U be a subset of S. We denote by $s(U)$ the sum of N and all members of U, i.e., $s(U) = N + \sum_{A \in U} A$. If V is a subset of U then clearly $s(V) \subset s(U)$ and besides if V runs over all finite subsets of U then $s(U)$ is the union of $s(V)$'s. This implies that if $s(V)$ is pure in M for all finite subsets V of U then $s(U)$ is pure in M too. We now call U *N-independent* (in M) if $s(U) = N \oplus (\sum_{A \in U} \oplus A)$ and $s(U)$ is pure in M. If V is a subset of an N-independent subset U then clearly $s(V) = N \oplus (\sum_{A \in V} \oplus A)$ and $s(V)$ is a direct summand of $s(U)$, so that $s(V)$ is also pure in M. Thus we know that a subset U of S is N-independent if and only if every finite subset of U is N-independent. Let $\mathbf{F}$ be the family of all N-independent subsets of M. Let $\mathbf{D}$ be any directed subfamily of $\mathbf{F}$, and let U be the union of all members of $\mathbf{D}$. Let U_0 be a finite subset of U. Since $\mathbf{D}$ is directed, there exists a member V of $\mathbf{D}$ such that $U_0 \subset V$. Since V is N-independent it follows that U_0 is also N-independent. Since this is true for all finite subsets U_0 of U, we know that U is N-independent, i.e., U is in $\mathbf{F}$. This implies that $\mathbf{F}$ has a maximal member, say W. Suppose now $s(W) = N \oplus \left(\sum_{A \in W} \oplus A\right) \neq M$. Since then $s(W)$ is a proper pure submodule of M, there exists an indecomposable submodule B of M such that $s(W) \cap B = 0$ and $s(W) + B = s(W) \oplus B = N \oplus \left(\sum_{A \in W} \oplus A\right) \oplus B$ is pure in M, as seen above. This shows that if we add B to the set W we have an N-independent subset $W \cup \{B\}$ of M, which properly contains W. But this contradicts the maximality of W in $\mathbf{F}$. Thus we know that $s(W) = M$. This shows that N is a direct summand of M and M/N is a direct sum of indecomposable submodules. If we take $N = 0$ then the latter fact means that M itself is a direct sum of indecomposable submodules. This completes the proof of our proposition.

2. PURE-INJECTIVE MODULES AND ALGEBRAIC COMPACTNESS

Let M be a left R-module. M is called *pure-injective* if, for every left R-module A and every pure submodule B of A, every homomorphism $B \to M$ can be extended to a homomorphism $A \to M$.

Proposition 2.1. *Let M be the direct product of left R-modules M_i's. Then M is pure-injective if and only if every M_i is pure-injective.*

Proof. Let A be a left R-module and B a pure submodule of A. Let, for each i, $p_i : M \to M_i$ be the projection and $e_i : M_i \to M$ be the canonical embedding. Then $p_i \circ e_i = 1_i$, the identity map of M_i. Suppose now that each M_i is pure-injective. Let $h : B \to M$ be a homomorphism. Then $p_i \circ h$ is a homomorphism $B \to M_i$, so it can be extended to a homomorphism $f_i : A \to M_i$. Since M is a direct product of M_i's the system of homomorphisms $[f_i]$ defines a homomorphism $f : A \to M$ such that $p_i \circ f = f_i$ for every i. Thus, for each $b \in B$ we have $p_i(f(b)) = f_i(b) = p_i(h(b))$ for every i, and therefore $f(b) = h(b)$. This shows that f is an extension of h, and so M is pure-injective. Suppose conversely that M is pure-injective. Choose any M_i and let $h : B \to M_i$ be a homomorphism. Then $e_i \circ h$ is a homomorphism $B \to M$ and so can be extended to a homomorphism $g : A \to M$. Let $f = p_i \circ g : A \to M_i$. Then clearly f is an extension of $p_i \circ e_i \circ h = 1_i \circ h = h$. Thus M_i is pure-injective.

Let $\mathbf{T}$ be the factor group $\mathbb{Q}/\mathbb{Z}$ of the additive group $\mathbb{Q}$ of rational numbers modulo its subgroup $\mathbb{Z}$ of integers. For any additive Abelian group A, we define $A^* = \mathrm{Hom}(A, \mathbf{T})$ and call it the *dual* group of A. Let B be another Abelian group and $\varphi : A \to B$ a homomorphism. Then we define its dual $\varphi^* = \mathrm{Hom}(\varphi, \mathbf{T})$, i.e., φ^* is the homomorphism $B^* \to A^*$ such that $\varphi^*(\beta) = \beta \circ \varphi$ for all $\beta \in B^*$. It is clear that if φ is an epimorphism then φ^* is a monomorphism. Since $\mathbf{T}$ is divisible and hence injective as a $\mathbb{Z}$-module, we also know that if φ is a monomorphism then φ^* is an epimorphism. Moreover, for any positive integer n, $\mathbf{T}$ has a cyclic subgroup of order n, which implies that $\mathbf{T}$ contains an isomorphic image of every simple $\mathbb{Z}$-module and thus $\mathbf{T}$ is an injective cogenerator. It follows then that if $A \neq 0$ then $A^* \neq 0$, or more precisely, if $a \in A$ and $a \neq 0$ then there exists an $\alpha \in A^*$ such that $\alpha(a) \neq 0$. From this we deduce that a homomorphism $\varphi : A \to B$ is an epimorphism or monomorphism whenever $\varphi^* : B^* \to A^*$ is a monomorphism or epimorphism respectively. For the proof, suppose φ^* is a monomorphism. Let C be the cokernel of φ and $\nu : B \to C$ be the natural epimorphism. Then $\nu \circ \varphi = 0$ and ν^* is a monomorphism. Therefore we have $\varphi^*(\nu^*(\gamma)) = (\gamma \circ \nu) \circ \varphi = \gamma \circ (\nu \circ \varphi) = 0$ for every $\gamma \in C^*$. Since however both φ^* and ν^* are monomorphisms, it follows that $\gamma = 0$ for all $\gamma \in C^*$,i.e., $C^* = 0$. This implies that $C = 0$, or equivalently, φ is an epimorphism. Next assume that φ^* is an epimorphism. Suppose that φ were not a monomorphism, i.e., the kernel of ϕ were non-zero. Take an $a \neq 0$ from the kernel. Then there would be an $\alpha \in A^*$ such that $\alpha(a) \neq 0$. By assumption, there exists a $\beta \in B^*$ such that $\alpha = \varphi^*(\beta) = \beta \circ \varphi$. We have then that $\alpha(a) = \beta(\varphi(a)) = 0$, a contradiction. Thus φ is a monomorphism. Finally, for any element a of A, we denote by $\bar{a}$ the mapping which associates each $\alpha \in A^*$ with $\alpha(a)$. Then clearly $\bar{a}$ is in $A^{**} = \mathrm{Hom}(A^*, \mathbf{T})$, the double dual of A, and moreover by associating a with $\bar{a}$ we have a monomorphism $\lambda_A : A \to A^{**}$, which we call the $*$-*canonical embedding* of A. Thus if we identify a with $\lambda_A(a) = \bar{a}$, A is regarded as a subgroup of A^{**}. symbol has been changed from A

We now consider the case where A is a left (or right) R-module. Then A^*

can be converted into a right (or left) R-module by setting αr (or $r\alpha$), for $\alpha \in A^*$ and $r \in R$, to be the mapping $a \mapsto \alpha(ra)$ (or $a \mapsto \alpha(ar)$) for $a \in A$. If B is another left (or right) R-module, and $\varphi : A \to B$ is a left (or right) R-homomorphism then it is easy to see that $\varphi^* : B^* \to A^*$ is a right (or left) R-homomorphism. Moreover, if A is a left R-module then A^{**} is also a left R-module and the $*$-canonical embedding $\lambda_A : A \to A^{**}$ becomes a left R-monomorphism, as we can easily check that for any $r \in R$ and $a \in A$ both $\lambda_A(ra)$ and $r\lambda_A(a)$ coincide with the mapping $\alpha \mapsto \alpha(ra)$ for $\alpha \in A^*$; of course, the same is true for every right R-module.

Proposition 2.2. *Let A be a left R-module, B is submodule of A and $\mathcal{E}$: $B \to A$ the inclusion map. Then the following conditions are equivalent:*

(1) *B is pure in A.*
(2) *The epimorphism $\mathcal{E}^* : A^* \to B^*$ splits.*
(3) *$B^* \otimes \mathcal{E} : B^* \otimes_R B \to B^* \otimes_R A$ is a monomorphism.*
(4) $\mathrm{Hom}(\mathcal{E}, M^*) : \mathrm{Hom}_R(A, M^*) \to \mathrm{Hom}_R(B, M^*)$ *is an epimorphism for all right R-modules M.*

Proof. Let M be a right R-module. Then we have a well-known natural isomorphism $\sigma(A) : \mathrm{Hom}_R(M, A^*) \to (M \otimes_R A)^*$, and thus we have the following commutative diagram:

$$\begin{array}{ccc} \mathrm{Hom}_R(M, A^*) & \xrightarrow{\sigma(A)} & (M \otimes_R A)^* \\ \downarrow{\scriptstyle \mathrm{Hom}(M,\mathcal{E}^*)} & & \downarrow{\scriptstyle (M\otimes\mathcal{E})^*} \\ \mathrm{Hom}_R(M, B^*) & \xrightarrow{\sigma(B)} & (M \otimes_R B)^* \end{array}$$

Assume (1), i.e., $M \otimes \mathcal{E} : M \otimes_R B \to M \otimes_R A$ is a monomorphism for all right R-modules M. Then in particular $B^* \otimes \mathcal{E} : B^* \otimes_R B \to B^* \otimes_R A$ is a monomorphism, i.e., we have (3). Assume next (3). Then it follows that $(B^* \otimes \mathcal{E})^* : (B^* \otimes_R A)^* \to (B^* \otimes_R B)^*$ is an epimorphism. By the commutativity of the above diagram for $M = B^*$, we know that $\mathrm{Hom}(B^*, \mathcal{E}^*)$: $\mathrm{Hom}_R(B^*, A^*) \to \mathrm{Hom}_R(B^*, B^*)$ is an epimorphism and so in particular the identity map 1 of B^* is the image of an element η of $\mathrm{Hom}_R(B^*, A^*)$, which means that $\mathcal{E}^* \circ \eta = 1$. Thus we have the condition (2). Assume now (2), i.e., the existence of a homomorphism $\eta : B^* \to A^*$ such that $\mathcal{E}^* \circ \eta = 1$. Then, for every right R-module M and any homomorphism $h : M \to B^*$, we have that $\mathcal{E}^* \circ \eta \circ h = h$, which means that $\mathrm{Hom}(M, \mathcal{E}^*) \circ \mathrm{Hom}(M, \eta) = 1$, the identity map of $\mathrm{Hom}_R(M, B^*)$ and so in particular we know that $\mathrm{Hom}(M, \mathcal{E}^*)$: $\mathrm{Hom}_R(M, A^*) \to \mathrm{Hom}_R(M, B^*)$ is an epimorphism. Again by the commutativity of the above diagram, $(M \otimes \mathcal{E})^* : (M \otimes_R A)^* \to (M \otimes_R B)^*$ is an epimorphism and hence its dual $M \otimes \mathcal{E} : M \otimes_R B \to M \otimes_R A$ is a monomorphism for all right R-modules M. Thus we have (1).

Finally by considering the situation $({}_RA, M_R, \mathrm{T})$ instead of $(M_R, {}_RA, \mathrm{T})$, we have another natural isomorphism $\tau(A) : \mathrm{Hom}_R(A, M^*) \to (M \otimes_R A)^*$, and this yields the following commutative diagram:

$$\begin{array}{ccc} \mathrm{Hom}_R(A, M^*) & \xrightarrow{\tau(A)} & (M \otimes_R A)^* \\ \downarrow{\scriptstyle \mathrm{Hom}(\mathcal{E}, M^*)} & & \downarrow{\scriptstyle (M \otimes \mathcal{E})^*} \\ \mathrm{Hom}_R(B, M^*) & \xrightarrow{\tau(B)} & (M \otimes_R B)^* \end{array}$$

Now the condition that $M \otimes \mathcal{E}$ is a monomorphism is equivalent to the condition that $(M \otimes \mathcal{E})^*$ is an epimorphism, and in view of the above commutative diagram this is equivalent to the condition that $\mathrm{Hom}(\mathcal{E}, M^*)$ is an epimorphism. This shows the equivalence of (1) and (4).

Corollary 2.3. *For every right (or left) R-module M, the left (or right) R-module M^* is pure-injective.*

Proof. Let M be a right R-module. Let A be a left R-module and B be a pure submodule of A. Then by the preceding proposition we see that $\mathrm{Hom}(\mathcal{E}, M^*) : \mathrm{Hom}_R(A, M^*) \to \mathrm{Hom}_R(B, M^*)$ is an epimorphism, where $\mathcal{E}$ is the inclusion map $B \to A$. But this means that every homomorphism $B \to M^*$ can be extended to a homomorphism $A \to M^*$. Thus M^* is pure-injective.

Proposition 2.4. *Let M be a left R-module. Then M is pure in M^{**} when M is regarded as a submodule of M^{**} in the natural manner.*

Proof. Let $\lambda : M \to M^{**}$ and $\mu : M^* \to M^{***}$ be the $*$-canonical embeddings, and let $\lambda^* : M^{***} \to M^*$ be the dual of λ. Take any χ from M^* and any x from M, and put $\omega = \mu(x)$ and $\delta = \lambda(x)$ for convenience. Since δ is in M^{**} and χ is in M^*, we have that $\omega(\delta) = \delta(\chi)$ and $\delta(\chi) = \chi(x)$. Thus we have that $\omega\big(\lambda(x)\big) = \omega(\delta) = \delta(\chi) = \chi(x)$. But since this is true for all $x \in M$ and $\omega = \mu(x)$ is independent of $x \in M$, we know that $\mu(\chi) \circ \lambda = \omega \circ \lambda = \chi$. On the other hand, since $\mu(x)$ is in $M^{***} = \mathrm{Hom}(M^{**}, \mathbf{T})$, we have that $\lambda^*\big(\mu(\chi)\big) = \mu(\chi) \circ \lambda$ and therefore is $= \chi$. This is true for all $\chi \in M^*$ and so we have that $\chi^* \circ \mu = 1$, the identity map of M, which implies that the epimorphism λ^* splits. Thus it follows from Proposition 2.2 that M is pure in M^{**}, since to regard M as a submodule of M^{**} means to identify λ with the inclusion map $M \to M^{**}$.

Proposition 2.5. *For a left R-module M, the following conditions are equivalent:*

(1) *M is pure-injective.*

(2) *If M is a pure submodule of a left R-module Q, M is a direct summand of Q.*

(3) *M is a direct summand of M^{**}.*

Proof. If M is pure-injective and is pure in Q then the identity map of M can be extended to a homomorphism $Q \longrightarrow M$ which means that M is a direct summand of Q. This proves (1)⇒(2). Since M is pure in M^{**} by the preceding proposition, (2)⇒(3) is clear. (3)⇒(1) follows from the fact that $M^{**} = (M^*)^*$ is pure-injective by Corollary 2.3 and every direct summand of a pure-injective module is pure-injective by Proposition 2.1.

Proposition 2.6 *Let M be a left R-module and N a submodule of M. Then N is pure in M if and only if the $*$-canonical embedding $\lambda : N \longrightarrow N^{**}$ can be extended to a homomorphism $M \longrightarrow N^{**}$.*

Proof. If N is pure in M then, since N^{**} is pure-injective by Corollary 2.3, λ can be extended to a homomorphism $M \longrightarrow N^{**}$. On the other hand, suppose that λ is extended to a homomorphism $\mu : M \longrightarrow N$. Let Φ be a finite, say $m \times n$-matrix over R. Let $[x_j] \in M^n$ and $[v_i] \in N^m$ satisfy $\Phi[x_j] = [v_i]$. By operating μ, we have then $\Phi[\mu(x_j)] = [\lambda(v_i)]$. Since $\lambda(N)$ is a pure submodule of N^{**} by Proposition 2.4, there exists $[y_j] \in N^n$ such that $\Phi[\lambda(y_j)] = [\lambda(v_i)]$ by Theorem 1.8. Since λ is a monomorphism, it follows that $\Phi[y_j] = [v_i]$. This is true for all finite matrices Φ over R, so that N is pure in M, again by Theorem 1.8.

Let $\Phi = [r_{ij}]$ be a row-finite $I \times J$-matrix over R, and let $g_i = [r_{ij} \mid j \in J]$ be the i-th row of Φ for every $i \in I$. Let M be a left R-module. A system of linear equations in M, $\sum_{j \in J} r_{ij}\xi_j = v_i \ i \in I$, where ξ_j's for $j \in J$ are indeterminates and v_i's for $i \in I$ are elements of M, is called *finitely soluble* if for any finite subset I_0 of I there exists a vector $[x_j] \in M^J$ such that $(g_i.[x_j] =) \sum_{j \in J} r_{ij}x_j = v_i$ for all $i \in I_0$. We now call M *Φ-compact* if every above type of system of linear equations has a whole solution in M, i.e., a vector $[x_j]$ in M^J which satisfies $\sum_{j \in J} r_{ij}x_j = v_i$ for all $i \in I$, or equivalently, $\Phi[x_j] = [v_i]$ whenever it is finitely solvable (for v_i). Next let A be a left R-module and B a submodule of A. A homomorphism $h : B \longrightarrow M$ is called *locally extendable* to A in M if for each finitely generated submodule B_0 of B the restriction of h to B_0 can be extended to a homomorphism $A \longrightarrow M$.

Proposition 2.7. *Let Φ be a row-finite $I \times J$-matrix over R, and let g_i be the i-th row of Φ for every $i \in I$. Let $F = R^{(J)}$ and let G be the submodule of (the free left R-module) F generated by all g_i's. Then a left R-module M is Φ-compact if and only if every homomorphism $G \longrightarrow M$ which is locally extendable to F in M can be extended to a homomorphism $F \longrightarrow M$.*

Proof. Let $\Phi = [r_{ij}]$. Then $g_i = [r_{ij} \mid j \in J]$ for each $i \in I$. Suppose that M is a Φ-compact left R-module. Let $h : G \to M$ be a homomorphism locally extendable to F in M. Let $v_i = h(g_i)$ for each $i \in I$. Then the system of linear equations $\sum_{j\in J} r_{ij}\xi_j = v_i$ $(i \in I)$ is finitely solvable. For, let I_0 be any finite subset of I. Then the restriction of h to the finitely generated submodule $\sum_{i\in I_0} Rg_i$ of G can be extended to a homomorphism $f_0 : F \to M$. If, as usual, we denote by e_j the vector in F whose i-th entry is 1 and other entries are 0 for each $j \in J$ and put $x_j^0 = f_0(e_j) \in M$ for each $j \in J$ then we have $\sum_j r_{ij}x_j^0 = \sum_j r_{ij}f_0(e_j) = f_0(\sum_j r_{ij}e_j) = f_0(g_i) = h(g_i) = v_i$ for all $i \in I_0$. Since M is Φ-compact, there exists a vector $[x_j]$ in M^J such that $\Phi[x_j] = [v_i]$. Let $f : F \to M$ be the homomorphism defined by associating each $[r_j] \in F = R^{(J)}$ with $\sum r_j x_j$. Then $f(g_i) = \sum_j r_{ij}x_j = v_i = h(g_i)$ for all $i \in I$.

Suppose conversely that every homomorphism $G \to M$ which is locally extendable to F in M can be extended to a homomorphism $F \to M$. Let $[v_i] \in M^I$ be such that the system of linear equations $\Phi[\xi_j] = [v_i]$ $(i \in I)$ is finitely solvable. Let I_0 be any finite subset of I. Then there exists a vector $[x_j^0] \in M^J$ such that $\sum_j r_{ij}x_j^0 = v_i$ for all $i \in I_0$. Now by associating each vector $[r_j] \in F = R^{(J)}$ with $\sum r_j x_j^0$ we have a homomorphism $f_0 : F \to M$, which therefore satisfies $f_0(g_i) = \sum_j r_{ij}x_i^0 = v_i$ for all $i \in I_0$. On the other hand, suppose that $\sum_{i\in I} r_i g_i = 0$ for some $[r_i] \in R^{(I)}$. Since r_i=0 for almost all i, there is a finite subset, say I_0 of I such that $r_i = 0$ if $i \in I - I_0$. Then there corresponds a homomorphism $f_0 : F \to M$ such that $f_0(g_i) = v_i$ for all $i \in I_0$, as seen just above. It follows then that $\sum_{i\in I} r_i v_i = \sum_{i\in I_0} r_i v_i = \sum_{i\in I_0} r_i f_0(g_i) = f_0(\sum_{i\in I_0} r_i g_i) = 0$. This shows that there is a well defined homomorphism $h : G = \sum Rg_i \to M$ such that for every finite subset I_0 of I there is a homomorphism $f_0 : F \to M$ satisfiying $f_0(g_i) = h(g_i)$ for all $i \in I_0$; if we observe the fact that every finitely generated submodule of G is contained in the submodule of G generated by a suitable finite number of g_i's we know that h is locally extendable to F in M. Therefore, by assumption, h can be extended to a homomorphism $f : F \to M$. Let now $x_j = f(e_j)$ for each $j \in J$. Then $x_j \in M$ and $v_i = h(g_i) = f(g_i) = f(\sum_j r_{ij}e_j) = \sum_j r_{ij}f(e_j) = \sum_j r_{ij}x_j$ for all $i \in I$, which shows that x_j satisfies $\Phi[x_j] = [v_i]$.

Now a left R-module M is called *algebraically compact* if M is Φ-compact for every row-finite matrix Φ over R.

Proposition 2.8. *Let M be a right R-module. Then its dual $M^* =$ Hom$(M, \mathbf{T})$ is an algebraically compact left R-module.*

Proof. Let A be a left R-module. Then we have a natural isomorphism $\tau(A)$: $\mathrm{Hom}_R(A, M^*) = \mathrm{Hom}_R(A, \mathrm{Hom}(M, \mathbf{T})) \to (M \otimes_R A) = \mathrm{Hom}(M \otimes_R A, \mathbf{T})$, as considered in the proof of Proposition 2.2; indeed, the isomorphism is such that if $f \in \mathrm{Hom}_R(A, M^*)$ corresponds to $\varphi = \tau(A)f \in (M \otimes_R A)^*$ then they satisfy $\big(f(a)\big)(x) = \varphi(x \otimes a)$ for all $a \in A$ and $x \in M$. Let B be a submodule of A and $\mathcal{E} : B \to A$ the inclusion map. Then we have the following commutative diagram:

$$\begin{array}{ccc} \mathrm{Hom}_R(A, M^*) & \xrightarrow{\tau(A)} & (M \otimes_R A)^* \\ \big\downarrow {\scriptstyle \mathrm{Hom}(\mathcal{E}, M^*)} & & \big\downarrow {\scriptstyle (M \otimes \mathcal{E})^*} \\ \mathrm{Hom}_R(B, M^*) & \xrightarrow{\tau(B)} & (M \otimes_R B)^* \end{array}$$

Let now $h \in \mathrm{Hom}_R(B, M^*)$ be locally extendable to A in M^*. Take an element $\sum x_i * b_i$ of $M \otimes_R B$ which is in the kernel of $M \otimes \mathcal{E}$, i.e., $\sum x_i \otimes b_i = 0$ in $M \otimes_R A$, where $x_1, x_2, \ldots, x_n$ are in M, $b_1, b_2, \ldots, b_n$ are in B, and $*$ denotes the tensor multiplication in $M \otimes_R B$. Then the restriction of h to the finitely generated submodule $Rb_1 + Rb_2 + \ldots + Rb_n$ can be extended to a homomorphism $f_0 : A \to M$. Let $\psi = \tau(B)h \in (M \otimes_R B)^*$ and $\varphi_0 = \tau(A)f_0 \in (M \otimes_R A)^*$. We have then $\psi(\sum x_i * b_i) = \sum \big(h(b_i)\big)(x_i) = \sum \big(f_0(b_i)\big)(x_i) = \varphi_0(\sum x_i \otimes b_i) = 0$. This shows that the kernel K of $M \otimes \mathcal{E} : M \otimes_R B \to M \otimes_R A$ is contained in the kernel of $\psi : M \otimes_R B \to \mathbf{T}$. Thus ψ induces naturally a homomorphism $\bar{\psi} : (M \otimes_R B)/K \to \mathbf{T}$. Since however $(M \otimes_R B)/K$ can be identified with a submodule of $M \otimes_R A$ in the natural manner and $\mathbf{T}$ is an injective $\mathbb{Z}$-module, $\bar{\psi}$ can be extended to a homomorphism $\varphi : M \otimes_R A \to \mathbf{T}$. It is then clear that $\psi = (M \otimes \mathcal{E})^* \varphi$. Let f be the element of $\mathrm{Hom}_R(A, M)$ which corresponds to $\varphi \in (M \otimes_R A)^*$ by the isomorphism $\tau(A)$, i.e., $\tau(A)f = \varphi$. Then, by the commutativity of the above diagram, we have that $\mathrm{Hom}(\mathcal{E}, M^*)f = h$, which means that f is an extension of h to A. Thus it follows from Proposition 2.6 that M is Φ-compact for all row-finite matrices Φ over R, i.e., M is algebraically compact.

Now we prove the following theorem of Warfield which characterizes pure-injective modules:

Theorem 2.9. *For a left R-module M, the following conditions are equivalent:*

(1) *M is pure-injective.*
(2) *M is algebraically compact.*

Proof. Assume (1). Then M is a direct summand of M^{**} by Proposition 2.5. Let Φ be a row-finite $I \times J$-matrix over R and $[v_i]$ a vector in M^I for which the system of linear equations $\Phi[\xi_j] = [v_i]$ is finitely solvable in M. Then, since M^{**} is Φ-compact by the preceding proposition, the equations have a solution in M^{**}, i.e., $\Phi[x_j] = [v_i]$ for some $[x_j] \in (M^{**})^J$. Let $p : M^{**} \to M$ be the projection. Then, since $p(v_i) = v_i$ for all $i \in I$, by operating p we have $\Phi[p(x_j)] = [v_i]$. This shows that M is Φ-compact. Since this is true for all row-finite matrices Φ, we have (2). Next assume (2). Let $\Phi = [r_{ij}]$ be again a row-finite $I \times J$-matrix over R, and let $[x_j]$ and $[v_i]$ be vectors in $(M^{**})^J$ and M^I respectively satisfying $\Phi[x_j] = [v_i]$. Let I_0 be any finite subset of I. Then, since Φ is row-finite, there exists a finite subset J_0 of J such that $r_{ij} = 0$ if $i \in I_0$ and $j \in J - J_0$. We have then that $\sum_{j \in J_0} r_{ij}x_j = v_i$ for all $i \in I_0$. Since however M is pure in M^{**} by Proposition 2.4, there exists a $y_j^0 \in M$ for each $j \in J_0$ such that $\sum_{j \in J_0} r_{ij}y_j^0 = v_i$ for all $i \in I_0$ due to Cohn's theorem (Theorem 1.8). This shows that the system of linear equations $\Phi[\xi_j] = [v_i]$ is finitely solvable in M. Since M is Φ-compact, the system is solvable in M, i.e., there exists a vector $[y_j] \in M^J$ such that $\Phi[y_j] = [v_i]$. Thus we know that M is Φ=pure in M^{**} for all row-finite matrices Φ over R. If we take Φ as a defining matrix of the factor module M^{**}/M and let f be the natural epimorphism $M^{**} \to M^{**}/M$ then it follows from Proposition 1.3 that M^{**}/M is f-projective, which implies that f splits, or equivalently, M is a direct summand of M^{**}. Since M^{**} is pure-injective by Corollary 2.3 its direct summand M is also pure-injective by Proposition 2.1. Thus (2)⇒(1) is proved.

Remark 1. Theorem 2.9 was established in Warfield [14], where he proved it by using the theory of compact topological Abelian groups. Our proof presented above is however by a method not using the topological concept, as given in [4].

Remark 2. The following two significant works are pointed out by the referee:

1. R. Kielpinski, *On Γ-pure injective modules*, [19].
2. M. Prest, *Model Theory and Modules*, [20]. The first one is a paper dealing with a notion of purity with respect to a certain family of submodules of a fixed free module, which amounts to generalizing the known theory of purity. In particular, it appears that another proof of the Warfield theorem (Theorem 2.9) is given there. The second one is a book centering around a detailed theory on pure-injective modules by a model theoretic approach. Moreover finite matrix subgroups, which are to be a subject of the next section 3, are discussed in terms of p.p. types and in connection with pure-injective modules.

3. FINITE MATRIX SUBGROUPS AND Σ-PURE INJECTIVITY.

We shall denote by R-Mod the collection of all left R-modules. A mapping $U : R\text{-Mod} \to \mathbf{Z}\text{-Mod}$ is called an *inner functor* of R-Mod if it satisfies the conditions that (i) $U(M) \subset M$ for all M in R-Mod, and (ii) if M, M' are in

R-Mod and $f : M \to M'$ is a homomorphism then $f(U(M)) \subset U(M')$. Let U be an inner functor of R-Mod. Let M be a left R-module. If N is a submodule of M then by taking f to be the inclusion map $N \to M$ we know that $U(N) = f(U(N)) \subset U(M)$. On the other hand, if f is an endomorphism of M, i.e., if $f : M \to M$ then $f(U(M)) \subset U(M)$; thus if S is the endomorphism ring of M then $U(M)$ is a submodule of the S-module M. Moreover, we can show that U commutes with direct sums, i.e., if $M = \oplus M_i$ with each M_i in R-Mod then it follows that $U(M) = \oplus U(M_i)$. For, if $p_i : M \to M_i$ is the projection for each i then we have that $p_i(U(M)) \subset U(M_i)$ for all i and so $U(M) \subset \oplus U(M_i)$, while if $e_i : M_i \to M$ is the canonical embedding for each i then we have that $e_i(U(M_i)) \subset U(M)$ for all i and this implies that $\oplus U(M_i) = \sum e_i(U(M_i)) \subset U(M)$. Let U and V be inner functors of R-Mod. We define the relation $U \subset V$ to be $U(M) \subset V(M)$ for all $M \in R$-Mod. Then clearly the relation is a partial order. On the other hand, if we associate each $M \in R$-Mod with the sum $U(M) + V(M)$ then we have clearly an inner functor, which we denote by $U + V$. Similarly, by associating each $M \in R$-Mod with the intersection $U(M) \cap V(M)$ we have another inner functor $U \cap V$. The concepts of sum and intersection of inner functors can be generalized for more than two (and even infinite numbers of) inner functors.

Let Φ be a row-finite $I \times J$-matrix over R. Let j_0 be an index in J. Let M be a left R-module. If we associate each vector $[x_j] \in M^J$ with its j_0-th entry x_{j_0}, we have a homomorphism $p_0 : M^J \to M$. Now a vector $[x_j] \in M^J$ is called a solution vector of Φ in M if $\Phi[x_j] = 0$. Then the set of those solution vectors forms a subgroup $S_M(\Phi)$ of M^J, and therefore its homomorphic image $p_0(S_M(\Phi))$ is a subgroup of M. If M' is another left R-module and $f : M \to M'$ a homomorphism then for any solution vector $[x_j]$ of Φ in M the image $[f(x_j)]$ is clearly a solution vector of Φ in M' and this implies that $f(p_0(S_M(\Phi))) \subset p_0(S_{M'}(\Phi))$. Thus we know that by associating each $M \in R$-Mod with $p_0(S_M(\Phi))$ we have an inner functor U of R-Mod. This functor U is usually denoted by $[\Phi; j_0]$ and called a *matrix functor* of R-Mod; moreover the subgroup $U(M)$ is usually called a *matrix subgroup* of M. Let next j_1 be another index in J. Then the group functor $[\Phi; j_1]$ is well defined. But we point out that if Ψ is the row-finite $I \times J$-matrix which is obtained from Φ by interchanging its j_0-th and j_1-columns then clearly $[\Psi; j_0] = [\Phi; j_1]$.

Let now Φ be a finite, say $m \times n$-matrix over R. Then, for each number j_0 such that $1 \leq j_0 \leq n$, the *finite matrix functor* $[\Phi; j_0]$ is defined. But if we interchange the first and the j_0-th column of Φ we have an $m \times n$-matrix Ψ such that $[\Psi; 1] = [\Phi; j_0]$. This shows that every finite matrix functor of R-Mod is of the form $[\Phi; 1]$ with a finite matrix Φ, which therefore we had better denote by $[\Phi]$. Thus x is in $[\Phi](M)$ if and only if there exist x_1, $x_2, \ldots x_n$ in M such that $x_1 = x$ and $\Phi[x_j] = 0$, where n is the number of columns of Φ. The subgroup of M of the form $[\Phi](M)$ is called a *finite matrix subgroup* of M.

Proposition 3.1. *The intersection of a finite number of finite matrix functors of R-Mod is a finite matrix functor too.*

Proof. We have only to prove the proposition for two finite matrix functors. Let namely Φ_1 and Φ_2 be finite matrices over R, whose sizes are $m_1 \times n_1$ and $m_2 \times n_2$ respectively. Let C be the first column of Φ_2 and let $\Phi_2 = [C \quad \Phi_2']$ with the $m_2 \times (n_2 - 1)$- matrix Φ_2'. Consider then the following $(m_1 + m_2) \times (n_1 + n_2 - 1)$- matrix:

$$\Phi = \begin{bmatrix} \Phi_1 & 0 \\ C\,0 & \Phi_2 \end{bmatrix}.$$

Let $X_1 = \begin{bmatrix} x_1 \\ x_2 \\ \vdots \\ x_{n_1} \end{bmatrix}$, $X_2 = \begin{bmatrix} x_{n_1+1} \\ x_{n_2+2} \\ \vdots \\ x_{n_1+n_2-1} \end{bmatrix}$ and $X = \begin{bmatrix} X_1 \\ X_2 \end{bmatrix} = \begin{bmatrix} x_1 \\ x_2 \\ \vdots \\ x_{n_1+n_2-1} \end{bmatrix}$ with elements $x_1, x_2, \ldots, x_{n_1+n_2-1}$ of a module $M \in R$-Mod. Then we have that $\Phi X = \begin{bmatrix} \Phi_1 X_1 \\ \Phi_2 X_2 \end{bmatrix}$ and therefore $\Phi X = 0$ if and only if $\Phi_1 X_1 = 0$ and $\Phi_2 X_2 = 0$. This shows that $x_1 \in [\Phi](M)$ if and only if $x_1 \in [\Phi_1](M)$ and $x_1 \in [\Phi_2](M)$, i.e., $[\Phi](M) = [\Phi_1](M) \cap [\Phi_2](M)$.

Now Zimmermann [16] gave another characterization of pure-injective modules in terms of finite matrix subgroups:

Theorem 3.2. *For a left R-module M, the following conditions are equivalent:*

(1) *M is pure-injective.*
(2) *M is algebraically compact.*
(3) *Every family of cosets modulo finite matrix subgroups of M with the finite intersection property has a non-empty intersection.*

Proof. Since the equivalence of (1) and (2) was proved in Theorem 2.9, we need to show that (2) and (3) are equivalent. Assume now (2). Let $\{U_\alpha \mid \alpha \in \Lambda\}$ be a set of finite matrix functors of R-Mod and $\{u_\alpha \mid \alpha \in \Lambda\}$ a subset of the module M such that the corresponding family of cosets $\{U_\alpha(M) + u_\alpha \mid \alpha \in \Lambda\}$ has the finite intersection property, i.e., for every (non-empty) finite subset Λ_0 of Λ the intersection $\bigcap_{\alpha \in \Lambda_0} (U_\alpha(M) + u_\alpha)$ is non-empty. Let, for each $\alpha \in \Lambda$, Φ_α be a finite matrix over R such that $[\Phi_\alpha] = U_\alpha$ and let Φ_α be an $m_\alpha \times n_\alpha$-matrix. Let, for each $\alpha \in \Lambda$, C_α be the first column of Φ_α and Φ_α' be the $m_\alpha \times (n_\alpha - 1)$-matrix such that $\Phi_\alpha = [C_\alpha \quad \Phi_\alpha']$; let further $r_{\alpha,i}$ be the i-th entry of C_α (i.e., the $(i,1)$-entry of Φ_α) for $1 \le i \le m_\alpha$. Let

I be the set of those pairs (α, i) for which $\alpha \in \Lambda$ and $1 \leq i \leq m_\alpha$, and let, for each $\alpha \in \Lambda$, I_α be the set of those pairs (α, i) with $1 \leq i \leq m_\alpha$. Then each I_α is a subset of I, and I is the disjoing union of these I_α's. Let next J' be the set of those pairs (α, j) for which $\alpha \in \Lambda$ and $2 \leq j \leq n_\alpha$, and J'_α be the set of those pairs (α, j) with $2 \leq j \leq n_\alpha$ for each $\alpha \in \Lambda$. Then J' is the disjoint union of these J'_α's. We next suppose the symbol 1 is not a member of J' and let J be the disjoint union of J' and $\{1\}$. We now consider the $I \times J$-matrix Φ whose $((\alpha, i), (\beta, j))$-entry is defined to be the (i, j)-entry of Φ_α or 0 according as $\alpha = \beta$ or $\alpha \neq \beta$ and whose $((\alpha, i), 1)$-entry is defined to be $r_{\alpha,i}$. Thus the $I_\alpha \times J'_\beta$-minor matrix of Φ is equal to Φ'_α or 0 according as $\alpha = \beta$ or $\alpha \neq \beta$ and the $I_\alpha \times 1$-minor matrix of Φ is C_α, and in particular Φ is row-finite. Next we let $v_{\alpha,i} = r_{\alpha,i} u_\alpha$ for each $(\alpha, i) \in I$; thus we have a vector $[v_{\alpha,i}]$ in M^I. Now let Λ_0 be a finite subset of Λ. Then by the finite intersection property there exists an $x \in M$ such that x is in the coset $U_\alpha(M) + u_\alpha$, or equivalently, $x - u_\alpha$ is in $U_\alpha(M)$ for every $\alpha \in \Lambda_0$. Since $[\Phi_\alpha] = U_\alpha$, the last condition implies that for each $\alpha \in \Lambda_0$ we can find $x_{\alpha,2}, \ldots, x_{\alpha,n}$ in M such that

$$\Phi_\alpha \begin{bmatrix} x - u_\alpha \\ x_{\alpha,2} \\ \vdots \\ x_{\alpha,n_\alpha} \end{bmatrix} = 0,$$

or equivalently,

$$\Phi_\alpha \begin{bmatrix} x \\ x_{\alpha,2} \\ \vdots \\ x_{\alpha,n_\alpha} \end{bmatrix} = \begin{bmatrix} v_{\alpha,1} \\ v_{\alpha,2} \\ \vdots \\ v_{\alpha,n_\alpha} \end{bmatrix}.$$

Define a vector X_0 in M^J such that its first entry is x and (α, j)-th entry, for $2 \leq j \leq n_\alpha$, is $x_{\alpha,j}$ or 0 according as $\alpha \in \Lambda_0$ or not. If we denote by $\Phi_{\alpha,i}$, for $\alpha \in \Lambda$ and $1 \leq i \leq m_\alpha$, the (α, i)-th row of Φ then we have that $\Phi_{\alpha,i} X_0 = v_{\alpha,i}$ whenever α is in Λ_0 and $1 \leq i \leq m_\alpha$. If we take into consideration the fact that every finite subset of I is contained in the union $\bigcup_{\alpha \in \lambda_0} I_\alpha$ for a suitable finite subset Λ_0 of Λ, we conclude that the system of linear equations $\Phi \cdot Y = [v_{\alpha,i}]$, with $J \times 1$-matrix Y whose entries are indeterminates, is finitely solvable in M. Since however M is algebraically compact by assumption, the system is solvable in M, i.e., there exists a vector X in M^J such that $\Phi \cdot X = [v_{\alpha,i}]$. This implies that

$$C_\alpha x_1 + \Phi'_\alpha \begin{bmatrix} x_{\alpha,2} \\ \vdots \\ x_{\alpha,n_\alpha} \end{bmatrix} = \begin{bmatrix} v_{\alpha,1} \\ v_{\alpha,2} \\ \vdots \\ v_{\alpha,m_\alpha} \end{bmatrix} = C_\alpha u_\alpha$$

and so

$$\Phi \cdot \begin{bmatrix} x_1 - u_\alpha \\ x_{\alpha,2} \\ \vdots \\ x_{\alpha,n_\alpha} \end{bmatrix} = C_\alpha(x_1 - u_\alpha) + \Phi'_\alpha \cdot \begin{bmatrix} x_{\alpha,2} \\ \vdots \\ x_{\alpha,n_\alpha} \end{bmatrix} = 0$$

for all $\alpha \in \Lambda$, where x_1 is the first entry of X and $x_{\alpha,j}$ $(1 \le j \le n_\alpha)$ is the (α, j)-entry of X. From this we know that $x_1 - u_\alpha$ is in $U_\alpha(M)$, or equivalently, x_1 is in the coset $U_\alpha(M) + u_\alpha$ for all $\alpha \in \Lambda$, and thus the intersection of all the cosets $U_\alpha(M) + u_\alpha$ $(\alpha \in \Lambda)$ is non-empty. This shows (2)⇒(3).

Assume conversely the condition (3). Let $\Phi = [r_{ij}]$ be a row-finite $I \times J$-matrix over R and v_i, $(i \in I)$, elements of M for which the system of linear equations $\Phi[\xi_j] = [v_i]$ with indeterminates ξ_j $(j \in J)$ is finitely solvable in M. Then we first prove that *if j_0 is any index in J and if Φ_0 is the $I \times (J - \{j_0\})$-matrix obtained from Φ by removing its j_0-th column then there exists an $x_0 \in M$ such that the system of linear equations $\Phi_0[\xi_j] = [v_1 - r_{ij_0}x_0]$ is finitely solvable in M.* For the proof, let $\{I_\alpha \mid \alpha \in \Lambda\}$ be the family of all finite subsets of I. Let, for each $\alpha \in \Lambda$, Φ_α be the $I_\alpha \times J$-matrix of Φ and let $U_\alpha = [\Phi_\alpha; j_0]$ be the matrix functor of R-Mod. Since Φ is row-finite, there exists a finte subset J_α of J which contains j_0 and such that $r_{ij} = 0$ if $i \in I_\alpha$ and $j \in J - J_\alpha$. Let Φ^α be the $I_\alpha \times J_\alpha$-minor matrix of Φ. Let N be a left R-module, and let $Y = [y_j \mid j \in J]$ be a vector in N^J and $Y_\alpha = [y_j \mid j \in J_\alpha]$ its restriction to J_α. Then clearly we have that $\Phi_\alpha Y = \Phi^\alpha Y_\alpha$, and this implies that $U_\alpha(N) = [\Phi^\alpha; j_0](N)$. Thus we know that $U_\alpha = [\Phi^\alpha; j_0]$ is a finite matrix functor. We also point out that if I_α and I_β are finite subsets of I such that $I_\alpha \subset I_\beta$ then it follows that $U_\alpha \supset U_\beta$. Now that $\Phi[\xi_j] = [v_i]$ is finitely solvable in M implies that the finite system of linear equations $\sum_j r_{ij}\xi_j = v_i$ $(i \in I_\alpha)$ is solvable in M for every $\alpha \in \Lambda$. We choose a solution vector $[x_i^\alpha] \in M^J$ for the system for each $\alpha \in \Lambda$. Let I_α and I_β be finite subsets of I such that $I_\alpha \subset I_\beta$. Then we have $\sum_j r_{ij}x_j^\alpha = v_i$ $(i \in I_\alpha)$ and $\sum_j r_{ij}x_j^\beta = v_i$ $(i \in I_\beta)$. From this follows that $\sum_j r_{ij}(x_j^\beta - x_j^\alpha) = 0$ $(i \in I_\alpha)$ and so $x_{j_0}^\beta - x_{j_0}^\alpha \in U_\alpha(M)$, or equivalently $x_{j_0}^\beta \in U_\alpha(M) + x_{j_0}^\alpha$; but since $U_\alpha \subset U_\beta$ it follows that $U_\beta(M) + x_{j_0}^\beta \subset U_\alpha(M) + x_{j_0}^\alpha$. This shows that the family $\left\{U_\alpha(M) + x_{j_0}^\alpha \mid \alpha \in \Lambda\right\}$ of cosets modulo $U_\alpha(M)$'s has the finite intersection property. Thus by assumption there exists an x_0 which is in the intersection of all cosets $U_\alpha(M) + x_{j_0}^\alpha$ $(\alpha \in \Lambda)$. Take any α from Λ. Since x_0 is in $U_\alpha(M) + x_{j_0}^\alpha$, $x_0 - x_{j_0}^\alpha$ is in $U_\alpha(M)$ and this means the existence of a vector $[z_j] \in M^J$ such that $\sum_j r_{ij}z_j = 0$ $(i \in I_\alpha)$ and $z_{j_0} = x_0 - x_{j_0}^\alpha$, i.e., $z_{j_0} + x_{j_0}^\alpha = x_0$. It follows now that

$$\sum_j r_{ij}(z_j + x_j^\alpha) = \sum_j r_{ij}z_j + \sum_j r_{ij}x_j^\alpha = v_i$$

and hence

$$\sum_{j \neq j_0} r_{ij}(z_j + x_j^\alpha) = v_i - r_{ij_0}x_0 \quad (i \in I_\alpha).$$

Thus the system of linear equations $\Phi_0[\xi_j] = [v_i - r_{ij_0}x_0]$ is finitely solvable in M.

Having proved the above proposition, we shall now complete the proof for (3)$\Rightarrow$(2), that is, from the assumptions of (3) and that $\Phi = [r_{ij}]$ is a row-finite $I \times J$-matrix over R and $[v_i]$ is a vector in M^I for which the system of linear equations $\sum_{j\in J} r_{ij}\xi_j = v_i$ $(i \in I)$ is finitely solvable in M we shall derive that the system of equations has a solution in M.

First we consider the case where J consists only of a single index. In this case our system of linear equations is of the from $R_i\xi = v_i$ with $r_i \in R$ and $v_i \in M$ $(i \in I)$. Let U_i be the finite matrix functor $[(r_i)]$ defined by the 1×1-matrix (r_i) for each $i \in I$. Then clearly $x \in U_i(M)$ if and only if $x \in M$ and $r_ix = 0$. Since the system of equations is finitely solvable, there exists an $x_i^0 \in M$ such that $r_ix_i^0 = v_i$ for each $i \in I$. It is then clear that an $x \in M$ satisfies $r_ix = v_i$ if and only if it is in $U_i(M) + x_i^0$ for each $i \in I$. Since the system of equations is finitely solvable, for every finite subset I_0 of I we can find an $x_0 \in M$ such that $r_ix_0 = v_i$ for all $i \in I_0$, which is equivalent to the condition that x_0 is in the intersection of $U_i(M) + x_i^0$ for all $i \in I_0$. Thus the family $\{U_i(M) + x_i^0 \mid i \in I\}$ of cosets has the finite intersection property. Therefore, by assumption (3), there exists an $x \in M$ which is in $U_i(M) + x_i^0$, or equivalently, satisfies $r_ix = v_i$ for all $i \in I$. This shows that our system of equations has a solution x in M.

Thus we may and shall assume henceforth that J contains at least two indices. Let S be the set of those pairs (K, Y) where K is a proper subset of J and $Y = [y_k]$ is a vector in M^K such that the system of linear equations $\sum_{j\in J-k} r_{ij}\xi_j = v_i - \sum_{k\in K} r_{ik}y_k$ $(i \in I)$ is finitely solvable in M. Let (K, Y) and (L, Z) be pairs in S. We define the relation $(K, Y) \leq (L, Z)$ by the condition that $K \subset L$ and $y_k = z_k$ for all $k \in K$ where $Z = [z_l] \in M^L$. Then $\leq$ is a partial order in S. Let $\{K_\nu, Y_\nu)\}$ be an ascending chain in S with respect to the partial order. (S is non-empty because of the proposition proved above.) Let $K = \bigcup K_\nu$, and let $Y = [y_k] \in M^K$ be a vector such that its restriction $[y_k \mid k \in K_\nu]$ to K_ν coincides with Y_ν for all ν; since $\{(K_\nu, Y_\nu)\}$ is an ascending chain such a vector Y is well-defined.

Now suppose $K = J$. Let h be any index from I. Since $\Phi = [r_{ij}]$ is row-finite, there exists a finite subset $J(h)$ of J such that $r_{hj} = 0$ if $j \in J - J(h)$. Since $J = K$ is the union of the ascending chain of K_ν's, $J(h) \subset K_\mu$ for some μ. The pair (K_μ, Y_μ) is in S and so K_μ is a proper subset of J and the system of linear equations

$$\sum_{j\in J-k_\mu} r_{ij}\xi_j = v_i - \sum_{k\in K_\mu} r_{ik}y_k \quad (i \in I)$$

is finitely solvable in M. Therefore the single equation

$$\sum_{j\in J-k_\mu} r_{hj}\xi_j = v_h - \sum_{k\in K_\mu} r_{hk}y_k$$

is solvable in M. Since however $j \in J - K_\mu$ implies that $j \in J - J(h)$ and so $r_{hj} = 0$, it follows that

$$0 = v_h - \sum_{k \in K_\mu} r_{hk} y_k = v_h - \sum_{j \in J} r_{hj} y_j,$$

i.e.,

$$\sum_{j \in J} r_{hj} y_j = v_h.$$

Since this is true for every h in I, this shows that the vector $[y_j] \in M^J$ is a solution of the system of linear equations $\Phi[\xi_j] = [v_i]$. Therefore we need only consider the case where $K = \bigcup K_\nu$ is a proper subset of J for every ascending chain $\{(K_\nu, Y_\nu)\}$ in S. Now let I_0 be a finite subset of I. Since $[r_{ij}]$ is a row-finite matrix, there exists a finite subset j_0 of J such that $r_{ij} = 0$ whenever $i \in I_0$ and $j \in J - J_0$. Let $\{(K_\nu, Y_\nu)\}$ be an ascending chain in S and let $K = \bigcup K_\nu$. Then the finite subset $K \cap J_0$ of K is contained in K_μ for some μ. It follows that $r_{ik} = 0$ if $i \in I_0$ and $k \in K - K_\mu$. Let $Y = [y_k] \in M^K$ be the vector defined above, so that Y_ν is the restriction of Y to K_ν for every ν. Since K is the disjoint union of $K - K_\mu$ and K_μ, we have that $\sum_{k \in K} r_{ik} y_k = \sum_{k \in K_\mu} r_{ik} y_k$ for $i \in I_0$. On the other hand, since (K_μ, F_μ) is in S, K_μ is a proper subset of J and the system of linear equations

$$\sum_{j \in J - K_\mu} r_{ij} \xi_j = v_i - \sum_{k \in K_\mu} r_{ik} y_k \quad (i \in I)$$

is finitely solvable in M, and therefore we can find a vector $[x_j^0]$ in M^{J-K} such that

$$\sum_{j \in J - K_\mu} r_{ij} x_j^0 = v_i - \sum_{k \in K_\mu} r_{ik} y_k$$

for all $i \in I_0$. Since however $J - K_\mu$ is the disjoint union of $J - K$ and $K - K_\mu$, we have that

$$\sum_{j \in J - K} r_{ij} x_j^0 = \sum_{j \in J - K_\mu} r_{ij} x_j^0.$$

This, together with the above equality $\sum_{k \in K} r_{ik} y_k = \sum_{k \in K_\mu} r_{ik} y_k$, implies that

$$\sum_{j \in J - K} r_{ij} x_j^0 = v_i - \sum_{k \in K} r_{ik} y_k$$

for all $i \in I_0$. Since this is true for every finite subset I_0 of I, we know that the system of linear equaions

$$\sum_{j \in J - K} r_{ij} \xi_j = v_i - \sum_{k \in K} r_{ik} y_k \quad (i \in I)$$

is finitely solvable in M. Thus it turns out that the pair (K, Y) is in S. Therefore, by Zorn's Lemma, S has a maximal member, which we denote again by (K, Y). Since (K, Y) is in S, the above system of linear equations is finitely solvable in M for this pair (K, Y) too. Suppose that $J - K$ contains at least two indices. Let j_0 be an index from $J - K$. Then, by applying the proposition proved above to this system of equations, we can find an $x_0 \in M$ such that the system of equations

$$\sum_{\substack{j \neq j_0 \\ j \in J-K}} r_{ij}\xi_j = v_i - \sum_{k \in K} r_{ik}y_k - r_{ij_0}x_0 \quad (i \in I)$$

is finitely solvable in M. This shows that if we put $K_0 = K \cup \{j_0\}$ and define Y_0 to be the vector in M^{K_0} such that the j-th entry of Y_0 is the j-th entry y_j of Y for $j \in K$ and the j_0-th entry of Y_0 is x_0 then (K_0, Y_0) is in S. Since $(K, Y) \subsetneqq (K_0, Y_0)$ this contradicts the maximality of (K, Y). Thus $J - K$ must consist of a single index, say j_0. Then the corresponding locally solvable system of equations is $r_{ij_0}\xi = v_i - \sum_{k \in K} r_{ik}y_k$ $(i \in I)$. But, as was shown above, such a type of system of equations has a solution $\xi = x$ in M (under the assumption of (3)) and this shows that x together with y_k's give a solution of the system of equations $\Phi[\xi_j] = [v_i]$. Thus the proof of our theorem is completed.

Remark. It is proved in [16, Satz 2.1] that the three conditions (1), (2) and (3) are equivalent to another condition which is defined by merely replacing the term 'finite matrix subgroups' with the term 'matrix subgroups' in (3). Also it is shown in [16, §1] that Proposition 3.1 remains true for any number of matrix functors, i.e., the intersection of arbitrary numbers of matrix functors of R-Mod is a matrix functor too.

Let M be a left R-module. M is called *Σ-pure-injective* if $M^{(I)}$ is pure-injective for every (infinite) set I. The following is a main theorem to characterize Σ-pure-injective modules obtained by Zimmermann [16], with some refinement by the author:

Theorem 3.3. *For a left R-module M, the following conditions are equivalent:*

(1) *M is Σ-pure injective.*
(2) *$M^{(\mathbb{N})}$ is pure-injective.*
(3) *$M^{(\mathbb{N})}$ is a direct summand of $M^{(\mathbb{N})}$.*
(4) *For every submodule L of $M^{(\mathbb{N})}$ containing $M^{(\mathbb{N})}$ such that the factor module $L/M^{(\mathbb{N})}$ is countably generated, $M^{(\mathbb{N})}$ is a direct summand of L.*
(5) *Descending chain condition for finite matrix subgroups of M.*

Proof. (1)⇒(2) is trivial. (2)⇒(3) follows from the fact that $M^{(\mathbb{N})}$ is a pure submodule of $M^{(\mathbf{N})}$. (Generally, given a family of left R-modules M_i, the direct sum $S = \oplus M_i$ is a pure submodule of the direct product $P = \prod M_i$. For, if Q is a right R-module and $x = \sum_{j=1}^{n} q_j \otimes s_j$ ($q_j \in Q$, $s_j \in S$) an element of the kernel of the homomorphism $Q \otimes \mathcal{E} : Q \otimes_R S \to Q \otimes_R P$, where $\mathcal{E}$ is the inclusion map $S \to P$, then $s_1, s_2, \ldots, s_n$ are contained in a finite partial sum S_0 of S; but S_0 is a direct summand of P and this means that there exists a homomorphism $\pi : P \to S$ which fixes each element of S_0 and hence each of $s_1, s_2, \ldots, s_n$. This also implies that $\pi \circ \mathcal{E} : S \to S$ fixes each of $s_1, s_2, \ldots, s_n$ and therefore $Q \otimes (\pi \circ \mathcal{E}) : Q \otimes_R S \to Q \otimes_R S$ fixes x. Since, however $Q \otimes (\pi \circ \mathcal{E}) = (Q \otimes \pi) \circ (Q \otimes \mathcal{E})$ and x is in the kernel of $Q \otimes \mathcal{E}$, it follows that $x = 0$. This shows that $Q \otimes \mathcal{E}$ is a monomorphism.) (3)⇒(4) is also clear; indeed, (3) implies that $M^{(\mathbf{N})}$ is a direct summand of every submodule L of $M^{(\mathbf{N})}$ containing $M^{(\mathbf{N})}$.

Now assume (4). Suppose that (5) is not true, i.e., there is a sequence $U_1, U_2, U_3, U_4, \ldots$ of finite matrix functors of R-Mod such that the corresponding sequence of finite matrix subgroups of M is a descending chain $U_1(M) \supset U_2(M) \supset \ldots$ which does not terminate. Then, by replacing the sequence $U_1, U_2, U_3, \ldots$ with a suitable subsequence, we may assume without loss of generality that $U_n(M) \neq U_{n+1}(M)$ for all n. Then choose an $x_n \in U_n(M) - U_{n+1}(M)$ for each n. If we set $u_1 = [x_1, x_2, x_3, \ldots]$, $u_2 = [0, x_2, x_3, \ldots]$ and generally $u_n = [0, 0, 0, \ldots, 0, x_n, x_{n+1}, \ldots]$, then they are vectors in $M^{\mathbf{N}}$ and $u_1 - u_{n+1} = [x_1, x_2, \ldots, x_n, 0, 0, \ldots]$ is in $M^{\mathbf{N}}$ for $n = 1, 2, 3, \ldots$. Let Φ_n be a finite, say $r_n \times s_n$-matrix over R, which defines U_n for each n.

Let n be in $\mathbf{N}$. Let m be any integer such that $n \le m$. Since $U_m(M) \subset U_n(N)$ and $x_m \in U_m(M)$, it follows that $x_m \in U_n(M)$ and this means that there exist $x^{(n)}_{1,m}$, $x^{(n)}_{2,m}, \ldots, x^{(n)}_{s_n,m}$ in M such that

$$\Phi(n) \cdot \begin{bmatrix} x^{(n)}_{1,m} \\ x^{(n)}_{2,m} \\ \vdots \\ x^{(n)}_{s_n,m} \end{bmatrix} = 0 \quad \text{and} \quad x_m = x^{(n)}_{1,m}.$$

Let

$$X_n = \begin{bmatrix} 0 & \ldots & 0 & x^{(n)}_{1,n} & x^{(n)}_{1,n+1} & \ldots \\ 0 & \ldots & 0 & x^{(n)}_{2,n} & x^{(n)}_{2,n+1} & \ldots \\ \vdots & & \vdots & \vdots & \vdots & \\ 0 & \ldots & 0 & x^{(n)}_{s_n,n} & x^{(n)}_{s_n,n+1} & \ldots \end{bmatrix}$$

be the $s_n \times \mathbf{N}$-matrix whose (i,j)-entry is 0 or $x^{(n)}_{1,j}$ according as $j < n$ or $j \ge n$. Then clearly we have that $\Phi_n \cdot X_n = 0$. Now let $u^{(n)}_i$ ($1 \le 1 \le s_n$) be the i-th row of X_n, i.e., $u^{(n)}_i = [0, \ldots, x^{(n)}_{i,n}, x^{(n)}_{i,n+1}, \ldots]$. Then each $u^{(n)}_i$ is in

$M^{(\mathbb{N})}$ and $u_1^{(n)} = u_n$. Let us consider now the countably generated submodule

$$C = \sum_{n=1}^{\infty} (Ru_1^{(n)} + Ru_2^{(n)} + \cdots + Ru_{s_n}^{(n)})$$

of $M^{(\mathbb{N})}$, and let $L = C + M^{(\mathbb{N})}$. Then L is a submodule of $M^{\mathbb{N}}$ containing $M^{(\mathbb{N})}$ such that $L/M^{(\mathbb{N})}$ is countably generated. Therefore, by our assumption (4), $M^{(\mathbb{N})}$ is a direct summand of L, i.e., there exists a homomorphism $p : L \to M^{(\mathbb{N})}$ such that $p(u) = u$ for all $u \in M^{(\mathbb{N})}$.

Now, since $p(u_1)$ is in $M^{(\mathbb{N})}$, there is a $k \in \mathbb{N}$ such that the k-th entry of $p(u_1)$ is 0. Denote by $p(X_{k+1})$ the $s_{k-1} \times \mathbb{N}$-matrix whose i-th row is $p(u_i^{(k+1)})$, i.e.,

$$p(X_{k+1}) = \begin{bmatrix} p(u_1^{(k+1)}) \\ p(u_2^{(k+1)}) \\ \vdots \\ p(u_{s_{k+1}}^{(k+1)}) \end{bmatrix}.$$

Since

$$X_{k+1} = \begin{bmatrix} u_1^{(k+1)} \\ u_2^{(k+2)} \\ \vdots \\ u_{s_{k+1}}^{k+1} \end{bmatrix}$$

and $\Phi_{k+1} \cdot X_{k+1} = 0$, we have that $\Phi_{k+1} \cdot p(X_{k+1}) = 0$. Let w_j be the j-th column of $p(X_{k+1})$ for each $j \in \mathbb{N}$, i.e., $p(X_{k+1}) = [w_1\ w_2\ w_3\ \ldots]$. Then clearly $\Phi_{k+1} \cdot w_j = 0$, which implies that the first entry of w_j is in $U_{k+1}(M)$. But the first entry of w_j is nothing else than the j-th entry of $p(u_1^{(k+1)}) = p(u_{k+1})$. Thus we know that every entry of $p(u_{k+1})$ is in $U_{k+1}(M)$. Since now u_1 and u_{k+1} are in L and $u_1 - u_{k+1} = [x_1\ x_2\ \ldots\ x_k\ 0\ 0\ \ldots]$ is in $M^{(\mathbb{N})}$, we have that

$$[x_1\ x_2\ \ldots\ x_k\ 0\ 0\ \ldots] = p(u_1 - u_{k+1}) = p(u_1) - p(u_{k+1}).$$

Since the k-th entry of $p(u_1)$ is 0, by taking k-th entries we have

$$x_k = -\big(k\text{-th entry of } p(u_{k+1})\big)$$

and so x_k is in $U_{k+1}(M)$. But this contradicts the condition $x_k \in U_k(M) - U_{k+1}(M)$. Thus we have proved (4)$\Rightarrow$(5).

Finally assume (5). We first prove that M is pure-injective, or that M satisfies the equivalent condition that every family of cosets modulo finite matrix subgroups of M with the finite intersection property has a non-empty intersection (Theorem 3.2). Let $\{M_\alpha + x_\alpha \mid \alpha \in \Lambda\}$ be a family of cosets modulo finite matrix subgroups M_α of M with the finite intersection property. If we take any finite number of M_α from the family $\{M_\alpha \mid \alpha \in \Lambda\}$ then their intersection is also a finite matrix subgroup of M by Proposition 3.1. Let F

be the family of all those finite intersections of M_α's. Then by the condition (5) F has a minimal member, say $M_0 = M_{\alpha_1} \cap M_{\alpha_2} \cap \ldots M_{\alpha_n}$. Let $\alpha \in \Lambda$. Then $M_0 \cap M_\alpha$ is in F and $M_0 \cap M_\alpha \subset M_0$, and so by the minimality of M_0 we have that $M_0 \cap M_\alpha = M_0$, i.e., $M_0 \subset M_\alpha$. This shows that M_0 is indeed the intersection of all M_α's. Now, by the finite intersection property of $\{M_\alpha + x_\alpha \mid \alpha \in \Lambda\}$ there exists an x_0 in $(M_{\alpha_1} + x_{\alpha-1}) \cap (M_{\alpha_2} + x_{\alpha_2}) \cap \ldots \cap (M_{\alpha_n} + x_{\alpha_n})$. It follows then that $M_{\alpha_i} + x_0 = M_{\alpha_i} + x_{\alpha_i}$ for $i = 1, 2, \ldots, n$ and therefore $M_0 + x_0 = (M_{\alpha_1} + x_{\alpha_1}) \cap (M_{\alpha_2} + x_{\alpha_2}) \cap \ldots \cap (M_{\alpha_n} + x_{\alpha_n})$. Let $\alpha \in \Lambda$. Then by the finite intersection property there exists an x in $(M_0 + x_0) \cap (M_\alpha + x_\alpha)$. It follows then that $M_0 + x_0 = M_0 + x \subset M_\alpha + x = M_\alpha + x_\alpha$. This shows that $M_0 + x_0$ is contained in (and indeed coincides with) the intersection of all $M_\alpha + x_\alpha$'s. Thus M is pure-injective. Next, let I be a set and let there be given a descending chain $U_1(M^{(I)} \supset U_2(M^{(I)} \supset U_3(M^{(I)}) \supset \ldots$ of finite matrix subgroups of $M^{(I)}$, where $U_1, U_2, U_3, \ldots$ are finite matrix functors of R-Mod. Since generally inner functors commute with direct sum, we have that $U_n(M^{(I)}) = (U_n(M))^{(I)}$ for $n = 1,2,3,\ldots$. Therefore, we have the descending chain $U_1(M) \supset U_2(M) \supset U_3(M) \supset \ldots$ of finite matrix subgroups of M. Then it follows from (5) tht the chain terminates, i.e., $U_n(M) = U_{n-1}(M) = U_{n-2}(M) = \ldots$ for some n, but this clearly implies that $U_n(M^{(I)}) = U_{n-1}(M^{(I)}) = U_{n-2}(M^{(I)}) =\ldots$. Thus we know that the condition (5) for M implies the condition (5) for $M^{(I)}$ and so the pure-injectivity of $M^{(I)}$; since this is true for every set I it follows that M is Σ-pure-injective. This shows (5)$\Rightarrow$(1).

Lemma 3.4. *Let M be a left R-module and N a pure submodule of M. Then for every finite matrix functor U of R-Mod we have that $U(M) \cap N = U(N)$.*

Proof. Let $\Phi = [r_{ij}]$ be a finite, say $m \times n$-matrix over R that defines U, and let y be any element of $U(M) \cap N$. If $n = 1$, i.e., if Φ is of the form $\begin{bmatrix} r_1 \\ r_2 \\ \vdots \\ r_m \end{bmatrix}$ then y satisfies $r_1 y = r_2 y = \cdots = r_m y = 0$, and, since $y \in N$, this means that $y \in U(N)$. So we may assume that $n > 1$. Then if $y \in U(M)$ then there exist $x_2, x_3, \ldots, x_n$ in M such that $r_{i1} y + r_{i2} x_2 + \cdots + r_{in} x_n = 0$, i.e., $r_{i2} x_2 + \cdots + r_{in} x_n = -r_{i1} y$ for $i = 1,2,\ldots,m$. Since N is pure in M and $y \in N$, this implies the existence of $y_2, y_3, \ldots, y_n$ in N such that $r_{i2} y_2 + \cdots + r_{in} y_n = -r_{i1} y$, i.e., $r_{i1} y + r_{i2} y_2 + \cdots + r_{in} y_n = 0$ for $i = 1,2,\ldots,m$ (because N is Φ_0-pure in M for the $m \times (n-1)$-matrix Φ_0 obtained from Φ by removing its first column), and this shows that y is in $U(N)$.

Proposition 3.5. *Let M be a Σ-pure-injective left R-module. Then every pure submodule of M is a direct summand of M and is Σ-pure-injective and*

M is a direct sum of indecomposable submodules.

Proof. Let N be a pure submodule of M. Let $N_1 \supset N_2 \supset N_3 \supset \ldots$ be a descending chain of finite matrix subgroups of N. Then there exist finite matrix functors $V_1, V_2, V_3, \ldots$ of R-Mod such that $V_1(N) = N_1$, $V_2(N) = N_2, \ldots$. Let now $U_1 = V_1$, $U_2 = V_1 \cap V_2$, $U_3 = V_1 \cap V_2 \cap V_3, \ldots$. Then they are finite matrix functors of R-Mod (Proposition 3.1) satisfying $U_1 \supset U_2 \supset U_3 \supset \ldots$ and $U_1(N) = V_1(N) = N_1$, $U_2(N) = V_1(N) \cap V_2(N) = N_1 \cap N_2 = N_2$, $U_3(N) = V_1(N) \cap V_2(N) \cap V_3(N) = N_1 \cap N_2 \cap N_3 = N_3, \ldots$. Since however M is Σ-pure-injective, the descending chain $U_1(N) \supset U_2(N) \supset U_3(N) \ldots$ of finite matrix subgroups of M terminates by Theorem 3.3, i.e., $U_n(M) = U_{n-1}(M) = \ldots$ for some n. On the other hand, we have $U_i(M) \cap N = U_i(N) = N_i$ for $i = 1, 2, 3, \ldots$ by Lemma 3.4, and therefore it follows that $N_n = N_{n-1} = \ldots$. Thus the descending chain condition holds fro finite matrix subgroups of N and so N is Σ-pure-injective, again by Theorem 3.3. Since the Σ-pure-injectivity implies the pure-injectivity, N must be a direct summand of M. That M is a direct sum of indecomposable submodules follows from Proposition 1.13.

Remark. An inner functor of R-Mod is called a *p-functor* if it commutes with direct products. As is easily seen, the finite sum of p-functors is a p-functor, and the intersection of an arbitrary number of p-functors is a p-functor too. Moreover, every matrix functor is a p-functor. If we replace the condition (4) in Theorem 3.3 by the condition that *for every descending chain $U_1 \supset U_2 \supset U_3 \supset \ldots$ of p-functors the corresponding descending chain $U_1(M) \supset U_2(M) \supset U_3(M) \supset \ldots$ of subgroups of M terminates*, then it gives exactly the theorem of Zimmermann [16, Folgerung 3.4]. In fact our proof for (4)$\Rightarrow$(5) is a modification of the original proof that (3) implies the above condition.

4. RINGS OF PURE GLOBAL DIMENSION ZERO

First of all, the following proposition can easily be seen if we notice the facts that a module M is pure-projective if and only if every pure epimorphism onto M splits (Proposition 1.12) and that M is pure-injective if and only if M is a direct summand of every module which contains M as a pure submodule (Proposition 2.5);

Proposition 4.1. *The following conditions are equivalent:*

(1) *Every left R-module is pure-projective.*

(2) *For any left R-module M, every pure submodule of M is a direct summand of M.*

(3) *Every left R-module is pure-injective.*

We call R a ring of *left pure global dimension zero* if R satisfies the equivalent conditions in Proposition 4.1.

Proposition 4.2. *R is of left pure global dimension zero if (and only if) every countably generated left R-module is pure-projective.*

Proof. Suppose every countably generated left R-module is pure-projective. Let M be any left R-module. Let L be a submodule of $M^{(\mathbf{N})}$ such that $M^{(\mathbf{N})} \subset L$ and $L/M^{\mathbf{N}}$ is countably generated. Since $M^{\mathbf{N}}$ is pure in $M^{(\mathbf{N})}$ whence in L the natural epimorphism $L \to L/M^{\mathbf{N}}$ is pure so splits, i.e., $M^{(\mathbf{N})}$ is a direct summand of L. Therefore, by Theorem 3.3, $M^{\mathbf{N}}$ and hence M is pure-injective. Thus R is of left pure global dimension zero.

Now we call R a *left Chase ring* if every left R-module is a direct sum of finitely generated submodules. According to the theorem of Faith-Walker (Anderson-Fuller [1,25.8]), R is left Noetherian if every injective left R-module is a direct sum of finitely generated submodules. Therefore it follows that every left Chase ring is left Noetherian. If however R is a left Noetherian ring then every submodule of a finitely generated left R-module is finitely generated too, and so every finitely generated left R-module is finitely presented. Hence if R is a left Chase ring then every left R-module is a direct sum of finitely presented submodules and therefore is pure-projective by Proposition 1.12. Thus we have:

Proposition 4.3. *Every left Chase ring is left Noetherian and of left pure global dimension zero.*

We now prove the following crucial proposition:

Proposition 4.4. *Let every countably generated left R-module be pure-injective. Then R is a left Chase ring and a left Artinian ring.*

Proof. Let M be a countably generated left R-module. Then $M^{(\mathbf{N})}$ is also countably generated and hence pure-injective, i.e., M is Σ-pure-projective. Consider now a countable number of finitely generated indecomposable left R-modules $M_1, M_2, \ldots$ Let M be their direct sum, i.e., $M = M_1 \oplus M_2 \oplus \cdots$; here we may and shall assume that each M_i is a submodule of M. Let N be a proper direct summand of M, and let $M = N \oplus N'$ with a submodule $N' \neq 0$ of M. As a finitely generated module, each indecomposable M_i is pure-injective and hence the endomorphism ring of M_i is a local ring by Zimmermann-Huisgen and Zimmermann [18, Theorem 9]. Therefore both the direct summands N and N' of M are direct sums of (indecomposable) submodules with local endomorphism rings by the theorem of Crawley-Jonsson-Warfield([1,26.5]). (That both N and N' are direct sums of indecomposable submodules is also a consequence of Proposition 3.5 and the fact that as direct summands of the countably generated module M both N and N' are countably generated and so Σ-pure-injective.) Let $N' = \sum \oplus N_j$ be a direct decomposition of N' into indecomposable submodules. Then we have the direct decomposition $M = N \oplus (\sum \oplus N_j)$. Choose a summand N_{j_0}, and let $p : M \to N_{j_0}$ be the projection with respect to this decomposition. According to Azumaya [3, Theorem 1], there exists an $i_0 \in \mathbf{N}$ such that the restriction of p to M_{i_0} gives an isomorphism $M_{i_0} \to N_{j_0}$. Since the kernel of p is $N \oplus (\sum_{j=j_0} \oplus N_j)$, we have then the direct decomposition $M = N \oplus M_{i_0} \oplus (\sum_{j=j_0} \oplus N_j)$ and thus $N \cap M_{i_0} = 0$ and $N \oplus M_{i_0}$ is a direct summand of M. Consider now the family of those non-empty subsets I of $\mathbf{N}$ for which $N \cap \sum_{i\in I} \oplus M_i = 0$ and $N \oplus (\sum_{i\in I} \oplus M_i)$ is a direct summand of M. The existence of such a non-empty I is assured just above. Let $\{I_\alpha\}$ be an ascending chain in the family. Let I be the union of all I_α's. Then clearly $N \cap \sum_{i\in I} \oplus M_i = 0$, and $N \oplus (\sum_{i\in I} \oplus M_i)$ is the union of the ascending chain $\{N \oplus (\sum_{i\in I_\alpha} \oplus M_i)\}$ of direct summands of M and hence is a pure submodule of M. Since M is countably generated whence Σ-pure-injective, it follows from Proposition 3.5 that $N \oplus (\sum_{i\in I} \oplus M_i)$ is a direct summand of M, and this shows that I is in our family. Thus, by Zorn's Lemma, there exists a maximal member, say I_0. Then the corresponding submodule $N \oplus (\sum_{i\in I_0} \oplus M_i)$ is a direct summand of M. Suppose that this were a proper submodule of M. Then by applying the above argument to this direct summand instead of N we know that there would be an $i_0 \in \mathbf{N}$ such that $(N \oplus (\sum_{i\in I_0} \oplus M_i)) \cap M_{i_0} = 0$ (whence $i_0 \neq I_0$) and $N \oplus (\sum_{i\in I_0} \oplus M_i) \oplus M_{i_0}$ is a direct summand of M. This clearly implies that by adjoining i_0 to I_0 we have a subset $I_0 \cup \{i_0\}$ in the family, which contradicts the maximality of I_0. Thus we have $N \oplus \sum_{i\in I_0} \oplus M_i = M$. This shows that the direct decomposition $M = \sum_{i\in\mathbf{N}} \oplus M_i$ complements N (in the sense of [1, p.141]).

Suppose, for each $i \in \mathbf{N}$, there is given a homomorphism $f_i : M_i \to M_{i+1}$ which is not an isomorphism. Ssince the decomposition $M = M_1 \oplus M_2 \oplus \cdots$ into finitely generated submodules M_i with local endomorphism rings complements direct summands, Harada and Sai [10, Lemma 9] implies that there exists an $n > 0$ such that $f_n \circ f_{n-1} \circ \cdots f_2 \circ f_1 = 0$, i.e., the sequence $f_1, f_2, f_3, \ldots$ is Noetherian in the sense of Auslander [2].

Now the left R-module R is trivially countably generated and so it is Σ-pure-injective. Therefore it satisfies the descending chain condition for finite matrix subgroups of R by Theorem 3.3. If however a is an element of R then it is obvious that the finite matrix subgroup of R corresponding to the 1×2-matrix $[1\ \ -a]$ is the principal right ideal aR. This implies that in particular R satisfies the descending chain condition for principal right ideals. Suppose that there were a non-zero idempotent e of R which cannot be a sum of a finite number of orthogonal primitive idempotents of R. Let E be the set of all such idempotents. Then the family of those principal right ideals eR with $e \in E$ has a minimal member e_0R with $e_0 \in E$. Of course, e_0 is not primitive, i.e., $e_0 = e_1 + e_2$ with orthogonal non-zero idempotents e_1 and e_2. Then we have that $e_0R = e_1R \oplus e_2R$ and so both e_1R and e_2R are properly contained in e_0R. It follows that $e_1 \notin E$ and $e_2 \notin E$, which means that both e_1 and e_2 are sums of finite numbers of orthogonal primitive idempotents. Obviously this implies that e_0 is also a sum of a finite number of orthogonal primitive idempotents, i.e., $e_0 \in E$, a contradiction. Thus we know that every non-zero idempotent and in particular the identity element 1 is a sum of a finite number of orthogonal primitive idemopotents of R. It follows now from Fuller [8, Theorem] that R is a left Chase ring.

Let $a_1, a_2, a_3, \ldots$ be an infinite sequence in J, the Jacobson radical of R. Then the descending chain $a_1R \supset a_1a_2R \supset a_1a_2a_3R \supset \ldots$ of principal right ideals of R terminates, i.e., we have that $a_1a_2 \ldots a_nR = a_1a_2 \ldots a_na_{n+1}R = \ldots$ for some n. This implies that there exists an $r \in R$ such that

$$a_1 \ldots a_n = a_1 \ldots a_na_{n+1}r$$

and so we have $a_1 \ldots a_n(1 - a_{n+1}r) = 0$. But since $a_{n+1} \in J$ it follows that $a_{n+1}r \in J$ and so $1 - a_{n+1}r$ is invertible. Therefore we have that $a_1 \ldots a_n = a_1 \ldots a_n(1 - a_{n+1}r)(1 - a_{n+1}r)^{-1} = 0$. Thus J is left T-nilpotent in the sense of Bass [6].

Let generally $a_1,a_2,a_3,\ldots$ be any infinite sequence in R. Then we have a descending chain $a_1R \supset a_1a_2R \supset a_1a_2a_3R \supset \ldots$ of principal right ideals of R. Suppose conversely that there is given a descending chain $c_1R \supset c_2R \supset \ldots$ of principal right ideals of R. Then we can find elements $a_1, a_2, \ldots$ of R such that $c_2 = c_1a_2$, $c_3 = c_2a_3, \ldots$, and if we put $c_1 = a_1$, we have $c_2 = a_1a_2$, $c_3 = a_1a_2a_3$, ... Thus we know that every descending chain of principal right ideals of R is given as $a_1R \supset a_1a_2R \supset a_1a_2a_3R \supset \ldots$ by a suitable infinite sequence $a_1,a_2,a_3, \ldots$ in R. This fact implies that the descending chain condition for principal right ideals of R is inherited to every ring which is a homomorphic image of R.Let $\bar{R}$ denote the factor ring R/J. Then in particular $\bar{R}$ satisfies the descending chain condition for principal right ideals. Let K be a right ideal of $\bar{R}$. Then among those principal right ideals P of $\bar{R}$ which satisfies $P + K = \bar{R}$ there exists a minimal one, which we denote by $\bar{a}\bar{R}$ with $\bar{a} \in \bar{R}$. Let Q be a right ideal of $\bar{R}$ such that $Q \in \bar{a}\bar{R}$ and $Q + (\bar{a}\bar{R} \cap K) = \bar{a}\bar{R}$. Then there are $\bar{b} \in Q$ and $\bar{c} \in \bar{a}\bar{R} \cap K$ such that $\bar{b} + \bar{c} = \bar{a}$ and hence $\bar{b}\bar{R} + (\bar{a}\bar{R} \cap K) = \bar{a}\bar{R}$. It follows then that $\bar{b}\bar{R} + K = \bar{a}\bar{R} + K = \bar{R}$. By the minimality of $\bar{a}\bar{R}$ we have that $\bar{b}\bar{R} = \bar{a}\bar{R}$ whence $Q = \bar{a}\bar{R}$. This shows that $\bar{a}\bar{R} \cap K$ is small in $\bar{a}\bar{R}$ whence in $\bar{R}$, i.e., $\bar{a}\bar{R} \cap K$ is in the Jacobson radical of $\bar{R}$, which however is zero as is well-known. Thus we have that $\bar{a}\bar{R} \cap K = 0$ and so $\bar{a}\bar{R} \oplus K = \bar{R}$. This shows that every right ideal of $\bar{R}$ is a direct summand of $\bar{R}$, or equivalently, $\bar{R}$ is semi-simple Artinian.

Now, for every power J^i of J, we denote by $l(J^i)$ the left annihilator of J^i in R. Then we have an ascending chain $l(J) \subset l(J^2) \subset l(J^3) \subset \ldots$ of two-sided ideals of R. But according to Proposition 4.3 the left Chase ring R is left Noetherian. Therefore the ascending chain terminates, so that there exists an $n > 0$ such that $l(J^n) = l(J^{n+1})$. Then it follows from $aJ \subset l(J^n)$, $a \in R$ that $a \in l(J^n)$, because $aJ \subset l(J^n)$ implies $aJ^{n+1} = 0$ whence $aJ^n = 0$. Suppose that $J^n \neq 0$, or equivalently $l(J^n) \neq R$. Then $J \not\subset l(J^n)$, because if $J = 1J \subset l(J^n)$ then $1 \in l(J^n)$ whence $l(J^n) = R$. Choose then an $a_1 \in J$ such that $a_1 \notin l(J^n)$. Then $a_1 J \not\subset l(J^n)$. Choose then an $a_2 \in J$ such that $a_1 a_2 \notin l(J^n)$. Then $a_1 a_2 J \not\subset l(J^n)$. Continuing in this way we obtain an infinite sequence $a_1, a_2, a_3, \ldots$ in J such that $a_1 a_2 \ldots a_k \notin J$ whence $a_1 a_2 \ldots a_k = 0$ for every $k \in \mathbb{N}$. This clealy contradicts the fact that J is left T-nilpotent, so that we have that $J^n = 0$. Let now i be an integer such that $1 \leq i \leq n$, and consider the factor module J^{i-1}/J^i. Since $J \cdot (J^{i-1}/J^i) = 0$, J^{i-1}/J^i is regarded as a left module over the factor ring $\bar{R} = R/J$, in the natural manner. But since $\bar{R}$ is an Artinian semi-simple ring, every left $\bar{R}$-module and so J^{i-1}/J^i is completely reducible, i.e., J^{i-1}/J^i is a direct sum of simple submodules. Since moreover R is left Noetherian, J^{i-1}/J^i is Noetherian too and this means that J^{i-1}/J^i is a finite direct sum of simple submodules, or equivalently, of finite length. Therefore it follows that J^{i-1}/J^i is Artinian. Since this is true for $i = 1, 2, \ldots, n$, we know that R is left Artinian. This completes the proof of our proposition.

We have now the following theorem:

Theorem 4.5. *The following conditions for a ring R are equivalent:*

(1) *R is a left Chase ring.*
(2) *R is a ring of left pure global dimension zero.*
(3) *Every indecomposable left R-module is finitely presented.*
(4) *Every countably generated left R-module is pure-projective.*
(5) *Every countably generated left R-module is pure-injective.*
(6) *For any countably generated left R-module M, every countably generated pure submodule of M is a direct summand of M.*
(7) *Every countably generated indecomposable left R-module is finitely presented.*

If R satisfies any of the equivalent conditions then R is a left Artinian ring.

Proof. The equivalence of (1), (2), (4), and (5) and the fact that these conditions imply that R is left Artinian have already been proved. In fact, the equivalence of (2) and (4) is proved in Proposition 4.2 and (1)⇒(2) is shown in Proposition 4.3, while (2)⇒(5) is trivial and (5)⇒(1), including that R is left Artinian., is proved in Proposition 4.4. Therefore we need only prove that these conditions are equivalent to each of the other conditions (3), (6) and (7). Assume (1). Then in particular every indecomposable left R-module is finitely generated. But since R is left Noetherian by Proposition 4.3, every finitely genearated left R-module is finitely presented. This implies (1)⇒(3). Conversely assume (3). Let M be a left R-module. Let M_0 be a pure submodule of M such that M/M_0 is indecomposable. Then M/M_0 is finitely presented. Since the natuaral epimorphism $M \to M/M_0$ is pure, this must split , i.e., M_0 is a direct summand of M. According to Proposition 1.13, M is a direct sum of indecomposable submodules, each of which is therefore finitely presented. Thus R is a left Chase ring. This shows (3)⇒(1). (3)⇒(7) is trivial. Assume (7). If we notice that every factor module as well as every direct summand of a countably generated module is countably generated too, we can see in exactly the same way as above that every countably generated left R-module is a direct sum of finitely presented submodules and so is pure-projective. Thus (7)⇒(4) is proved. (2)⇒(6) is trivial. Assume (6). Let M be a countably generated left R-module. Let L be a submodule of $M^{\mathbf{N}}$ such that $M^{(\mathbf{N})} \subset L$ and $L/M^{(\mathbf{N})}$ is countably generated. Since $M^{(\mathbf{N})}$ is countably generated, it follows that L itself is countably generated. Moreover $M^{(\mathbf{N})}$ is pure in L, because $M^{(\mathbf{N})}$ is pure in $M^{\mathbf{N}}$. Thus $M^{(\mathbf{N})}$ is a direct summand of L, and so by Theorem 3.3 $M^{(\mathbf{N})}$ and hence M is pure-injective. This shows (6)⇒(5), and the proof of our theorem is completed.

Remark 1. In the proof of Proposition 4.4, we derived from the assumption that the left R-module $R^{(\mathbf{N})}$ is pure-injective (i.e., R is Σ-pure-injective) the descending chain condition for principal right ideals of R by making use of Theorem 3.3. We can also prove this by using the method given in Bass [6] and Anderson-Fuller [1] as follows. Let F be a free left R-module with a countable free basis $u_1, u_2, \ldots$, and let $a_1, a_2, \ldots$ be a sequence in R. Let $v_i = u_i - a_i u_{i-1}$ for each $i \in \mathbf{N}$, and let G be the submodule of F generated by $v_1, v_2, \ldots$. Then, as is shown in [1, p.313], $v_1, v_2, \ldots$ are linearly independent and so form a free basis of G, and more generally for every $k \in \mathbf{N}$, $v_1, \ldots, v_k, u_{k+1}, u_{k+2}, \ldots$ form a free basis of F. Thus $Rv_1 \oplus Rv_2 \oplus \cdots \oplus Rv_k$ is a direct summand of F, and let p_k be the corresponding projection. Let now Φ be a finite, say $m \times n$-matrix over R, and let $[x_j] \in F^n$ and $[g_i] \in G^m$ be such that $\Phi[x_j] = [g_i]$. Choose a sufficiently large k such that all g_j's are in $rv_1 \oplus Rv_2 \oplus \cdots \oplus Rv_k$ and apply p_k. Then we have $\Phi[p_k(x_j)] = [p_k(g_i)]$. But that $p_k(x_j)$ is in G for all j and $p_k(g_i) = g_i$ for all i implies that G is Φ-pure in F. Since this is true for all finite matrices Φ, G is pure in F by Theorem 1.8. Since moreover G is isomorphic to $R^{(\mathbf{N})}$, G is pure-injective and so G is a direct summand of F. It follows then by [6, Lemma 1.3] (or [1, Lemma 28.2]) that the descending chain $a_1R \supset a_1a_2R \supset \ldots$ terminates. Thus R satisfies the descening chain condition for principal right ideals. In this connection, this condition is equivalent to the left perfectness of R, as shown in [6].

Remark 2. As another characterization of rings of pure global dimnersion zero, Zimmermann-Huisgen [17] gives the following theorem: *R is a ring or left pure global dimension zero if and only if left R-module is a direct sum of indecomposable submodules.* Here the 'only if' part is an immediate consequence of Proposition 3.5, because every left module over a ring of left pure global dimension zero is Σ-pure-injective. The proof for the 'if' part is rather difficult, but we point out that this can be proved by her other theorem that a pure-injective module M is Σ-pure-injective if and only if $M^{\mathbf{N}}$ is a direct sum of indecomposable submodules.

Remark 3.

In Azumaya-Facchini [5, Theorem 8], it is proved that a ring R is of left pure global dimension zero if and only if every left R-module is a Mittag-Leffler module in the sense of Raynaud-Gruson[11], where the 'only if' part follows immediately from the known fact that every pure-projective module is a Mittag-Leffler module. But if we use Proposition 4.2 then the 'if' part can easily be seen, because every countably generated Mittag-Leffler module is pure-projective by [11, Corollaire 2.2.2]. It is to be mentioned in this connection that our theorem was also obtained by Simson in [12, Theorem 6.3].

REFERENCES.

1. F.W. Anderson and K.R. Fuller, *Rings and Categories of Modules*, Springer-Verlag, New York / Heidelberg / Berlin, 1973.

2. M. Auslander, *Representation theory of artin algebras. II*, Comm. Algebra **1** (1974), 269–310.

3. G. Azumaya, *Corrections and supplementaries to my paper concerning Krull-Remak-Schimdt's theorem*, Nagoya Math. J. **1** (1950), 117–124.

4. G. Azumaya, *An algebraic proof of a theorem of Warfield on algebraically compact modules*, Math. J. Okayama Univ. **28** (1986), 53–60.

5. G. Azumaya and A. Facchini, *Rings of pure global dimension zero and Mittag-Leffler modules*, J. Pure. Appl. Algebra **62** (1989), 109–122.

6. H. Bass, *Finitistic dimension and a homological generalization of semi-primary rings*, Trans. Amer. Math. Soc. **95** (1960), 466–488.

7. P.M. Cohn, *On the free product of associative rings*, Math. Z. **71** (1959) 380–398.

8. K.R. Fuller, *On rings whose left modules are direct sums of finitely generated modules*, Proc. Amer. Math. Soc. **54** (1976), 39–44.

9. L. Gruson and C.U. Jensen, *Deux applications de la notion de L-dimension*, C. R. Acad. Sci. Paris. Sér A **282** (1976), 23–24.

10. M. Harada and Y. Sai, *On categories of indecomposable modules, I*, Osaka J. Math. **7** (1970), 323–344.

11. M. Raymaud and L. Gruson, *Critéres de platitude et de projectivé*, Invent. Math. **13**, 1–89.

12. D. Simson, *On pure global dimension of locally finitely presented Grothendieck categories*, Fundamenta Math. **96** (1977), 91–116.

13. B. Strenström, *Direct sum decompositions in Grothendieck categories*, Arkiv Math. **7** (1968), 427–432.

14. R.B. Warfield, Jr., *Purity and algebraic compactness for modules*, Pacific J. Math. **28** (1969), 699–719.

15. W. Zimmermann, *Einige Charakterisierungen der Ringe, über denen reine Untermoduln direkte Summanden sind*, Bayer. Akad. Wiss. Math.-Natur. Kl. Sitzungsber. (1972), 77–79.

16. W. Zimmermann, *Rein injektive direkte Summen von Moduln*, Comm. Algebra **5** (1977), 1083–1117.

17. B. Zimmermann-Huisgen, *Rings whose right modules are direct sums of indecomposable modules*, Proc. Amer. Math. Soc **77** (1979), 191–197.

18. B. Zimmermann-Huisgen and W. Zimmermann, *Algebraically compact rings and modules*, Math. Z. **161** (1978), 81–93.

Additional References

19. R. Kielpinski, *On Γ-pure modules*, Bull. Acad. Polon. Sci. Sér. Sci. Math. Astr. Phys. **15** (1967), 127–131

20 M. Prest, *Model Theory and Modules*, London Math. Soc. Lecture Note Series **130**, Cambridge Univ. Press, 1988.

Projective Resolutions and Degree Shifting for Cohomology and Group Rings

by Jon F. Carlson[1]
Department of Mathematics
University of Georgia
Athens, Georgia 30602

1. Introduction.

One of the interesting features about the cohomology of finite groups is that it can be and - for many reasons - should be indexed on the entire set of integers rather than just on the nonnegative integers. In the negative degrees the cohomology is usually known as Tate cohomology. Although it is not as widely studied as ordinary cohomology it is no less natural. From the standpoint of modular or integral representation theory, every module is a member of a family all of whose elements share many of the same homological characteristics. For example, suppose that $k = \mathbb{Z}$ or k is a field of finite characteristic. Let G be a finite group and M a kG-module which is k-projective. Then there is a doubly infinite exact sequence

$$\cdots \longrightarrow F_2 \xrightarrow{\partial_2} F_1 \xrightarrow{\partial_1} F_0 \xrightarrow{\partial_0} F_{-1} \longrightarrow \cdots$$

of projective kG-modules such that $\partial(F_0) \cong M$. The members of the family $\{\partial(F_i) \mid i \in \mathbb{Z}\}$ all have the same cohomology. Only the degrees are shifted. Specifically we have that

$$\hat{H}^n(G, \partial(F_i)) = \hat{H}^{n+j-i}(G, \partial(F_j))$$

for all i, j and n. The members also share several other properties such as having the same varieties.

As a consequence, we encounter a curious question when investigating the homological algebra of modules over finite groups. The

[1]Partially supported by a grant from NSF.

question is: where is the natural starting point (middle point?) for the degree shifting process? Degree can be shifted in either direction and by any amount. Yet, in the above example, the modules $\partial(F_i)$ are all different if M is not periodic. Moreover the differences are manifested in some subtle ways. Hence we would like to have a benchmark for the degree shift if only to serve as hypothesis of future theorems.

At first we might be tempted to take as the middle point of the family, $\{\partial(F_i)\}$, that member which has least k-dimension or k-rank. The idea seems most natural in the case of an irreducible kG-module or kG-lattice. For here, the obvious middle point is the irreducible member of the family which usually also has the minimal dimension. However there may not be a unique member with least dimension. Even worse, experience indicates that the homological characteristics of modules are only loosely correlated with dimension or rank.

In this paper we define a new invariant which we call the index of a module. It is offered as a possible benchmark for the degree shifting process. We also present several associated invariants which in different contexts may serve the purpose better. The index is defined in terms of homogeneous systems of parameters for cohomology rings and modules. The idea of the index of a module grew out of the investigation of the spectral sequences associated to the complexes in [7]. In that setting the concept seems very natural. The main problem that we address here is the proof of the existence.

David Benson and I first started looking at homogeneous systems of parameters for cohomology rings some five years ago, and our first results were on the structure of projective resolutions [6]. Our most recent work on the subject discusses some of the complexes associated to the systems of parameters. That paper, [7] , should really be viewed as a companion to the one here presented. In it we present and discuss some of the many open problems and attempt to show the relationships among some of the unanswered questions. A lot of the motivation for asking the questions is given in [7] and not repeated here. None the less, we have tried to include here enough of the background information to make the paper independently readable. Also, we should state that none of the conjectures of [7] are actually settled here. What is accomplished is a presentation of some of the same ideas in a entirely new way.

We begin in Section 2 with some preliminary results which are necessary for the definition of the index. The theorems which we need are not very difficult to prove and may even be known in some other context. The only background required is some well known commutative algebra. Section 3 is mostly a discussion of notation and background for varieties of modules. Some material from [6] and [7] is reviewed for later use. A major purpose is also to give some perspective on the work.

Although Section 4 gives some review of Tate cohomology and duality, its main aim is the proof of Theorem 4.1. This is the last major obstacle in the path toward defining the index in Section 5. In that section we also establish the relationship between the index and its properties and the results of [7]. Particularly stated is the connection to the structure of the module and its projective resolution. This connection leads to one of the several possible variations on the index in the next section. Section 6 also discusses the behavior of the index under induction and restriction to subgroups and presents several examples.

Except for notation and background the last two sections are independent of the rest of the paper. Section 7 focuses on quasi-regular sequences, which were introduced in [7] as a device for finding the (not then defined) index of a module. We prove a strong converse to a theorem of the earlier paper. Section 8 discusses two questions about contravariantly finite subcategories that arose at the Tsukuba conference.

Finally, I want to thank the many people who have helped me to focus my thinking through helpful discussions. Particularly I want to mention Maurice Auslander, Dave Benson, Dieter Happel, Helmut Lenzing and Robert Varley. Some of the ideas for this paper were developed during a month's visit to the Universität Bielefeld. I am grateful to Claus Ringel and the many people there who made my stay so pleasant, and to the Deutsche Forschungsgemeinschaft for financial support.

2. **Homology of Koszul complexes.**

Throughout this section we assume that k is a Noetherian ring with unit and that $\mathcal{P} = k[x_1, \cdots, x_n]$ is a polynomial ring over k in n variables. Let A be a finitely generated $\mathcal{P}$-module. In this section we consider the Koszul complex associated to A. The reader will recognize that our notation is not standard even if our construction is. Among other things, the Koszul complexes of this treatment are cochain complexes. The reason for this indexing scheme should be obvious in later sections. The Koszul complex of A, $\mathcal{K}(A)$ can be defined as follows.

For each $i = 0, \cdots, n$ let $\mathcal{K}_i(\mathcal{P})$ denote the free $\mathcal{P}$-module with basis consisting of the set of symbols b(S) indexed on the set of all subsets $S \subseteq T = \{1, \cdots, n\}$ with $|S| = i$. That is, $\mathcal{K}_i(\mathcal{P})$ is the direct sum of $\binom{n}{i}$ copies of $\mathcal{P}$, and the symbols b( ) index the summands. Now for $0 \leq i < n$, define

$$\delta_i\colon \mathcal{K}_i(\mathcal{P}) \longrightarrow \mathcal{K}_{i+1}(\mathcal{P})$$

by

$$\delta_i(a\, b(S)) = \sum_{j \notin S} (-1)^{u(S,j)}\, x_j\, a\, b(S \cup \{j\})$$

where $u(S, j) = |\{ i \in S | i < j\}|$. It is easy to check that $\delta_{i+1}\delta_i = 0$ and hence

$$\mathcal{K}(\mathcal{P})\colon\ 0 \longrightarrow \mathcal{K}_0(\mathcal{P}) \xrightarrow{\delta_0} \mathcal{K}_1(\mathcal{P}) \xrightarrow{\delta_1} \cdots \longrightarrow \mathcal{K}_n(\mathcal{P}) \longrightarrow 0$$

is a complex. In fact, $\mathcal{K}(\mathcal{P})$ is the standard projective resolution of k as a $\mathcal{P}$-module [16]. In particular we have the following.

Lemma 2.1 The homology of $\mathcal{K}(\mathcal{P})$ is $H^*(\mathcal{K}(\mathcal{P})) = H^n(\mathcal{K}(\mathcal{P})) = k$.

Now let A be a $\mathcal{P}$-module. Define the Koszul complex of A to be

$$\mathcal{K}(A) = A \otimes_{\mathcal{P}} \mathcal{K}(\mathcal{P}).$$

For convenience we write $ab(S)$ for $a \otimes b(S)$, $a \in A$. Then $\mathcal{K}_i(A) = \sum_{|S|=i} Ab(S) \cong A^{\binom{n}{i}}$ and the coboundary homomorphism is given

by the

$$\delta_i(ab(S)) = \sum_{j \notin S} (-1)^{u(S,j)} x_j \, ab(S \cup \{j\})$$

for $a \in A$. Notice that if $\phi\colon A \longrightarrow A'$ is a homomorphism of $\mathcal{P}$-modules then there is a chain map $\hat{\phi}\colon \mathcal{K}(A) \longrightarrow \mathcal{K}(A')$ given by $\hat{\phi}(a \cdot b(S)) = \phi(a) \cdot b(S)$ for $a \in A$, $S \subseteq T$. From this viewpoint it can be seen that $\mathcal{K}(\ )$ is an exact functor from the category of $\mathcal{P}$-modules to the category of complexes of $\mathcal{P}$-modules and chain maps.

Theorem 2.2 Suppose that A is a finitely generated $\mathcal{P}$-module. Then the homology of $\mathcal{K}(A)$,

$$H^*(\mathcal{K}(A)) = \sum_{j=0}^{n} H^j(\mathcal{K}(A)) = \sum_{j=0}^{n} \mathrm{Tor}_j^{\mathcal{P}}(k, A)$$

is finitely generated as a k-module.

Proof The equalities follow from Lemma 2.1. Notice that $\mathcal{K}_i(A)$ is a finitely generated $\mathcal{P}$-module, since A is , and $\ker\delta_i$ is a finitely generated $\mathcal{P}$-module because $\mathcal{P}$ is noetherian. It follows that

$$H^i(\mathcal{K}(A)) = \ker\delta_i / \delta_{i-1}(\mathcal{K}_{i-1}(A))$$

is a finitely generated $\mathcal{P}$-module. However the elements $x_1, \cdots, x_n$ annihilate $H^i(\mathcal{K}(A))$ for all i. That is, the action of $\mathcal{P}$ on $H^i(\mathcal{K}(A))$ factors through the map $\mathcal{P} \longrightarrow k$ which takes each x_j to zero. Therefore $H^i(\mathcal{K}(A))$ is finitely generated as a k-module.

From now on let's assume that $A = \sum_{i \geq 0} A_i$ is a graded $\mathcal{P}$-module. However we should not restrict ourselves to the usual grading on $\mathcal{P} = k[x_1, \cdots, x_n]$. Instead we suppose only that $\mathcal{P} = \sum_{i \geq 0} \mathcal{P}_i$ has a grading in which, for each i, $x_i \in \mathcal{P}_{n_i}$ is homogeneous of some degree $n_i > 0$. Of course, $\mathcal{P}_r \cdot \mathcal{P}_s \subseteq \mathcal{P}_{r+s}$ and $\mathcal{P}_r \cdot A_s \subseteq A_{r+s}$. In this case the Koszul complex $\mathcal{K}(A)$ is a direct sum of k-subcomplexes

$$\mathcal{K}(A, t): \ 0 \longrightarrow \mathcal{K}_0(A, t) \xrightarrow{\delta_0} \mathcal{K}_1(A, t) \xrightarrow{\delta_1} \cdots \longrightarrow \mathcal{K}_n(A, t) \longrightarrow 0$$

where $\mathcal{K}_i(A, t) = \sum_{|S|=i} A_{t + deg(S)} \cdot b(S)$ for $deg(S) = \sum_{i \in S} deg(x_i) = \sum_{i \in S} n_i$. It is easy to see that for $a \in A_{t + deg(S)}$,

$$\delta_i(a \cdot b(S)) = \sum_{j \notin S} (-1)^{u(S,j)} x_j \, a \, b(S \cup \{j\})$$

is in $\mathcal{K}_{i+1}(A, t)$ if $|S| = i$. Thus we have that

$$\mathcal{K}(A) = \sum_{t \geq -deg(T)} \mathcal{K}(A, t)$$

as complexes of k-modules. However as a complex of k-modules $\mathcal{K}(A)$ has finitely generated cohomology by Theorem 2.2. Moreover

$$H^*(\mathcal{K}(A)) = \oplus \sum_{t \geq -deg(T)} H^*(\mathcal{K}(A, t)).$$

Consequently we have proved the following.

Theorem 2.3. For all but a finite number of t, the complex $\mathcal{K}(A, t)$ is an exact sequence (i.e. $H^*(\mathcal{K}(A, t)) = 0$).

The theorem shows that the complex $\mathcal{K}(A, t)$ is an exact sequence for t sufficiently large. Hence we may make the following definition.

<u>Definition</u> Let A be a nonzero graded $\mathcal{P}$-module. The index of A is the least integer r such that $\mathcal{K}(A, t)$ is exact whenever $t \geq r$. Denote it by $\mathrm{Index}_{\mathcal{P}}(A)$ or by $\mathrm{Index}(A)$ if there can be no confusion.

At this point we need to expand the scope of our discussion. For the remainder of the section we let R be a ring having a unit element and having the following properties.

(1) R is a k-algebra. By this we mean only that there is a ring homomorphism of k into the center of R which sends unit element to unit element and makes R into a k-module.

(2) R is finitely generated as a module over its center, $\mathcal{C}(R)$.

(3) The center, $\mathcal{C}(R)$, is finitely generated as a ring over k. That is, there is a finite set of generators such that every element of $\mathcal{C}(R)$ is a polynomial in these elements with coefficients in k.

(4) $R = \sum_{i \geq 0} R_i$ is a graded ring with $k \cdot 1 = R_0$.

Of course conditions (2) and (3) imply that R is also finitely generated as a ring. Note that by (4), $\mathcal{C}(R)$ must also be a graded ring.

<u>Definition</u> Let A be a finitely generated, graded R-module. We say that $\zeta_1, \cdots, \zeta_n \in \mathcal{C}(R)$ is a homogeneous set of parameters for A if

(i) each ζ_i is homogeneous of positive degree, i.e. $\zeta_i \in R_{n_i}$ for some $n_i = \deg(\zeta_i) > 0$, and

(ii) if $\mathcal{P} = k[\zeta_1, \cdots, \zeta_n]$ is the polynomial ring in the symbols $\zeta_1, \cdots, \zeta_n$ then A is a finitely generated module over $\mathcal{P}$.

It should be emphasized that we are not assuming that $\zeta_1, \cdots, \zeta_n$ are algebraically independent in R, nor are we supposing that the map $\mathcal{P} \longrightarrow R$ is injective. That is, $\mathcal{P}$ is meant to be the formal polynomial ring in those symbols. It has an obvious action on A and on R, but it is not equated with its image in R.

Now suppose that A is a finitely generated, graded R-module, and $\zeta_1, \cdots, \zeta_n$ is a homogeneous set of parameters for A. Let $\mathcal{P} = k[\zeta_1, \cdots, \zeta_n]$. It makes sense to talk about the index of A, meaning $\text{Index}_{\mathcal{P}}(A)$. A major result of this section is that the index does not depend on the choice of the parameters $\zeta_1, \cdots, \zeta_n$. Specifically, we show the following.

Theorem 2.4 Let A be a finitely generated, graded R-module. Let $\zeta_1, \cdots, \zeta_n$ and $\gamma_1, \cdots, \gamma_m$ be two homogeneous sets of parameters for A. If $\mathcal{P} = k[\zeta_1, \cdots, \zeta_n]$ and $\mathcal{P}' = k[\gamma_1, \cdots, \gamma_m]$, then

$$\text{Index}_{\mathcal{P}}(A) = \text{Index}_{\mathcal{P}'}(A).$$

Proof In order to prove the theorem we need first some notation. For $\eta_1, \cdots, \eta_\ell \in \mathcal{C}(R)$, let

$$\mathcal{K}(A; \eta_1, \cdots, \eta_\ell) = \mathcal{K}(A)$$

where $\mathcal{K}(A)$ is the Koszul complex defined previously with $x_1 = \eta_1, \cdots, x_\ell = \eta_\ell$, $n = \ell$. Notice that this is defined even if $\eta_1, \cdots, \eta_\ell$ is not a homogeneous set of parameters. If in addition, $\eta_1, \cdots, \eta_\ell$ are homogeneous then we may define

$$\mathcal{K}(A; \eta_1, \cdots, \eta_\ell ; t) = \mathcal{K}(A, t).$$

Lemma 2.5 Suppose that $\alpha_1, \cdots, \alpha_\ell$ are units (invertible elements) of k. Then the complexes $\mathcal{K}(A; \eta_1, \cdots, \eta_\ell ; t)$ and $\mathcal{K}(A; \alpha_1\eta_1, \cdots, \alpha_\ell\eta_\ell ; t)$ are isomorphic as complexes of k-modules.

Proof Define a chain map

$$\mu: \mathcal{K}(A; \eta_1, \cdots, \eta_\ell ; t) \longrightarrow \mathcal{K}(A; \alpha_1\eta_1, \cdots, \alpha_\ell\eta_\ell ; t)$$

by

$$\mu(a \cdot b(S)) = (\prod_{j \notin S} \alpha_j)\, a \cdot b(S)$$

for all $a \in A_{t+deg(S)}$, $S \subseteq \{1, \cdots, \ell\} = T$. It is now straight forward to check that $\mu\delta = \delta\mu$. Clearly μ is an isomorphism.

Now suppose that $U \subseteq T = \{1, \cdots, \ell\}$. Let $\mathcal{K}_U(A; \eta_1, \cdots, \eta_\ell; t)$ be the subcomplex of $\mathcal{K}(A; \eta_1, \cdots, \eta_\ell ; t)$ given as

$$0 \longrightarrow \mathcal{K}_{U,0} \longrightarrow \mathcal{K}_{U,1} \longrightarrow \cdots \longrightarrow \mathcal{K}_{U,n} \longrightarrow 0$$

where $\mathcal{K}_{U,i} = \sum_{U \subseteq S \subseteq T} A_{t+deg(S)} \cdot b(S)$. Obviously, $\mathcal{K}_{U,i} = \{0\}$ if $i < |U|$. Now observe that for $U \leq S, j \notin S$

$$u(S, j) = u(U, j) + u(S \backslash U , j)$$

where $u(S, j) = |\{i \in S | \ i < j\}|$ by definition. Then for $a \in A_{t+deg(S)}$, $U \subseteq S$

$$\begin{aligned}\delta(a \cdot b(S)) &= \sum_{j \notin S} (-1)^{u(S,j)} \eta_j\, a \cdot b(S \cup \{j\}) \\ &= \sum_{j \notin S} (-1)^{u(S \backslash U, j)}((-1)^{u(U,j)} \eta_j) a \cdot b(S \cup \{j\})\end{aligned}$$

Now suppose that $T = U = \{i(1), \cdots, i(r)\}$, where $i(1) < i(2) < \cdots < i(r)$.

<u>Lemma 2.6</u> The subcomplex $\mathcal{K}_U(A; \eta_1, \cdots, \eta_\ell ; t)$ is isomorphic to the shift of the complex $\mathcal{K}(A; \eta_{i(1)} , \cdots , \eta_{i(r)} ; t + \deg U)$ by $\ell - r = |U|$ places.

<u>Proof</u> From the preceeding calculation it is clear that $\mathcal{K}_U(A; \eta_1, \cdots , \eta_\ell ; t)$ is isomorphic to the shift of the complex $\mathcal{K}(A; \alpha_1\eta_{i(1)}, \cdots , \alpha_r\eta_{i(r)} ; t)$ where $\alpha_j = (-1)^{u(U,j)}$. The result is now a consequence of Lemma 2.5.

<u>Lemma 2.7</u> If $U = V \cup \{j\}$, then $\mathcal{K}_U(A; \eta_1, \cdots , \eta_\ell ; t)$ is a subcomplex of $\mathcal{K}_V(A; \eta_1, \cdots , \eta_\ell ; t)$ and the quotient complex is isomorphic to

$$\mathcal{K}_U(A; \eta_1, \cdots , \eta_\ell ; t - \deg \eta_j)$$

shifted (-1) places (one place to the left).

Proof The first statement is obvious. To prove the second define the chain map

$$\nu\colon \mathcal{K}_V(A;\, \eta_1, \cdots, \eta_\ell\, ;\, t) \longrightarrow \mathcal{K}_U(A;\, \eta_1, \cdots, \eta_\ell\, ;\, t - \deg \eta_j)$$

by

$$\nu(a \cdot b(S)) = \begin{cases} 0 & \text{if } U \subseteq S \\ (-1)^{(S \backslash V,\, j)} ab(S \cup \{j\}) & \text{if } U \not\subseteq S \end{cases}$$

for $a \in A_{t + degS}$. The formula assumes that $V \subseteq S$, and hence $U \subseteq S$ if and only if $j \in S$. It can be checked that ν is a chain map. Clearly its kernel is $\mathcal{K}_U(A;\, \eta_1, \cdots, \eta_\ell\, ;\, t)$. Hence we are done.

We can now complete the proof of Theorem 2.4. Let $r = \mathrm{Index}_{\mathcal{P}}(A)$ and $s = \mathrm{Index}_{\mathcal{P}'}(A)$. We assume that $s < r$ so that $s \le r - 1 < r$. Our object then is to get a contradiction. For U a subset of $T = \{1, \cdots, m{+}n\}$, let $\mathcal{K}_U(A, t)$ be an abbreviation for the complex $\mathcal{K}_U(A;\, \zeta_1, \cdots, \zeta_n, \gamma_1, \cdots, \gamma_m;\, t)$. Also $\mathcal{K}_\emptyset(A, t) = \mathcal{K}(A, t)$. Our claim is that $\mathcal{K}(A, r-1)$ is an exact sequence. We have a nested set of subsets of T

$$\emptyset = U_0 \subseteq U_1 \subseteq \cdots \subseteq U_m$$

where $U_i = \{m{+}n,\ m{+}n-1, \cdots, m{+}n{+}1-i\}$ if $i > 0$. Hence $T - U_i = \{1, \cdots, n{+}m-i\}$. By Lemma 2.6 we have exact sequences

$$0 \longrightarrow \mathcal{K}_{U_{i+1}}(A, t) \longrightarrow \mathcal{K}_{U_i}(A, t) \longrightarrow \mathcal{K}^{[-1]}_{U_{i+1}}(A, t - \deg \gamma_{n-i}) \longrightarrow 0$$

where the exponent $[-1]$ indicates the shift. By Lemma 2.5, $\mathcal{K}_{U_i}(A, t)$ is isomorphic to the complex

$$\mathcal{K}^{[i]}(A;\, \zeta_1, \cdots, \zeta_n, \gamma_1, \cdots, \gamma_{m-i}\, ;\, t + \deg(U_i))$$

where $\deg U_i = \sum_{j=m-i+1}^{m} \deg(\gamma_j)$. Therefore our claim is proved as follows. Notice that $\mathcal{K}(A, r-1)$ is exact provided $\mathcal{K}_{U_1}(A, r-1)$ and $\mathcal{K}_{U_1}(A, r - 1 - \deg\gamma_m) = \mathcal{K}^{[1]}(A;\, \zeta_1, \cdots, \zeta_n, \gamma_1, \cdots, \gamma_{m-1}\, ;\, r-1)$ are exact. In turn these two are exact provided $\mathcal{K}_{U_2}(A, t)$ is exact for $t = r-1, r-1-\deg\gamma_m, r-1-\deg\gamma_{m-1}$ and $r-1-\deg\gamma_m-\deg\gamma_{m-1}$. So ultimately, the exactness of $\mathcal{K}(A, r-1)$ is implied by the exactness of $\mathcal{K}_{U_n}(A, t)$ for all $t = r-1-\deg(S)$, $S \subseteq \{m+1, \cdots, n+m\} = U_n$. But finally

$$\mathcal{K}_{U_n}(A, t) \cong \mathcal{K}(A; \zeta_1, \cdots, \zeta_n ; t + \deg(U_n))$$

which is exact in these cases because

$$r-1-\deg(S) + \deg(U_n) = r-1+\deg(U_n - S) \geq s,$$

by our assumption.

Before finishing the proof we should notice that we have proved a more general statement than was stated in our claim. Namely we have the following.

Corollary 2.8 Suppose that $\zeta_1, \cdots, \zeta_n \in R$ is a homogeneous set of parameters for a finitely generated graded R-module A. Let $\zeta_{n+1}, \cdots, \zeta_t$ be homogeneous elements in R. Let $\mathcal{P} = k[\zeta_1, \cdots, \zeta_n]$, $\mathcal{P}' = k[\zeta_1, \cdots, \zeta_t]$. Then

$$\text{Index}_{\mathcal{P}'}(A) \leq \text{Index}_{\mathcal{P}}(A).$$

Continuing with the proof, we know that $\mathcal{K}(A, r-1)$ is exact. Now consider the nested sets

$$\emptyset = V_0 \subseteq V_1 \subseteq \cdots \subseteq V_n$$

where $V_i = \{1, \cdots, i\}$ for $i > 0$. Again by Lemma 2.7 we have exact sequences

$$0 \longrightarrow \mathcal{K}_{V_{i+1}}(A, t) \longrightarrow \mathcal{K}_{V_i}(A, t) \longrightarrow \mathcal{K}_{V_{i+1}}^{[-1]}(A, t-\deg\zeta_{i+1}) \longrightarrow 0 .$$

By lemma 2.6, and Corollary 2.8 we know that $\mathcal{K}_{V_1}(A, r-1)$ is exact. Since $\mathcal{K}_{V_0}(A, r-1)$ is exact we have that $\mathcal{K}_{V_1}(A, r-1-\deg\zeta_1)$ is exact. Continuing in this fashion we get that $\mathcal{K}_{V_j}(A, r-1-\deg V_j)$ is exact for $j = 1, \cdots, n$. It follows then that

$$\mathcal{K}_{V_k}(A, r-1-\deg V_n) \cong \mathcal{K}(A; \gamma_1, \cdots, \gamma_m ; r-1)$$

is exact. But this contradicts the assumption that $\text{Index}_{\mathcal{P}}(A) = r$.

The theorem allows us to generalize our definition.

Definition 2.9 Let A be a finitely generated graded R-module. The index of A is

$$\text{Index}_R(A) = \text{Index}_{\mathcal{P}}(A)$$

for $\mathcal{P} = k[\zeta_1, \cdots, \zeta_n]$ where $\zeta_1, \cdots, \zeta_n$ is some homogeneous set of parameters for A.

3. Cohomology, varieties and complexity.

A major aim of this paper is to exploit the results of the last section in the context of the cohomology of modules over group algebras. We will want to consider the ramifications of the results in the case where $R = H^*(G, k)$ and $A = H^*(G, M)$ for G a finite group, k a suitable coefficient ring and M a kG-module. In this section we prepare for the applications. We recall notations and facts established in previous works and address a few technical issues which have some importance in what follows. A principal theme is the usage of Tate cohomology. That is, our grading runs to both positive and negative degrees and the index of a cohomology module can be either negative or positive depending on the chosen orientation. While this allows for an expanded definition of the index, it also presents us with some problems which must be carefully addressed.

Throughout this section and those to follow we assume that G is a finite group and that k is a Dedekind domain. We reserve the notation K for the special case of a field of characteristic $p > 0$. Unless otherwise noted the term kG-module will be used only for finitely generated kG-lattices, i.e., those which are projective as k-modules.

The assumptions on the ring k are intended to ensure that projective kG-modules (lattices) are also weakly injective. Otherwise put, this means that kG is a Gorenstein order (see [14]). So if L is any kG-lattice then there exists a projective kG-module Q and an injective homomorphism ϕ: L $\longrightarrow$ Q which is split over k. In particular, we have that the cokernel $Q/\phi(L) = \Omega_{kG}^{-1}(L)$ is again a kG-lattice.

In general the module $\Omega_{kG}^{-1}(L)$ will depend on the choice of the weakly injective module, Q , and the injection ϕ. However it can be shown that any two such modules are stably isomorphic, and we shall use this notation to denote any appropriate module with the understanding that it is not uniquely determined. There are some cases in which it can be unique, for example where k = K is a field and we add the assumption that Q is an injective hull of L (which is also projective). The uniqueness is only up to isomorphism, but we assume it wherever possible.

In the other direction, suppose that L is a kG-lattice and that ψ: P $\longrightarrow$ L is surjective with P a projective kG-module. Then we let $\Omega(L)$ denote the kernel of ψ except that again we assume that $\Omega(L)$ is the unique minimal such module whenever possible. For example, if k = K is a field and if ψ_1: $P_1 \longrightarrow L$ and ψ_2: $P_2 \longrightarrow L$ are two such maps, then Schanuel's Lemma says that $\ker \psi_2 \oplus P_1 \cong \ker \psi_1 \oplus P_2$. In particular we can assume that P is a projective cover and $\Omega(L)$ has no projective submodules in this case.

Inductively, define $\Omega^n(L) = \Omega(\Omega^{n-1}(L))$ if $n > 1$ and]$\Omega^n(L) = \Omega^{-1}(\Omega^{n+1}(L))$ if $n < -1$. If k = K then $\Omega^0(L) \cong \Omega(\Omega^{-1}(L))$, so that $\Omega^i(\Omega^j(L)) \cong \Omega^{i+j}(L)$ for all i and j. In general we have only that $\Omega^i(\Omega^j(L))$ and $\Omega^{i+j}(L)$ are stably isomorphic.

For K a field of characteristic $p > 0$, the cohomology ring $H^*(G, K)$ is a finitely generated, graded K-algebra. It is graded commutative in the sense that $\zeta\gamma = (-1)^{(deg\zeta)(deg\gamma)}\ \gamma\zeta$ for ζ, γ homogeneous elements. If p = 2 then it is actually commutative. If p > 2, then all elements of odd degree are nilpotent and hence are in the radical, Rad $H^*(G,K)$. So in either case $H^*(G,K)/\text{Rad } H^*(G,K)$ is a commutative ring. Hence the maximal ideal spectrum $V_G(K)$ of $H^*(G,K)$ is a homogeneous affine variety. If M is a KG-module then $\text{Ext}^*_{KG}(M, M) \cong H^*(G, \text{Hom}_K(M, M))$ is a finitely generated module over $H^*(G, K)$. Let J(M) denote the annihilator in $H^*(G,K)$ of $\text{Ext}^*_{KG}(M, M)$. Let $V_G(M) \subseteq V_G(K)$ denote the subvariety corresponding to J(M). It is the set of all maximal ideals which contain J(M).

The varieties of modules have had several uses in the representation theory. Given sufficient information on the structure of the group G, the varieties can be computed with some difficulty. For G an elementary abelian p-group $V_G(M)$ can be calculated directly from the module structure of M . In more complicated groups the varieties can (in theory) be pieced together from the varieties of the restrictions to the maximal elementary abelian p-subgroups. The reader is referred to the book by Benson [5] or the survey article [12] for more information. For the purposes of this paper, we will list some properties of the varieties without proof.

Proposition 3.1. i) $V_G(M) = \{0\}$ if and only if M is projective.

ii) If $0 \longrightarrow M_1 \longrightarrow M_2 \longrightarrow M_3 \longrightarrow 0$ is an exact sequence of KG-module then $V_G(M_i) \subseteq V_G(M_j) \cup V_G(M_k)$ for {i, j, k} = {1,2,3}. In particular, $V_G(M \oplus N) = V_G(M) \cup V_G(N)$.

iii) $V_G(M \otimes N) = V_G(M) \cap V_G(N)$. (Here the tensor product is over K and $M \otimes_K N$ is given the diagonal G-action $g(m \otimes n) = gm \otimes gn$.)

iv) If $W \subseteq V_G(K)$ is any closed homogeneous subvariety, then there exists a KG-module M with $V_G(M) = W$.

v) If $\zeta \in H^n(G, M)$, $\zeta \neq 0$, and if ζ is represented by $\hat{\zeta}: \Omega^n(K) \longrightarrow K$, then the kernel L_ζ of ζ has $V_G(L_\zeta) = V_G(\zeta)$, the subvariety of all maximal ideals which contain ζ.

vi) $V_G(\Omega^n(M)) = V_G(M) = V_G(M^*)$ for any KG-module M and integer n. Here $M^* = \mathrm{Hom}_K(M, K)$.

vii) Let $t = \dim V_G(M)$, then t is the least nonnegative integer such that

$$\lim_{n \to \infty} \frac{\mathrm{Dim}\Omega^n(M)}{n^t} = 0.$$

It is worth noting that (v) can be proved by restriction to elementary abelian p-subgroups and looking at the rank varieties [11]. As for the notation, observe that if

$$\cdots \longrightarrow P_{n+1} \xrightarrow{\partial} P_n \xrightarrow{\partial} P_{n-1} \longrightarrow \cdots \longrightarrow P_0 \xrightarrow{\partial} K \longrightarrow 0$$

is a minimal projective resolution of K and if $\tilde{\zeta}: P_n \longrightarrow K$ is a cocycle representing ζ, then $\tilde{\zeta}$ induces the homomorphism $\hat{\zeta}: P_n/\partial(p^{n+1}) \longrightarrow K$ since $\tilde{\zeta}\partial = 0$. Because $P_n/\partial(P_{n+1}) \cong \Omega^n(K)$, the map $\hat{\zeta}$ is defined. Statement (iv) can be obtained directly from statements (iii) and (v). That is, choose any set of homogeneous generators for the ideal of W and take the tensor product of the corresponding modules to get M. Statement (vii) says that the dimension of $V_G(M)$ is, in fact, the complexity of M. The complexity is defined as the polynomial rate of growth of $\mathrm{Dim}_K(\Omega^n(M))$ as expressed in (vii). It is equal to the polynomial rate of growth of $\mathrm{Ext}^n_{KG}(M, M)$.

It should be noted that properties (i) and (iii) in combination provide a method predicting when the tensor product of two modules will be projective without requiring that the tensor product be computed. This has been one of the most useful tools to come out of the theory. In fact the study of homogeneous systems of parameters was initiated by a theorem on projective resolutions whose proof employs this technique [6]. In what follows we sketch the result. Some of the notation and ideas in the development are needed later.

Suppose that $\zeta_1, \cdots, \zeta_n \in H^*(G, K)$ is a homogeneous system of parameters for the module $\mathrm{Ext}^*_{KG}(M, M)$. Thus $\mathrm{Ext}^*_{KG}(M, M)$ is a finitely generated module over $K[\zeta_1, \cdots, \zeta_n]$. This is equivalent to the fact that $(\bigcap_{i=1}^{n} V_G(\zeta_i)) \cap V_G(M) = \{0\}$. Let $n_i = \deg \zeta_i$. So if (P_*, ϵ) is a minimal projective resolution, then we have the following diagram

$$\begin{array}{ccccccccccccc} \cdots \longrightarrow & P_{n_i} & \longrightarrow & P_{n_i-1} & \longrightarrow & P_{n_i-2} & \longrightarrow & \cdots & \longrightarrow & P_0 & \xrightarrow{\epsilon} & K & \longrightarrow 0 \\ & \zeta_i\downarrow & & \downarrow & & \| & & & & \| & & \| & \qquad (3.2) \\ 0 \longrightarrow & K & \xrightarrow{j} & L_i & \longrightarrow & P_{n_i-2} & \longrightarrow & \cdots & \longrightarrow & P_0 & \xrightarrow{\epsilon} & K & \longrightarrow 0. \end{array}$$

Here ζ_i denotes the cocycle representing the cohomology class of the same name, and the bottom row is the pushout of the diagram. Hence the bottom row is exact. It can be checked that $L_i \cong \Omega^{-1}(L_{\zeta_i})$ and so $V_G(L_i) = V_G(\zeta_i)$ by (v). Now let $X^{(i)}$ be the infinite splice of the bottom row with itself:

$$X^{(i)}: \quad \cdots \longrightarrow P_0 \xrightarrow{j\epsilon} L_i \longrightarrow \cdots \longrightarrow P_0 \xrightarrow{j\epsilon} L_i \longrightarrow \cdots \longrightarrow P_0 \longrightarrow 0.$$

So $H_*(X^{(i)}) = H_0(X^{(i)}) \cong K$.

Theorem 3.3 [6] With $X^{(i)}$ defined as above we have that $X_* = X^{(1)}_* \otimes \cdots \otimes X^{(n)}_* \otimes M$ is a projective resolution of M.

To prove the theorem we need only notice that $H_*(X) = H_0(X) \cong M$ by the Künneth formula, and that each tensor product

$$X^{(n)}_{v_1} \otimes X^{(2)}_{v_2} \otimes \cdots \otimes X^{(n)}_{v_n} \otimes M$$

is projective by properties (i), (iii) and the assumption that $V_G(M) \cap (\bigcap_{i=1}^{n} V_G(\zeta_i)) = \{0\}$.

For $i = 1, \cdots, n$, let $C^{(i)}$ denote the complex

$$C^{(i)}: \; 0 \longrightarrow L_i \longrightarrow P_{n_i-2} \longrightarrow \cdots \longrightarrow P_0 \longrightarrow 0$$

obtained by truncating the ends of the sequence in (3.2). In a later section we will need the following. The proof requires only an extension of the above argument.

Lemma 3.4 [7] Suppose that $\zeta_1, \cdots, \zeta_n$ is a homogeneous system of parameters for $H^*(G, K)$. Then $C = C^{(1)} \otimes C^{(2)} \otimes \cdots \otimes C^{(n)}$ is a finite complex of projective kG-modules and

$$H^*(C) = \Lambda_k(\hat{\zeta}_1, \cdots, \hat{\zeta}_n)$$

where Λ_k denotes the exterior algebra over k. Also deg $\hat{\zeta}_i = n_i - 1$.

The idea behind the lemma is that

$$H^*(C) = H^*(C^{(1)}) \otimes \cdots \otimes H^*(C^{(n)}) .$$

So $\hat{\zeta}_i = 1 \otimes \cdots \otimes \tilde{\zeta}_i \otimes \cdots \otimes 1$ where $\tilde{\zeta}_i$ is a generator for $H^{n_i - 1}(C^{(i)})$.

4. Tate cohomology, duality and rates of growth.

As before, G is a finite group and k is a Dedekind domain. We use K to designate a field of characteristic $p > 0$. As in the last section kG-modules will be assumed to be projective as k-modules. If M is a kG-module the Tate cohomology with coefficients in M is defined as follows. Let F_* be a doubly infinite projective resolution of the trivial kG-module k. That is, F_* has the form of an exact sequence of projective modules:

$$\cdots \longrightarrow F_2 \longrightarrow F_1 \xrightarrow{\partial_1} F_0 \xrightarrow{\partial_0} F_{-1} \xrightarrow{\partial_{-1}} F_{-2} \longrightarrow \cdots$$

where $\partial(F_0) = \ker \partial_{-1} \cong k$. Then the Tate cohomology is defined to be the homology of the complex $\mathrm{Hom}_{KG}(F_*, M)$. In particular we have that $\partial_i(F_i) = \ker\partial_{i-1} = \Omega^i(k)$. So that the nth Tate cohomology is given as

$$\hat{H}^n(G, M) = \mathrm{Hom}_{kG}(\Omega^n(k), M)/\mathrm{PHom}_{kG}(\Omega^n(k), M)$$

where $\mathrm{PHom}_{kG}(\Omega^n(k), M)$ is the submodule consisting of those homomorphisms θ: $\Omega^n(k) \longrightarrow M$ which factor through a projective kG-module. Of course, if $n > 0$ then $\hat{H}^n(G, M) = H^n(G, M)$. If M is a projective kG-module then M is cohomologically trivial, meaning that $\hat{H}^n(G, M) = 0$. In fact, defining $\hat{\mathrm{Ext}}^n$ by the usual relation

$$\hat{\mathrm{Ext}}^n_{KG}(L, M) \cong \hat{H}^n(G, \mathrm{Hom}_K(L, M)) \cong \hat{H}^n(G, L^* \otimes M)$$

($L^* = \mathrm{Hom}_k(L, k)$), we have that $\hat{\mathrm{Ext}}^n_{KG}(L, M) = 0$ whenever L or M is a projective kG-module. Hence if

$$0 \longrightarrow \Omega(M) \longrightarrow P \longrightarrow M \longrightarrow 0$$

is exact with P projective then the long exact sequence for cohomology says that

$$\hat{H}^n(G, M) \cong \hat{H}^{n+1}(G, \Omega(M))$$

for all n. More generally, for any n, j

$$\hat{H}^n(G, \Omega^j(M)) = \hat{H}^{n-j}(G, M).$$

One of the most useful aspects of Tate cohomology is the Tate duality. For this we consider mostly the case in which $k = K$ is a field of characteristic $p > 0$. The duality theorem states that the composition

$$\hat{H}^m(G, M^*) \otimes_K \hat{H}^{-m-1}(G,M) \longrightarrow \hat{H}^{-1}(G, M^* \otimes M) \longrightarrow \hat{H}^{-1}(G, K) \cong K$$

is a duality pairing. In this notation, $M^* = \mathrm{Hom}_K(M, K)$. The first map in the composition is cup product. Notice that $M^* \otimes_K M \cong \mathrm{Hom}_K(M, M)$ under the map $\lambda \otimes m \longrightarrow f$ where $f(m') = \lambda(m') \cdot m$. Consequently we have a KG-module homomorphism $M^* \otimes M \cong \mathrm{Hom}_K(M, M) \longrightarrow K$ which takes $f \in \mathrm{Hom}_K(M, M)$ to the trace of f. Alternatively, if $\lambda \in M^*$ and $m \in M$ then $\lambda \otimes m$ goes to $\lambda(m) \in K$. This induces the second part of the composition from $\hat{H}^{-1}(G, M^* \otimes M)$ to $\hat{H}^{-1}(G, K)$. The statement that this is a duality pairing means basically that it is nondegenerate. Consequently $\hat{H}^i(G, M)$ is isomorphic to the K-dual of $\hat{H}^{-i-1}(G, M^*)$.

A major aim of this section is to prove the following.

Theorem 4.1 Let M be a KG-module. Let r be the least nonnegative integer such that

$$\lim_{n \to \infty} \frac{\operatorname{Dim} H^n(G, M)}{n^t} = 0 \text{ , for } t \geq r \text{ .}$$

Let s be the least nonnegative integer such that

$$\lim_{n \to \infty} \frac{\operatorname{Dim} \hat{H}^{-n}(G, M)}{n^t} = 0, \text{ for } t \geq s.$$

Then $r = s$.

Proof We may assume that $r \leq s$, as otherwise we may exchange M^* for M, and the statement becomes true by duality. Before proceeding further we need the following.

Lemma 4.2 Let k be a commutative noetherian ring and let M be a graded module over the polynomial ring $k[\zeta_1, \cdots, \zeta_n]$. We assume here that $\zeta_1, \cdots, \zeta_n$ are homogeneous but do not assume that they all have the same degree. Then exactly one of the following statements is true.

i) There exists an element $m \in M$, $m \neq 0$, such that $\zeta_i m = 0$ for all $i = 1, \cdots, n$.

ii) There exists a homogeneous element $\gamma \in k[\zeta_1, \cdots, \zeta_n]$ which has positive degree and is regular for M (i.e. $\gamma m = 0$ for $m \in M$ implies $m = 0$).

Proof of Lemma For $m \in M$, let Ann(m) be the annihilator of m in $k[\zeta_1, \cdots, \zeta_n]$. Recall that the associated primes of M are the maximal elements in the set $\{\text{ann}(m) \mid m \in M , m \neq 0\}$. (see [17]). The associated primes are prime ideals. Statement (i) says that the ideal $(\zeta_1, \cdots, \zeta_n)$ is contained in some associated prime. Clearly if (i) is true then (ii) is false. So suppose that (i) is false and the ideal $(\zeta_1, \cdots, \zeta_n)$ is in no associated prime. Because there is only a finite number of associated prime ideals, it is not possible for $(\zeta_1, \cdots, \zeta_n)$ to be in the union of all the associated primes. Hence there is a homogeneous element γ of positive degree such that γ is in no associated prime and annihilates no element.

Returning to the proof of the theorem, we choose a homogeneous set of parameters $\zeta_1, \cdots, \zeta_n \in H^*(G, K)$ for $H^*(G, M)$. Now consider the (graded) Koszul complex $\mathcal{K}(A, t)$, where $A = H^*(G, M)$ regarded as a module over $K[\zeta_1, \cdots, \zeta_n]$. By Theorem 2.3, for t sufficiently large $\mathcal{K}(A, t)$ is exact. This implies that for t sufficiently large the homomorphism

$$\hat{H}^t(G, M) \longrightarrow \sum_{i=1}^{n} \hat{H}^{t+deg(\zeta_i)}(G, M)$$

given by $\theta \longrightarrow \sum \zeta_i\theta$ is injective. Therefore for some index ℓ, there exists no element $m \in \sum_{t \geq \ell} \hat{H}^t(G, M)$ such that $\zeta_i m = 0$ for all i. By Lemma 4.2 there exists a homogeneous element $\gamma \in H^*(G, K)$, having positive degree, such that γ is a regular element for $\sum_{t \geq \ell} \hat{H}^t(G, M)$.

Now γ is represented by a cocycle $\hat{\gamma}: \Omega^n(K) \longrightarrow K$. So the sequence

$$0 \longrightarrow L \longrightarrow \Omega^n(K) \xrightarrow{\hat{\gamma}} K \longrightarrow 0$$

is exact. The long exact sequence on cohomology gives us an exact sequence

$$(4.3) \longrightarrow \hat{H}^m(G, L \otimes M) \longrightarrow \hat{H}^m(G, \Omega^n(K) \otimes M) \xrightarrow{\bar{\gamma}} \hat{H}^m(G, M) \longrightarrow \cdots$$

Note that $\hat{H}^m(G, \Omega^n(K) \otimes M) \cong \hat{H}^{m-n}(G, M)$ and $\bar{\gamma}$ is cup product with γ. Hence we have that for $m - n - 1 > \ell$,

$$\mathrm{Dim}\, \hat{H}^m(G, L \otimes M) = \mathrm{Dim}\, \hat{H}^{m-1}(G, M) - \mathrm{Dim}\, \hat{H}^{m-n-1}(G, M).$$

In sufficiently high degrees the dimensions $\mathrm{Dim}\hat{H}^m(G, M)$ are given by PORC functions [20]. That is, there exist an integer $u > 0$

and polynomial functions $f_0(t), \cdots, f_{u-1}(t)$ such that if $m = tu + v$, $0 \leq v < u$ and m sufficiently large then

$$\text{Dim } H^m(G, M) = f_v(t).$$

Of course, f_v must be a polynomial of degree $r - 1$ for at least one (actually all) value of v. In no case can it have degree greater than $r - 1$.

By replacing γ by γ^a for some a we can assume that $n = \deg \gamma = xu$ for some x. Then for $m = tu + v$,

$$\begin{aligned} \text{Dim } H^{m+n+1}(G, L \otimes M) &= \text{Dim } H^{(t+x)u+v}(G, M) - \text{Dim } H^m(G, M) \\ &= f_v(t + x) - f_v(t). \end{aligned}$$

So the PORC functions describing the dimensions of the cohomology of $L \otimes M$ have degree at most $r - 2$. Therefore

$$\lim_{m \to \infty} \frac{\text{Dim } H^m(G, L \otimes M)}{m^{r-1}} = 0.$$

Now by the duality $\text{Dim } \hat{H}^m(G, M)$ is the same as $\text{Dim } \hat{H}^{-m-1}(G, M^*)$. Consequently there are an integer $u > 0$ and polynomials $g_0, \cdots, g_{u-1}$ such that if $m = tu + v$, $0 \leq v < u$ and m is sufficiently large then

$$\text{Dim } \hat{H}^{-m}(G, M) = g_v(t).$$

The degree of the g_v's must be $s - 1$ in this case. Again we may assume that $n = xu$ for some x. From the long exact sequence on cohomology we conclude that $\text{Dim } \hat{H}^{-m}(G, L \otimes M)$ is at least as large as

$$\begin{aligned} &\text{Dim } \hat{H}^{-m-n}(G, M) - \text{Dim } \hat{H}^{-m}(G, M) \\ &= g_v(t + x) - g_v(t) \end{aligned}$$

if $m = tu + v$. But this is a polynomial in t of degree $s - 2$.

Therefore we have shown that, in the statement of the theorem, if M is replaced by $L \otimes_K M$, then r and s are replaced by $r - 1$ and $s - 1$ respectively. This allows us to prove the theorem by induction. The case $r = 0$ is proved in Theorem 1.1 of [8]. That is, $s = 0$ if $r = 0$. We noted earlier that we may assume $r \leq s$. By induction on r and considering the module $L \otimes M$ we see that $r - 1 = s - 1$. Hence $r=s$.

Remark 4.4 In theory there is nothing to prevent us from extending Theorem 4.1 to other coefficient rings, provided we can make proper sense of the statement. That is, we need a suitable notion to replace that of Dim $\hat{H}^m(G, M)$. For example, if $k = \mathbb{Z}$ or if k is a rank-one discrete valuation ring of characteristic 0 whose maximal ideal contains a prime p, then we can replace dimension by composition length. That is, for any kG-lattice M, then $\hat{H}^m(G, M)$ will have finite k-composition length, the necessary PORC functions will exist, etc. All that is really necessary is that $k/|G|k$ have finite composition length. There will be a few minor differences in the proof. For example, we must be certain that the cocycle $\hat{\gamma}: \Omega^n(k) \longrightarrow k$ is surjective. This can be accomplished by replacing $\Omega^n(k)$ by $\Omega^n(k) \oplus kG$ and $\hat{\gamma}$ by $\hat{\gamma}'$ where $\hat{\gamma}'(a, b) = \hat{\gamma}(a) + \epsilon(b)$, ϵ being the augmentation map. Then $\hat{\gamma}$ and $\hat{\gamma}'$ differ by a map which factors through a projective (see [6]).

Notice also that the Tate duality is different. For a coefficient ring k of characteristic zero, as described above, the duality pairing has the form

$$\hat{H}^m(G, M) \otimes_k \hat{H}^{-m}(G, M^*) \longrightarrow k/|G|k$$

for any finitely generated kG-lattice M.

5. Marking the degree shift.

The purpose of this section is to apply the results of Section 2 to the case in which $R = H^*(G, k)$ and $A = \hat{H}^*(G, M)$ for M a finitely generated kG-module. It has already been noted that the index of $H^*(G, M)$ can be defined. That is, $H^*(G, k)$ is a finitely generated ring and $H^*(G, M)$ is a finitely generated module over $H^*(G, k)$ by classical theorems of Evens [15]. However even more can be done if we use Tate cohomology. In particular, the index gives us a benchmark for the degree shifting process.

Some added hypothesis may be required here. As usual, G is a finite group and k is a commutative noetherian ring. We shall need Tate duality and also the conclusion of Theorem 4.1 (see also Remark 4.4). Hence to be safe we assume that $k = K$, a field of characteristic $p > 0$, $k = \mathbb{Z}$ or k is a rank-one discrete valuation ring of characteristic zero and whose maximal ideal contains $|G|$. We remind the reader that kG-modules are assumed to be finitely generated and projective as k-modules. The index, Index($H^*(G, M)$), means the index of $H^*(G, M)$ as an $H^*(G, k)$-module.

We say that the cohomology $H^*(G, M)$ is periodic if there exists a positive integer $s > 0$ such that $H^n(G, M) = H^{n+s}(G, M)$ for all n, sufficiently large. Recall that $\hat{H}^m(G, M) = \hat{H}^{m+n}(G, \Omega^n(M))$ for all n, m.

Proposition 5.1 Let M be a kG-module.

(1) If the cohomology $H^*(G, M)$ is periodic, then there exists a positive integer s such that $\hat{H}^n(G, M) \cong \hat{H}^{n+s}(G, M)$ for all n, and Index($H^*(G, \Omega^n(M))$) is either 0 or 1 for all n.

(2) If the cohomology $H^*(G, M)$ is not periodic, then there exists an integer s such that Index($H^*(G, \Omega^n(M))$) $= n - s$ for all $n \geq s + 1$.

Proof Consider first the periodic case. Then, in the notation of Theorem 4.1, r is either 0 or 1. If $r = 0$ then by Theorem 1.1 of [8], $\hat{H}^n(G, M) = 0$ for all n. So we may consider the case $r = 1$. In the proof of Theorem 4.1, it was shown that there exists an element $\gamma \in H^s(G, k)$ for some s, such that cup product with γ is injective, and hence is an isomorphism in large enough degrees. Therefore, in the notation of the proof, the module $L \otimes M$ has $H^n(G, L \otimes M) = 0$ for n sufficiently large. By Theorem 1.1 of [8], $H^n(G, L \otimes M) = 0$ for all n and cup product with γ induces an isomorphism of $H^n(G, M)$ to $H^{n+s}(G, M)$ for all n by (4.3). This proves the first statement. The second follows from the fact that $H^m(G, \Omega^n(M)) \cong \hat{H}^{m-n}(G, M)$ for

$m > 0$ and the set $\{\gamma\}$ is a homogeneous set of parameters for $H^*(G, \Omega^n(M))$.

Now suppose that the cohomology $H^*(G, M)$ is not periodic. We claim that in the notation of Theorem 4.1, $r > 1$. That is, the finitude of the index of $H^*(G, M)$ and Lemma 4.2 require that there exists $\gamma \in H^t(G, k)$ for some t such that cup product with γ is an injection $\gamma' : H^n(G, M) \longrightarrow H^{n+t}(G, M)$ for n sufficiently large. If there were some bound on the dimension or composition length of $H^n(G, M)$ for n large then γ' would be an isomorphism in high degrees. Hence there is no such bound. We have then, by Theorem 4.1 that $s > 1$ also.

Now choose any homogeneous set of parameters $\zeta_1, \cdots, \zeta_n$ in $H^*(G, k)$ for $H^*(G, M)$. Note that for any set of positive integers $a_1, \cdots, a_n$, the set of elements $\zeta_1^{a_1}, \cdots, \zeta_n^{a_n}$ is also a homogeneous set of parameters for $H^*(G, M)$. Hence we may assume that $\zeta_1, \cdots, \zeta_n$ all have the same degree t. Because $s > 1$ there is a positive integer ℓ such that

$$\mathrm{Dim}\, \hat{H}^m(G, M) > n \,\mathrm{Dim}\, \hat{H}^{m+\ell t}(G, M)$$

for some m (negative). Now $\zeta_1^\ell, \cdots, \zeta_n^\ell$ is a homogeneous set of parameters and if V_i is the kernel of cup product with ζ_i^ℓ on $\hat{H}^m(G, M)$ then $\mathrm{Dim}(V_1 \cap \cdots \cap V_n) > 0$. Consequently there exists $\gamma \in \hat{H}^m(G, M)$ such that $\zeta_i^\ell \cdot \gamma = 0$ for all i and $\mathcal{K}(H^*(G, \Omega^{1-m}(M)); \zeta_1^\ell, \cdots, \zeta_n^\ell ; 1)$ is not exact. That is, $H^1(G, \Omega^{1-m}(M)) \cong \hat{H}^m(G, M)$. Therefore the second part of the proposition is a consequence of the following.

__Lemma 5.2__ If $\mathrm{Index}(H^*(G, M)) = s \geq 2$ then $\mathrm{Index}(H^*(G, \Omega^j(M)) = s + j$, for all $j > 0$.

__Proof__ Note again that $H^\ell(G, M) = H^{\ell+j}(G, \Omega^j(M))$ for $\ell, j > 0$. Therefore

$$\mathcal{K}(H^*(G, M); \zeta_1, \cdots, \zeta_n ; \ell) \cong \mathcal{K}(H^*(G, \Omega^j(M)) ; \zeta_1, \cdots, \zeta_n ; \ell + j)$$

for ℓ, $j > 0$. Hence the lemma is proved.

We are now prepared to give a definition of the index of a kG-module. This generalizes the idea of section 2.

<u>Definition</u> The index of a kG-module M, denoted Ind(M) or $\mathrm{Ind}_G(M)$ is the least integer such that $\mathcal{K}(\hat{H}^*(G, M); \zeta_1, \cdots, \zeta_n; r)$ is exact for all $r \geq \mathrm{Ind}(M)$ and any set $\zeta_1, \cdots, \zeta_n \in H^*(G, k)$ of homogeneous parameters for $H^*(G, M)$. Of course Ind(M) does not depend on the choice of parameters. We say that $\mathrm{Ind}(M) = -\infty$ if $H^*(G, M)$ is periodic. If $H^*(G, M)$ is not periodic then $\mathrm{Ind}(M) = \mathrm{Index}(H^*(G, \Omega^j(M))) - j$ for j sufficiently large. (That is, $\mathrm{Ind}(M) = -s$ where s is defined in Prop. 5.1(2).)

By generalizing Lemma 5.2 we get the following.

<u>Lemma 5.3</u> $\mathrm{Ind}(\Omega^j(M)) = \mathrm{Ind}(M) + j$, for all j.

Several of the results in [7] are concerned with the index of special modules specifically the trivial module and irreducible modules. In every example which has been computed, where K is a field of characteristic $p > 0$ and M and N are simple KG-modules, it has been shown that $\mathrm{Ind}(M \otimes N) = 0$ unless $H^*(G, M \otimes N)$ is periodic. The problem, of course, is that very few examples are actually known.

When the Sylow p-subgroup of G is cyclic or generalized quaternion $(p = 2)$ then the cohomology of any KG-module is periodic, and there is nothing else to say in this regard. For more complicated groups the calculation are very difficult and have been done only in special cases. The following is a sampling of the actual results.

(5.4). The trivial module. The index Ind(K) is zero in the following cases.

(a) G has p-rank 2 [7].

(b) The Sylow p-subgroup of G is elementary abelian.

(c) $p = 2, G = M_{12}$ [2].

(d) $p = 2$ and the Sylow 2-subgroup of G is an extra special group [18].

(e) $G = GL(n, q^r)$, q a prime, $q \neq p$ [19].

All but the first case is a consequence of the result that either $H^*(G, K)$, the cohomology ring of G or that of its Sylow p-subgroup, is Cohen-Macaulay [7]. In Section 7 we will expand further on this idea.

(5.5) Simple modules. Suppose that M and N are irreducible KG-modules. Then $\mathrm{Ind}(M \otimes N)$ is either zero or $-\infty$ in the following cases.

(a) $p = 2$ and $G = M_{11}, L_3(2), A_4, A_5, A_6, A_7$.

(b) $p = 3$ and $G = A_6$.

In [7] it is shown that there is another direct connection between the notion of the index and the representation theory. We concentrate on the case in which $k = K$, is a field. Let $\zeta_1, \cdots, \zeta_n$ be a homogeneous set of parameters for $\mathrm{Ext}^*_{KG}(M,M) \cong H^*(G,\mathrm{Hom}_K(M, M))$. Then each ζ_i is represented by a cocycle $\hat{\zeta}_i$ in $\mathrm{Hom}_{KG}(\Omega^{n_i}(K), K)$, $n_i = \deg(\zeta_i)$. Let $D^{(i)}$ be the complex

$$D^{(i)}\colon\ \Omega^{n_i}(K) \xrightarrow{\hat{\zeta}_i} K$$

which has homology $H_*(D^{(i)}) = H_1(D^{(i)}) = L_{\zeta_i}$ as in Proposition 3.1 (ν). Let

$$D(r) = D^{(1)} \otimes \cdots \otimes D^{(n)} \otimes \Omega^r(M).$$

Then by the Künneth formula,

$$H_*(D(r)) \cong H_n(D(r)) = L_{\zeta_1} \otimes \cdots \otimes L_{\zeta_n} \otimes \Omega^r(M)$$

which is a projective KG-module. So as in [7] it is possible to write

$D(r) = \tilde{D}(r) \oplus Q(r)$ where $\tilde{D}(r)$ is an (n+1)-term exact sequence and $Q(r)$ is a totally split complex of projective modules. The questions is whether all of the projective modules in $\tilde{D}(r)$ can be stripped away leaving an exact sequence. The results on the index imply an affirmative answer for large enough r. In particular, we have the following.

Theorem 5.6 Suppose that $r \geq \mathrm{Ind}(M^* \otimes N)$ for all irreducible KG-modules N. Let $\zeta_1, \cdots, \zeta_n \in H^*(G, k)$ be a homogeneous set of parameters for $\mathrm{Ext}^*_{KG}(M, M)$. Then there exists an exact sequence of the form

$$V(r) : 0 \longrightarrow V_n \xrightarrow{\partial} V_{n-1} \xrightarrow{\partial} \cdots \xrightarrow{\partial} V_0 \longrightarrow 0$$

where $V_i = \oplus \sum_{|S|=i} \Omega^{r+deg(S)}(M) \cdot b(S)$. The sum is taken over all subsets $S \subseteq T = \{1, \cdots, n\}$. Here $\deg(S) = \sum_{j \in S} \deg(\zeta_j) = \sum_{j \in S} n_j$.

(The elements b(S) are meant to be place-keepers. That is, we may regard V_i as the set of all formal symbols of the form $\sum_{|S|=i} \alpha_S \cdot b(S)$, $\alpha_S \in \Omega^{r+deg(S)}(M)$ with the

obvious addition and KG-action).

The boundary maps are defined by

$$\partial(\alpha \cdot b(S)) = \sum (-1)^{u_{S,j}} \gamma_{S,j}(\alpha) \cdot b(S \setminus \{j\})$$

where $u_{S,j} = \{i \in S \mid i < j \notin S$ and

$$\gamma_{S,j} : \Omega^{t+n_j}(M) \longrightarrow \Omega^t(M),$$

$t = t + \deg(S) - n_j$, is a cocycle representing ζ_j.

It should be recalled that $\Omega^a(M) \otimes \Omega^b(N) \cong \Omega^{a+b}(M \otimes N) \oplus (\mathrm{proj})$ for any a, b and modules M, N. Hence the exact sequence V(r) is precisely what $\tilde{D}(r)$ would be if it had no projective submodules in any of its terms. The connection with the index is involved with the fact that if N is irreducible then

$$\hat{H}^t(G, M^* \otimes N) \cong \hat{Ext}^t_{KG}(M, N) \cong Hom_{KG}(\Omega^t(M), N).$$

This is because $\Omega^t(M)$ has no projective submodules, and hence no element of $Hom_{KG}(\Omega^t(M), N)$ factors through a projective. So if the functor $Hom_{KG}(-, N)$ is applied to the sequence $V(r)$, then the result is precisely the complex $\mathcal{K}(\hat{H}^*(G, M^* \otimes N), \zeta_1, \cdots, \zeta_n; t)$. Likewise if the functor $Hom_{KG}(-,N)/PHom_{KG}(-,N)$ is applied to $\tilde{D}(r)$.

The proof of the theorem is a straightforward generalization of the proof of Proposition 3.3 in [7]. Actually we should note that, as in [7], the theorem has a strong converse. That is, if there exists a sequence $V(r)$ for all $r \geq s$, then $s \geq Ind(M^* \otimes N)$ for all simple modules N.

We end this section by considering connection between the index of modules and topics in the cohomology of groups. Let C be the complex described in Lemma 3.4. Then C is defined by a set of homogeneous parameters, $\zeta_1, \cdots, \zeta_n$, for $H^*(G, K)$. In [7] it was shown that C is a Poincaré duality complex and that a certain spectral sequence associated to C is related to the Koszul complexes for the cohomology of a module. Suppose that

$$P_*: \quad \cdots \longrightarrow P_1 \xrightarrow{\partial_1} P_0 \xrightarrow{\partial_0} P_{-1} \longrightarrow \cdots$$

is a complete projective resolution of K. Thus $\partial_0 = j \circ \epsilon$ where $\epsilon: P_0 \longrightarrow K$ is surjective and $j: K \longrightarrow P_{-1}$ is injective. Now consider the double complex $E_0 = \{E_0^{r,s}\}$ where

$$E_0^{r,s} = Hom_{KG}(P_r \otimes C_s, M).$$

This is the E_0-term of a spectral sequence which converges to zero because C_s is projective for all s. The E_1-term is obtained by taking the cohomology with respect to the boundary on C. Then it is not difficult to see that the E_2-page has the terms

$$E_2^{r,s} = \hat{Ext}^r_{KG}(H_s(C), M) \cong \hat{H}^r(G, H^s(C) \otimes M).$$

Remember that r can be negative as well as positive. So

$$E_2^{*,*} = \hat{H}^*(G, M) \otimes \Lambda(\hat{\zeta}_1, \cdots, \hat{\zeta}_n)$$

by Lemma 3.4 . In [7] we showed that the early differentials are given by the formula

$$d_{n_i-1}(\alpha\hat{\zeta}_i) = \zeta_i\alpha \cdot \hat{1}$$

for $\alpha \in \hat{H}^*(G, M)$. Hence the spectral sequence resembles a piecemeal version of the complex $\mathcal{K}(H^*(G, M); \zeta_1, \cdots, \zeta_n; t)$, at least in the early stages. The higher differentials in the spectral sequence are given by a form of twisted Massey product. However the existence of the index suggests that there are no higher differentials. It is not clear exactly what this means for the Massey product. This should be the subject of further study.

6. Variations on the theme, induction and some examples.

We have seen in the previous sections that the index can serve as a benchmark for the degree shift in Tate cohomology and for the translation operation Ω in the representation theory. However the question remains as to whether Ind($-$) is the most natural or best such marker. Ind(M) certainly does not reveal some properties of M. For example, Ind(M) might be $-\infty$, indicating that the cohomology $H^*(G, M)$ is periodic, even when M itself is not periodic. On the other hand, if M is not periodic then there must exist some module N such that $\text{Ind}(M \otimes N) > -\infty$ ($N = M^*$ will do [10]).

Some of the defects of the index can be corrected by some simple variations on the theme. In this section we explore a few of the variations with some discussion and examples. As before, K is a field of characteristic $p > 0$ and k is suitable Dedekind domain over which we can define Tate duality and the results of Section 4 hold (see Remark 4.4).

Definition 6.1 a) Let M be a kG-module. The global index of M is

$$\mathrm{GInd}(M) = \max\{\mathrm{ind}(M \otimes V)\}$$

where the maximum is taken over the set of all irreducible modules (lattices if char k = 0).

(b) Let the dual index of M be defined as the greatest integer s such that $\mathcal{K}(\hat{H}^*(G, M); \zeta_1, \cdots, \zeta_n; r - \sum \deg \zeta_i)$ is exact for all $r \leq s$. The dual index of M is denoted by Dnd(M). Let $\mathrm{Dnd}(M) = \infty$ if $H^*(G, M)$ is periodic.

(c) The width of M is

$$\mathrm{Wid}(M) = \mathrm{Ind}(M) - \mathrm{Dnd}(M).$$

We say that Wid(M) = 0 in case $H^*(G, M)$ is periodic.

Several remarks are in order. It is clear that the notion of the global index is motivated by Theorem 5.6. It was noted before that GInd(M) = 0 in all known examples where k = K and M is a simple KG-module. We recall that $(\Omega^s(M))^* \cong \Omega^{-s}(M^*)$ for any s and for any KG-module M. In addition we have a double dual theorem: $M^{**} \cong M$. Hence we can show the following.

Lemma 6.2 Suppose that $\zeta_1, \cdots, \zeta_n$ is a homogeneous set of parameters for $\mathrm{Ext}^*_{KG}(M, M)$ where M is a KG-module. If $r \leq -\mathrm{GInd}(M^*) - \sum_{i=1}^{n} \deg \zeta_i$, then there exists an exact sequence V(r) as in Theorem 5.6.

The proof is very easy. If $V(t, M^*)$ is the sequence defined in (5.6) for M^* then since $-r - \sum \deg \zeta_i \geq \mathrm{GInd}(M^*)$,

$$V(-r - \textstyle\sum \deg(\zeta_i), M^*) = V(r) = V(r, M)$$

is exact. The corollary of the lemma is that there exists a sequence of the form V(r) for all but a finite number of integers r . From the remark following Theorem 5.6 we see that a similar statement is true for the graded Koszul complexes. The better proof uses only Tate duality. So for $k = K$, $\hat{H}^i(G, M)$ is the dual of $\hat{H}^{-i-1}(G, M^*)$. Hence in this case

$$\mathcal{K}(\hat{H}^*(G, M^*); \zeta_1, \cdots, \zeta_n; r)$$
$$= \mathcal{K}(\hat{H}^*(G, M); \zeta_1, \cdots, \zeta_n; -r - \sum \deg(\zeta_i) - 1).$$

If k has characteristic zero then $H^i(G, M)$ is the dual of $\hat{H}^{-i}(G, M^*)$ and so we get a similar relationship. Therefore we have the following.

Lemma 6.3 The graded Koszul complex $\mathcal{K}(\hat{H}^*(G, M); \zeta_1, \cdots, \zeta_n; r)$ is an exact sequence if $r \leq -\mathrm{Ind}(M^*) - \sum_{i=1}^{n} \deg(\zeta_i) - 1$ when $k = K$ or $r \leq -\mathrm{Ind}(M^*) - \sum \deg(\zeta_i)$ when k has characteristic zero.

The result of the above is that $\mathrm{Dnd}(M) = -\mathrm{Ind}(M^*) - 1$ if $k = K$ and $\mathrm{Dnd}(M) = -\mathrm{Ind}(M^*)$ if k has characteristic zero.

The width of a module merely measures the number of times that the graded Koszul complex is not exact. Unlike the other invariants which we have defined, the width is not affected by translation. That is, $\mathrm{Dnd}(\Omega(M)) = \mathrm{Dnd}(M) + 1$ and similarly for $\mathrm{Glnd}(-)$. However $\mathrm{Wid}(\Omega(M)) = \mathrm{Wid}(M)$ for all M.

More generally we may view all of these invariants as functions on the set $\mathfrak{X}(kG)$ of isomorphic classes of indecomposable kG-modules. That is, M defines a function $f_M\colon \mathfrak{X}(kG) \longrightarrow \mathbb{Z} \cup \{-\infty\}$ given by $f_M(U) = \mathrm{Ind}(M \otimes_k U)$. The question may be asked as to when the functions distinguish between nonisomorphic nonperiodic modules. In some of the examples given below we can see that the answer may depend on the index of the trivial module and other questions. It can be checked that if $G \cong (\mathbb{Z}/2)^2$ and K has characteristic 2, then two nonperiodic indecomposable KG-modules have the property that $f_M = f_N$ if and only if M = N. This however is a very special case.

Before looking at examples we consider the relation of induction and restriction to the index. First note that if M and N are kG-modules then $\mathrm{Ind}(M \oplus N) = \max\{\mathrm{Ind}(M), \mathrm{Ind}(N)\}$. Therefore if M is a direct summand of N, then $\mathrm{Ind}(M) \leq \mathrm{Ind}(N)$. Suppose that H is a subgroup of G and that M is a kH-module. Then Frobenius reciprocity or the Eckmann-Shapiro Lemma says that $\hat{H}^t(G, M^{\uparrow G}) \cong \hat{H}^t(H, M)$ where $M^{\uparrow G} = kG \otimes_{kH} M$ and t is any integer. Therefore we have that $\mathrm{Ind}_G(M^{\uparrow G}) = \mathrm{Ind}_H(M)$, since it is easy enough to choose a homogeneous set of parameters for $H^*(G, M^{\uparrow G})$ whose restriction to H is a set of parameters for $H^*(H, M)$. That is, the diagram

$$\begin{array}{ccc} H^m(G, M^{\uparrow G}) & \xrightarrow{\ \gamma\ } & H^{m+n}(G, M^{\uparrow G}) \\ \| & & \| \\ H^m(H, M) & \xrightarrow{\ \mathrm{res}(\gamma)\ } & H^{m+n}(H, M), \end{array}$$

where γ means cup product with $\gamma \in H^n(G, k)$, commutes.

If k is a field of characteristic $p > 0$ or a complete discrete valuation ring whose residue class field has characteristic p, then vertices, sources and Green correspondents for kG-modules are defined. Suppose that the kG-module M has vertex Q. Let $H = N_G(Q)$ and let N be the kH-module which is the Green correspondent of M. Let S be a kQ-module which is a source for N and for M. Then N is a direct summand of $S^{\uparrow H}$ and M is a direct summand of $N^{\uparrow G}$. Therefore we have that

<u>Proposition 6.4</u> In the above notation

$$\mathrm{Ind}_G(M) \leq \mathrm{Ind}_H(N) \leq \mathrm{Ind}_Q(S).$$

Also $\mathrm{Dnd}_G(M) \geq \mathrm{Dnd}_H(N) \geq \mathrm{Dnd}_Q(S)$, so that

$$\mathrm{Wid}_G(M) \leq \mathrm{Wid}_H(N) \leq \mathrm{Wid}_Q(S).$$

We had stated earlier that in all known examples $\mathrm{Ind}_G(K) = 0$. The proposition, among other things, says that to prove the statement for all groups it is really only necessary to prove it for p-groups. Before going farther we should note the following.

Lemma 6.5 If $k = K$ is a field of characteristic $p > 0$ and if $H^*(G, K)$ is not periodic then $\mathrm{Ind}(K) \geq 0$. If k has characteristic 0, and if $H^*(G, k)$ is not periodic then $\mathrm{Ind}_G(k) \geq 1$.

Proof In the field case, $\hat{H}^{-1}(G, k)$ has dimension one and is generated by the class of the almost split sequence. But then for any $\zeta \in H^n(G, K)$, $n > 0$ we have $\zeta \cdot \hat{H}^{-1}(G, K) = 0$. The argument for the statement goes as follows. Let

$$E:\ 0 \longrightarrow \Omega^2(K) \longrightarrow B \longrightarrow K \longrightarrow 0$$

be the almost split sequence. Let ζ be represented by a cocycle

$$\hat{\zeta} \in \mathrm{Hom}_{KG}(\Omega^2(K), \Omega^{2-n}(K)).$$

Then $\mathrm{cls}(\hat{\zeta}E) = \zeta \cdot \mathrm{cls}(E) = \mathrm{cls}(E')$ as in the diagram

$$\begin{array}{ccccccccc}
E: & 0 & \longrightarrow & \Omega^2(K) & \longrightarrow & B & \longrightarrow & K & \longrightarrow 0 \\
 & & & \downarrow \hat{\zeta} & & \downarrow & & \| & \\
E': & 0 & \longrightarrow & \Omega^{2-n}(K) & \longrightarrow & U & \longrightarrow & K & \longrightarrow 0.
\end{array}$$

Here E' is the pushout along $\hat{\zeta}$. Because E is the almost split sequence and $\hat{\zeta}$ is not a splittable injection, E' is split. The result of all this is that $\mathcal{K}(\hat{H}^*(G, K); \zeta_1, \cdots, \zeta_n; -1)$ can not be exact for any choice of parameters.

For the case in which k has characteristic 0, the same argument can be made if the category has almost split sequences. However a better argument can be made using exponents. That is, we know that $\hat{H}^0(G, k) = k/|G|k$ and if the cohomology is not periodic then we can argue as in [1] that $\exp(H^n(G, k)) < \exp(H^0(G, k))$ for $n > 0$. So cup product with any $\zeta \in H^n(G, k)$ will not be injective on $\hat{H}^0(G, k)$. In particular it will not be injective on the p-part of $\hat{H}^0(G, k)$ for some prime p. Therefore, as before $\mathcal{K}(\hat{H}^*(G, k); \zeta_1, \cdots, \zeta_n ; 0)$ can not be exact.

Several very interesting examples can be constructed using almost split sequences. The relevance of using the almost split sequences is that the connecting homomorphisms in the long exact sequence for cohomology are almost all zero. Hence we get exact sequences of graded Koszul complexes. An illustration of the technique is given by the following.

Lemma 6.6 Suppose that M is an indecomposable KG-module and that

$$0 \longrightarrow \Omega^2(M) \xrightarrow{\alpha} U \xrightarrow{\beta} M \longrightarrow 0$$

is an almost split sequence. If $\Omega^n(M) \ncong K$ for all n, then $\mathrm{Ind}(U) = \mathrm{Ind}(\Omega^2(M)) = \mathrm{Ind}(M) + 2$. Likewise $\mathrm{Dnd}(U) = \mathrm{Dnd}(M)$ and hence $\mathrm{Wid}(U) = \mathrm{Wid}(M) + 2$.

Proof The condition in the hypothesis insures that

$$0 \longrightarrow \hat{H}^r(G, \Omega^2(M)) \xrightarrow{\alpha_*} \hat{H}^r(G, U) \xrightarrow{\beta_*} \hat{H}^r(G, M) \longrightarrow 0$$

is exact for all r. That is, every $\zeta \in H^r(G, M)$ is represented by a cocycle $\hat{\zeta}: \Omega^r(K) \longrightarrow M$ which is not a splittable epimorphism. So $\hat{\zeta} = \beta\mu$ for some $\mu: \Omega^r(K) \longrightarrow U$. Hence $\beta^*(\mathrm{cls}(\mu)) = \zeta$. So we have an exact sequence of complexes

$$0 \longrightarrow \mathcal{K}(\hat{H}^*(G, \Omega^2(M)); \zeta_1, \cdots, \zeta_n; r) \xrightarrow{\alpha_*} \mathcal{K}(\hat{H}^*(G, U); \zeta_1, \cdots, \zeta_n; r)$$

$$\xrightarrow{\beta_*} \mathcal{K}(\hat{H}^*(G, M); \zeta_1, \cdots, \zeta_n; r) \longrightarrow 0 .$$

Clearly the middle is exact if the two ends are exact. If $r = \text{Ind}(M)+1$, then the right end is exact while the left end is not. So the middle is not exact.

It is interesting to see what happens for the case in which $M \cong \Omega^n(K)$. Let P be the projective cover of K, and $\epsilon: P \longrightarrow K$ be surjective with kernel $\text{rad}(P) \cong \Omega(K)$. Similarly we have an injection $j: K \longrightarrow P$ with cokernel isomorphic to $\Omega^{-1}(K)$. The heart of P is the module $E = \text{rad}\, P/j(K)$. This gives us an exact sequence

$$A:\ 0 \longrightarrow \Omega(K) \xrightarrow{\ \alpha\ } E \otimes P \xrightarrow{\ \beta\ } \Omega^{-1}(K) \longrightarrow 0$$

which is known to be almost split. It should be pointed out that E is indecomposable except in a few tame cases such as occur when $p = 2$ and the Sylow 2-subgroup of G is $(\mathbb{Z}/2)^2$ or a dihedral group [22].

Lemma 6.7 Suppose that $\text{Ind}(K) > 0$. Then $\text{Ind}(E) = \text{Ind}(K) + 1 = \text{Ind}(\Omega(K))$.

The proof is the same as for the previous lemma because the map $H^r(G, E) \longrightarrow H^r(G, \Omega^{-1}(K))$ is surjective as long as $r \geq 0$.

On the other hand, if $\text{Ind}(K) = 0$ then a very interesting phenomenon occurs. Specifically, it is true that $\mathcal{K}(\hat{H}^*(G, \Omega(K));$ $\zeta_1, \cdots, \zeta_n; 0)$ is not exact. But its homology can be seen to be concentrated in degree zero and is isomorphic to K. That is, the homology comes from the element in the degree zero term which is $\hat{H}^0(G, \Omega(K)) \cong \hat{H}^{-1}(G, K)$. As was pointed out in the proof of (6.5), this cohomology group is generated by the class of the almost split sequence and this class is annihilated by $\alpha_*: \hat{H}^0(G, \Omega(K)) \longrightarrow \hat{H}^0(G, E)$. Therefore, by considering the exact sequence of complexes as in (6.6) we get that $\mathcal{K}(\hat{H}^*(G, E), \zeta_1, \cdots, \zeta_n; 0)$ is exact. Therefore we have

Lemma 6.8 If $\mathrm{Ind}(K) = 0$ then $\mathrm{Ind}(E) = 0$.

We mentioned in (5.4) that $\mathrm{Ind}(K) = 0$ if G is an elementary abelian p-group or a p-group of rank 2. But if G is a p-group and H is a proper subgroup then the restriction to H of the almost split sequence (A) is split. That is,

$$E|_H \cong \Omega^{-1}(K_H) \oplus \Omega(K_H) \oplus Q$$

where Q is a projective KH-module. Therefore

$$\mathrm{Ind}_H(E_H) = 1 > \mathrm{Ind}_G(E)$$

if $\mathrm{Ind}_G(K) = 0$, and H has p-rank at least 2. If G is a dihedral 2-group, $p = 2$, then $E = E_1 \oplus E_2$ where E_1 and E_2 are endo-trivial modules. That is $E_1 \otimes E_2 \cong K \oplus (\mathrm{proj})$ and $E_2 = E_1^*$. If H is an elementary abelian subgroup of order 4 then

$$E_i|_H \cong \Omega^{-1}(H) \oplus (\mathrm{proj})$$
$$E_j|_H \cong \Omega(H) \oplus (\mathrm{proj})$$

for some i, j with $\{i, j\} = \{1, 2\}$. But then

$$\mathrm{Ind}_H(E_i|_H) = -1 < \mathrm{Ind}_G(E_i).$$

We know that $\mathrm{Ind}_G(E_i) = 0$ because there is an automorphism of G which interchanges E_1 and E_2. See [3] for a general reference. The consequence of all of this is that the index seems not to commute with restriction in any way. This is in spite of the fact that it does commute with induction.

In the same direction we can recall that if M is any indecomposable KG-module then the tensor product sequence

$$M \otimes A: \quad 0 \longrightarrow M \otimes \Omega(K) \longrightarrow M \otimes E \oplus (\mathrm{proj}) \longrightarrow M \otimes \Omega^{-1}(K) \longrightarrow 0$$

is either split or almost split [3]. If split, then $\mathrm{Ind}(M \otimes E) = \mathrm{Ind}(M \otimes \Omega(K)) = \mathrm{Ind}(\Omega(M)) = \mathrm{Ind}(M) + 1$. If it is almost split, then the same result holds as long as $M \not\cong \Omega^t(K)$ for any t, by Lemma 6.6. By Lemma 6.7 we have that $\mathrm{Ind}(M \otimes E) = \mathrm{Ind}(M \otimes \Omega(K))$ for all M if $\mathrm{Ind}(K) > 0$. On the other hand if $\mathrm{Ind}(K) = 0$ then $\mathrm{Ind}(\Omega^n(K) \otimes E) = \mathrm{Ind}(\Omega^n(K))) = n$ for all n.

It is tempting to try to define a multiplicative index of a module M . Such a definition might go as:

$$\mathrm{MuInd}(M) = \max_{L}\{\mathrm{Ind}(M \otimes L) - \mathrm{Ind}(L)\}.$$

It is clear that $\mathrm{MuInd}(\Omega^n(K)) = n$ for all n. Also $\mathrm{MuInd}(E) = 1$ by the preceding discussion. However, in general, it is not clear that MuInd(M) is finite.

Suppose now that $\zeta \in H^n(G, K)$ is a regular element, a nondivisor of zero. Such elements always exist. Then we have an exact sequence

$$0 \longrightarrow K \xrightarrow{\hat{\zeta}} \Omega^{-n}(K) \longrightarrow L \longrightarrow 0$$

where $\hat{\zeta}$ represents ζ . Now $H^m(G, \Omega^{-n}(K)) \cong H^{m+n}(G, K)$ and $\zeta\colon H^m(G, K) \longrightarrow H^{n+m}(G, K)$ is injective for $m \geq 0$. Therefore we have exact sequences

$$0 \longrightarrow H^m(G, K) \xrightarrow{\zeta} H^{m+n}(G, K) \longrightarrow H^m(G, L) \longrightarrow 0$$

for all $m > 0$. If L is not periodic then we have, as before, that $\mathrm{Ind}(L) \leq \mathrm{Ind}(K)$. Moreover we have equality if $\mathrm{Ind}(K) > 0$. If $\mathrm{Ind}(K) = 0$ then it is more difficult to determine Ind(L) but it seems likely that $\mathrm{Ind}(L) = -1$ for the same reasons as in the case of the module E. Assuming that this is the case then $\mathrm{Ind}(L^* \otimes L) = \mathrm{Ind}(\mathrm{Hom}_K(L, L))$ is almost certain to be positive. This is the index of the cohomology ring of the module L. If in addition, ζ annihilates $\mathrm{Ext}^*_{KG}(L, L)$ then $L^* \otimes L \cong \Omega^n(L) \oplus \Omega(L)$ and we have that $\mathrm{Ind}(L^* \otimes L) = n + \mathrm{Ind}(L)$. All of this speculation can be confirmed with examples even if general proofs are not yet available.

7. The index and quasi-regular sequences.

In this section we assume that the coefficient ring k is noetherian and that $k/|G|k$ has finite composition length as a k-module. Hence if M is a finitely generated kG-module, then $\hat{H}^n(G,M)$, being finitely generated, will have finite composition length as a k-module, for all n. In Section 9 of [7] we introduced the notion of a quasi-regular sequence. It is a device for proving that the index of a module is zero. The idea can certainly be defined in a general setting even if we only use it in the context of group cohomology. However it should be emphasized that the definition given below is <u>not</u> equivalent to the one in [6].

<u>Definition 7.1</u> Let R be a graded commutative ring and let A be a finitely generated graded R-module. Let $\zeta_1, \cdots, \zeta_n$ be homogeneous elements of R such that $\deg\zeta_i = n_i > 0$. Let $A_m^{(i)} = A_m/A_m \cap (\zeta_1, \cdots, \zeta_i)A$. We say that $\zeta_1, \cdots, \zeta_n$ is a quasi-regular sequence for A if for each i, the map

$$\hat{\zeta}_i\colon A_m^{(i-1)} \longrightarrow A_{m+n_i}^{(i-1)},$$

given as multiplication by ζ_i is injective whenever $m \geq \sum_{j=1}^{i-1} n_j$. We say that

$\zeta_1, \cdots, \zeta_n$ is a complete quasi-regular sequence if, in addition, $\hat{\zeta}_n$ is an isomorphism.

Of course if each $\hat{\zeta}_i$ is injective for all m then $\zeta_1, \cdots, \zeta_n$ is a regular sequence for A. The aim of this section is to prove the following.

<u>Theorem 7.2</u> Let $R = \sum_{m \geq 0} H^{2m}(G, k)$ and let $A = \sum_{j \geq t} H^j(G, M)$, where M is a kG-module. Then $\mathrm{Ind}(M) \leq t$ if and only if there exists a complete quasi-regular sequence for A in R.

The "if" part of the theorem is a consequence of Proposition 9.10 of [7]. Consequently we may assume that $\mathrm{Ind}(M) \leq t$ and find a quasi-regular sequence. For the proof we need the following lemma. If X is a k-module, let $\Gamma(X)$ denote the composition length when it exists.

Lemma 7.3 Suppose that $\mathrm{Ind}(M) \leq t$ and that $\zeta_1, \cdots, \zeta_r$ is a quasi-regular sequence for $A = \sum_{j \geq t} \hat{H}^j(G, M)$. Suppose that $\eta_1, \cdots, \eta_r$ is a homogeneous set of

parameters for A; i.e. A is finitely generated over $k[\eta_1, \cdots, \eta_r]$. Assume also that $\deg(\zeta_i) = \deg(\eta_i)$ for all $i = 1, \cdots, r$. Then $\zeta_1, \cdots, \zeta_r$ is a complete quasi-regular sequence for A . That is, $\zeta_1, \cdots, \zeta_r$ is a system of parameters for A.

Proof We have that for any $m \geq t$, the complex $\mathcal{K}(A; \eta_1, \cdots, \eta_r; m)$ is an exact sequence. This implies that

$$\sum_{S \leq T} (-1)^{|S|}\, \Gamma(H^{m+deg(S)}(G, M)) = 0 \tag{7.4}$$

for $T = \{1, \cdots, r\}$, and $\deg S = \sum_{i \in S} \deg(\eta_i)$.

Now we claim that if $m \geq t$ then

$$H^*(\mathcal{K}(A; \zeta_1, \cdots, \zeta_i; m) = H^i(\mathcal{K}(A; \zeta_1, \cdots, \zeta_i; m))$$

$$\cong A^{(i)}_{m + \sum deg(\zeta_i)} \tag{7.5}$$

for every i. That is, we want to show that all of the cohomology is concentrated in degree i. For $i = 1$, the statement is clear because the complex has only the two terms:

$$\hat{H}^m(G, M) \xrightarrow{\zeta_1} \hat{H}^{m+n_1}(G, M),$$

and ζ_1 is an injection by hypothesis. The cokernel clearly is $A^{(1)}_{m+n_1}$. The proof of (7.5) is by induction on i. For $i > 1$ we have an exact sequence of cochain complexes

$$0 \longrightarrow \mathcal{K}^{[-1]}(A; \zeta_1, \cdots, \zeta_{i-1}; m + \deg(\zeta_i))$$

$$\longrightarrow \mathcal{K}(A; \zeta_1, \cdots, \zeta_i; m)$$

$$\longrightarrow \mathcal{K}(A;\, \zeta_1, \cdots, \zeta_{i-1}, m) \longrightarrow 0$$

where the "$[-1]$" is meant to indicate a degree shift. This sequence is similar to the one in Lemma 2.7. This gives us a long exact sequence on cohomology

$$\cdots \longrightarrow H^{j-1}(\mathcal{K}(A;\, \zeta_1, \cdots, \zeta_{i-1}; m))$$

$$\longrightarrow H^j(\mathcal{K}^{[-1]}(A;\, \zeta_1, \cdots, \zeta_{i-1}, m + \deg \zeta_i))$$

$$\longrightarrow H^j(\mathcal{K}(A;\, \zeta_1, \cdots, \zeta_i;\, m)) \longrightarrow \cdots$$

By induction, it is only necessary to look at the portion in which $j = i$. The connecting homomorphism ζ must be cup product with $\pm\, \zeta_i$. So the sequence looks like

$$0 \longrightarrow A_\ell^{(i-1)} \xrightarrow{\ \zeta_i\ } A^{(i)}_{\ell + deg(\zeta_i)} \longrightarrow A^{(i)}_{\ell + n_i} \longrightarrow 0$$

when the left end is zero because ζ_i is injective and $\ell = \sum_{j=1}^{i-1} \deg\zeta_j + m$. This proves (7.5).

Now by (7.4), for $m \geq t$,

$$0 = \sum_{i=0}^{r} (-1)^i \Gamma(\mathcal{K}_i(A;\, \zeta_1, \cdots, \zeta_r;\, m))$$

$$= \sum_{i=0}^{r} (-1)^i\, \Gamma(H^i(\mathcal{K}(A;\, \zeta_1, \cdots, \zeta_r;\, m)))$$

$$= (-1)^r\, \Gamma(H^r(\mathcal{K}(A;\, \zeta_1, \cdots, \zeta_r;\, m)))\, .$$

Therefore, $H^r(\mathcal{K}(A;\, \zeta_1, \cdots, \zeta_r;\, m))$ is exact. It is then necessary that

$$\hat{H}^{m+\sum n_j}(G, M) \subseteq (\zeta_1, \cdots, \zeta_r)A.$$

That $\zeta_1, \cdots, \zeta_r$ is a homogeneous system of parameters is now obvious. The fact that the quasi-regular sequence is complete follows from Proposition 9.10 of [7].

Proof of Theorem 7.2 Let $\theta_1 \cdots, \theta_n$ be a homogeneous set of parameters for $A = \sum_{j \geq t} \hat{H}^j(G, M)$. Here $\theta_i \in H^{n_i}(G, k)$. Because $t \geq \mathrm{Ind}(M)$ the map

$$A \longrightarrow A^n$$

given by $\gamma \longrightarrow (\theta_1\gamma, \cdots, \theta_n\gamma)$ is injective. By Lemma 4.2 there exists a homogeneous element $\zeta_1 \in k[\theta_1, \cdots, \theta_n]$ such that the cup product

$$\zeta_1\colon A \longrightarrow A$$

is injective. Note here that we can replace ζ_1 by any power of ζ_1, if necessary, and the statement remains true. So now we can make an exact sequence of complexes, for $m \geq t$,

$$0 \longrightarrow \mathcal{K}(A; \theta_1, \cdots, \theta_n; m)$$

$$\longrightarrow \mathcal{K}(A; \theta_1, \cdots, \theta_n; m + \deg \zeta_1)$$

$$\longrightarrow \mathcal{K}(A/\zeta_1 A; \theta_1, \cdots, \theta_n; m + \deg \zeta_1) \longrightarrow 0$$

because $\zeta_1\colon H^{m+deg(S)}(G, M) \longrightarrow H^{m+deg(S)+deg(\zeta_1)}(G, M)$ is injective. Now the first two terms are exact sequences and so, by the long exact sequence on homology, the third is also exact. This says that $\mathrm{Index}(A/\zeta_1 A) \leq t + \deg \zeta_1$.

This argument can now be repeated on $A/\zeta_1 A$ to yield an element $\zeta_2 \in H^*(G,k)$ with the property that

$$\zeta_2\colon A_{m+deg\zeta_1}/\zeta_1 A_m \longrightarrow A_{m+deg\zeta_1+deg\zeta_2}/\zeta_1 A_{m+deg\zeta_2}$$

is injective for $m \geq t$. Also we get that $\mathcal{K}(A/(\zeta_1, \zeta_2)A; \theta_1, \cdots, \theta_n; m + \deg\zeta_1 + \deg\zeta_2)$ is exact for all $m \geq t$. Inductively we are constructing a quasi-regular sequence of arbitrarily long length. But note that we can replace any ζ_i by a power of itself. So we can be certain that the degrees of $\zeta_1, \cdots, \zeta_n$ match with the degrees of a homogeneous system of parameters. Therefore Lemma 7.3 finishes the proof.

8. Contravariantly finite subcategories.

We end the paper with a brief discussion of two questions which were raised at the conference in Tsukuba. Both of them were concerned with Auslander's lecture on contravariantly finite subcategories and, in particular, those subcategories which are closed under extensions. Questions were asked about two subcategories of mod(KG). Both are closed under extensions and under translations by Ω and Ω^{-1}. Both fail to be contravariantly finite. The proofs of the failure given at the conference were both indirect, using Wakamatsu's Lemma or some variation thereon. Subsequent study [13] has uncovered more general conditions which are necessary for a subcategory to be contravariantly finite. Nonetheless it is interesting to see how the module theory can be used. See [4] for a general reference on this subject.

Throughout this section let K be a field of characteristic $p > 0$, and G a finite group. Let mod(KG) denote the category of finitely generated left KG-modules. Recall that a subcategory $\mathcal{C}$ of mod(KG) is contravariantly finite if every KG-module M has a right $\mathcal{C}$-approximation. A right $\mathcal{C}$-approximation is a homomorphism $f: U \longrightarrow M$ where U is in $\mathcal{C}$, such that if V is in $\mathcal{C}$ and $\sigma: V \longrightarrow M$ is a homomorphism then there exists $\tau: V \longrightarrow U$ such that $f\tau = \sigma$. In the discussion to follow, we also assume that subcategories are full and are closed under isomorphisms, direct sums and summands. We state

Wakamatsu's Lemma without proof.

<u>Lemma 8.1</u> [21] Let $\mathcal{C}$ be a subcategory of mod(KG) which is closed under extensions. Let M be a KG-module and $f: U \longrightarrow M$ a right minimal $\mathcal{C}$-approximation of M . Let X be the kernel of f. Then $\mathrm{Ext}^1_{KG}(V, X) = 0$ for all V in $\mathcal{C}$.

(8.2) Let $\mathcal{C}$ be the full subcategory of mod(KG) consisting of all modules which are direct summands of trivial homology modules (see [8] for a definition). This subcategory is closed under extensions, duals, tensor products and translations by Ω and Ω^{-1} . It is not contravariantly finite unless every module in the principal block is a direct summand of a trivial homology module. The reason can be sketched as follows. Suppose that M is an indecomposable module in the principal block of KG, such that M is not a summand of a trivial homology module. Let $f: U \longrightarrow M$ be the minimal right $\mathcal{C}$-approximation. Then f must be surjective since the projective modules are in $\mathcal{C}$. Consequently we have an exact sequence

$$0 \longrightarrow X \longrightarrow U \xrightarrow{f} M \longrightarrow 0$$

where $\mathrm{Ext}^1_{KG}(V, X) = 0$ for all V in $\mathcal{C}$ by Wakamatsu's Lemma. By the translation invariance we have that $\hat{\mathrm{Ext}}^m_{KG}(V, X) = 0$ for all m and V in $\mathcal{C}$. Then by Tate duality $\hat{\mathrm{Ext}}^m_{KG}(V, X^*) = 0$ for all m , V. Let I be the injective hull of U. If we add I to M we get an exact sequence

(8.3) $$0 \longrightarrow U \longrightarrow M \oplus I \longrightarrow \Omega^{-1}(X) \longrightarrow 0.$$

But since $\mathrm{Ext}^1_{KG}(\Omega^{-1}(X), U) = \hat{\mathrm{Ext}}^2_{KG}(U^*, X^*) = \{0\}$, the sequence (8.3) must split. Then by the Krull-Schmidt theorem, M is the direct sum of a module in $\mathcal{C}$ and a module in the orthogonal category. However we now have a contradiction to the fact that the principal block is Ext-connected.

(8.4) Let H be a subgroup of G. Let $\mathfrak{C}_H$ be the full subcategory of mod(KG) consisting of all modules whose restrictions to H are projective. Then $\mathfrak{C}_H$ is also closed under extensions, duals, tensor products, and translations by Ω and Ω^{-1}. It is contravariantly finite only in the trivial cases.

<u>Proposition 8.5</u> The subcategory $\mathfrak{C}_H$ is contravariantly finite if either

(i) H contains no element of order p.

or (ii) H contains a conjugate of every elementary abelian p-subgroup.

In all other cases $\mathfrak{C}_H$ is not contravariantly finite.

<u>Proof</u> In case (i) $\mathfrak{C}_H = \text{mod(KG)}$. In case (ii) $\mathfrak{C}_H$ is the full subcategory of projective KG-modules by Chouinard's Theorem (see (2.24.5) of [5]). Clearly both of these are contravariantly finite.

In the notation of Section 3, let $V_H = V_G(K_H^{\uparrow G})$ be the subvariety of $V_G(K)$ corresponding to the induced module. A KG-module M is projective on restriction to H if and only if $(M_H)^{\uparrow G} = K_H^{\uparrow G} \otimes M$ is projective as a KG-module. By Proposition 3.1, M is projective on restriction to H if and only if $V_G(M) \cap V_H = \{0\}$. The cases (i) and (ii) are precisely the cases in which $V_H = \{0\}$ and $V_H = V_G(K)$ respectively. So suppose that H satisfies neither of the conditions (i) or (ii). Then $\{0\} \subset V_H \subset V_G(K)$.

We assume that $\mathfrak{C}_H$ is contravariantly finite and work towards a contradiction. There exists a module U in $\mathfrak{C}_H$ and a map

$$\theta\colon U \longrightarrow K$$

which is a $\mathfrak{C}_H$-approximation of K. Now $V_G(U) \cap V_H = \{0\}$ and the projectivized variety of $V_G(K)$ is connected (see [11]). So $V_H \cup V_G(U) \neq V_G(K)$. So there must exist a nonzero closed homogeneous subvariety W with $W \cap V_H = \{0\}$ and

$W \cap V_G(U) = \{0\}$. By Proposition 3.1, there is a KG-module M with $V_G(M) = W$. Because $V_G(M) \cap V_H = \{0\}$, M is in $\mathfrak{C}_H$. Moreover, $M^* \otimes M = \mathrm{Hom}_K(M, M)$ is also in $\mathfrak{C}_H$. Let Tr denote the trace map from $\mathrm{Hom}_K(M, M)$ to K. This is a KG-homomorphism and hence there is a map σ such that the diagram

$$\begin{array}{ccc} & & \mathrm{Hom}_K(M,M) \\ & \overset{\sigma}{\swarrow} & \downarrow{\scriptstyle \mathrm{Tr}} \\ U & \xrightarrow{\ \ \theta\ \ } & K \end{array}$$

commutes.

But we have also that $V_G(U) \cap V_G(M) = \{0\}$ so that $M^* \otimes M \otimes U = \mathrm{Hom}_K(\mathrm{Hom}_K(M, M), U)$ is projective. So any map from $\mathrm{Hom}_K(M, M)$ to U factors through a projective module. It follows that Tr must factor through a projective implying that M is a projective module (c.f. [3]). The contradiction completes the proof.

The contradiction can also be obtained from Wakamatsu's Lemma (8.1) by looking at the long exact sequence

$$\cdots \longrightarrow \mathrm{Hom}_{KG}(M^* \otimes M, U) \xrightarrow{\theta_*} \mathrm{Hom}_{KG}(M^* \otimes M, K)$$
$$\longrightarrow \mathrm{Ext}^1_{KG}(M^* \otimes M, \ker\theta) \longrightarrow \cdots.$$

That is, Tr can not be in the image of θ_* and so $\mathrm{Ext}^1_{KG}(M^* \otimes M, \mathrm{Ker}\theta) \neq 0$ as claimed.

References

[1] A. Adem, Cohomological exponents of ZG-lattices, J. Pure and Appl. Algebra 58 (1989), 1-5.

[2] A. Adem, J. Maginnis and R. J. Milgram, The geometry and cohomology of the Mathieu group M12, J. Algebra, 139 (1991), 90-133.

[3] M. Auslander and J. F. Carlson, Almost split sequences and group algebras, J. Algebra 103 (1986), 122-140.

[4] M. Auslander and I. Reiten, Applications of contravariantly finite subcategories, Advances in Math. (to appear)

[5] D. Benson, Modular Representation Theory: New Trends and Methods, Lecture Notes in Math. 1081, Springer-Verlag, New York, 1984.

[6] D. J. Benson and J. F. Carlson, Complexity and multiple complexes, Math. Zeit. 195 (1987), 221-238.

[7] D. J. Benson, J. F. Carlson, Projective resolutions and Poincaré duality complexes, Trans. Amer. Math. Soc. (to appear).

[8] D. J. Benson, J. F. Carlson, and G. R. Robinson, On the vanishing of cohomology, J. Algebra 131 (1990), 40-73.

[9] K. S. Brown, Cohomology of Groups, Springer-Verlag, New York, 1982.

[10] J. F. Carlson, Complexity and Krull dimension, Representations of Algebras, Lecture Notes in Math. 903, Springer-Verlag, New York, 1981, pp. 62-67.

[11] J. F. Carlson, The variety of an indecomposable module is connected, Invent. Math. 77 (1984), 291-299.

[12] J. F. Carlson, Varieties for modules, The Arcata Conference on Representations of Finite Groups, Proc. Sym. Pure Math. 47, Amer. Math. Soc., Providence, 1987, pp. 37-44.

[13] J. F. Carlson and D. Happel, Contravariantly finite subcategories and irreducible maps, Proc. Amer. Math. Soc. (to appear).

[14] C. W. Curtis and I. Reiner, Methods of Representation Theory, Vol. 1, Wiley-Interscience, New York, 1981.

[15] L. Evens, The cohomology ring of a finite group, Trans. Amer. Math. Soc. 101 (1961), 224-239.

[16] S. Mac Lane, Homology, Springer-Verlag, New York, 1967.

[17] H. Matsumura, Commutative Algebra, W. A. Benjamin, New York, 1970.

[18] D. Quillen, The mod-2 cohomology ring of extra-special 2-groups and the spinor groups, Math. Ann. 194 (1971), 197-212.

[19] D. Quillen, On the cohomology and K-theory of the general linear groups over a finite field, Ann. of Math. 96 (1972), 552-556.

[20] R. G. Swan, Groups with no odd dimensional cohomology, J. Algebra 17 (1971), 401-403.

[21] T. Wakamatsu, On modules with trivial self extensions, J. Algebra 114 (1988), 106-114.

[22] P. J. Webb, The Auslander-Reiten quiver of a finite group, Math. Zeit. 179 (1982), 97-121.

MODULES OF FINITE LENGTH OVER THEIR ENDOMORPHISM RINGS

WILLIAM CRAWLEY-BOEVEY

Mathematical Institute, Oxford University,
24-29 St. Giles, Oxford OX1 3LB, England

Given a ring R (associative, with 1) one can define the *endolength* of an R-module M to be its length when it is regarded in the natural way as an $\text{End}_R(M)$-module, and thus one can consider the class of modules of finite endolength. The aim of this paper is to show that this is a useful concept. Briefly, the contents are as follows. In §§1-3 we cover some background machinery, in §§4-6 we discuss the modules of finite endolength for a general ring, and in §§7-9 we show how these modules control the behaviour of the finite length modules for noetherian and artin algebras. Although much of this paper has a survey nature, there are some new results proved here, the main ones being the characterization of the pure-injective modules which occur as the source of a left almost split map in §2, the character theory for modules of finite endolength in §5, and the characterization of the artin algebras with an indecomposable module of infinite length and finite endolength (a *generic* module) proved in §§8-9.

The author is supported by an SERC Advanced Research Fellowship. I would like to thank M. Prest and A. H. Schofield for some useful discussions.

1 THE FUNCTOR CATEGORY

If R is a ring, we denote by R-Mod the category of left R-modules, and by mod-R be the category of finitely presented (f.p.) right R-modules. We denote by D(R) the category of additive functors from mod-R to $\mathbb{Z}$-Mod (the category of abelian groups). This category has a very rich structure, and is an invaluable tool for the study of R-modules. First of all, it is an abelian category, with kernels, images and cokernels computed "pointwise". For example, a morphism of functors $f:\mathcal{F}\longrightarrow\mathcal{G}$ is by definition a natural transformation, so that for each f.p. module X there is a homomorphism $f_X:\mathcal{F}(X)\longrightarrow\mathcal{G}(X)$ of abelian groups. The kernel of f is then the functor which is defined on objects by $(\mathrm{Ker}\ f)(X) = \mathrm{Ker}(f_X)$, and similarly for the image and cokernel of f. It follows that a sequence $\mathcal{F}\longrightarrow\mathcal{G}\longrightarrow\mathcal{H}$ of functors is exact if and only if the sequence $\mathcal{F}(X)\longrightarrow\mathcal{G}(X)\longrightarrow\mathcal{H}(X)$ is exact for all f.p. modules X.

We shall also need to use the fact that D(R) is a Grothendieck category, which as far as we are concerned means that it has injective envelopes [G], and in particular, that we can do homological algebra in D(R). There are several very important classes of functors which we now list.

1.1 The *representable functors* are the functors $(X,-) = \mathrm{Hom}_R(X,-)$ with X a f.p. right R-module. By Yoneda's lemma, $\mathrm{Hom}_{D(R)}((X,-),\mathcal{F})$ is isomorphic to $\mathcal{F}(X)$ for any functor $\mathcal{F}$, and it follows that (X,-) is a projective object in D(R).

Given a functor $\mathcal{F}$, a family of finitely presented modules $(X_\lambda)_{\lambda\in\Lambda}$, and elements $f_\lambda \in \mathcal{F}(X_\lambda)$, one can consider the smallest subfunctor $\mathcal{G}$ of $\mathcal{F}$ such that $f_\lambda \in \mathcal{G}(X_\lambda)$ for all $\lambda\in\Lambda$. This is the subfunctor of $\mathcal{F}$ *generated by* the elements f_λ. Now the f_λ determine maps $(X_\lambda,-)\longrightarrow\mathcal{F}$, and the functor $(X_\lambda,-)$ is generated by the identity endomorphism of X_λ, so that $\mathcal{G}$ is the image of the map $\coprod_{\lambda\in\Lambda}(X_\lambda,-)\longrightarrow\mathcal{F}$. Defining the notion of a *finitely generated* (f.g.) functor in the obvious way, the isomorphism

$$(X_1,-) \oplus \ldots \oplus (X_n,-) \cong (X_1\oplus\ldots\oplus X_n,-)$$

shows that a functor is finitely generated if and only if it is a quotient of a representable functor.

Finally let us observe that the representable functors are precisely the f.g. projective functors. Namely, any f.g. projective functor $\mathcal{F}$ is a summand of a representable functor $(X,-)$, and then the isomorphism $\mathrm{End}_{D(R)}((X,-)) \cong \mathrm{End}_R(X)^{op}$ shows that $\mathcal{F} \cong (Y,-)$ with Y a summand of X (so that it is f.p.).

1.2 A functor $\mathcal{F}$ is said to be *coherent* if it is a quotient of a representable functor by a finitely generated subfunctor, or in other words, if it is finitely presented. If $\mathcal{F}$ has projective presentation

$$(Y,-) \xrightarrow{f} (X,-) \longrightarrow \mathcal{F} \longrightarrow 0, \qquad (\dagger)$$

then f is determined by a homomorphism $\alpha: X \longrightarrow Y$, and if Z is the cokernel of α, then $\mathcal{F}$ actually has a projective resolution

$$0 \longrightarrow (Z,-) \longrightarrow (Y,-) \xrightarrow{f} (X,-) \longrightarrow \mathcal{F} \longrightarrow 0. \qquad (\ddagger)$$

The name "coherent" (rather than "finitely presented") is used because of the following well-known and important property, which implies that the category of coherent functors is closed under kernels, cokernels, images and extensions.

LEMMA. A f.g. subfunctor of a coherent functor is coherent.

Proof. It suffices to prove this for representable functors. Now a f.g. subfunctor of $(X,-)$ is the image of a map f as in $(\dagger)$, and so the projective resolution $(\ddagger)$ shows that Im(f) is coherent.

1.3 The simple functors in D(R) have been determined by Auslander [A3]. If X is a f.p. right R-module and J is a left ideal in $\mathrm{End}_R(X)$ one can define a subfunctor $(X,-)_J$ of $(X,-)$ via

$$(X,Y)_J = \{\theta \in \mathrm{Hom}_R(X,Y) \mid \phi \circ \theta \in J \text{ for all } \phi \in \mathrm{Hom}_R(Y,X)\}$$

If $\mathfrak{m}$ is a maximal left ideal in $\mathrm{End}_R(X)$ we set $\mathcal{S}_{X,\mathfrak{m}}=(X,-)/(X,-)_{\mathfrak{m}}$.

LEMMA. The functors $\mathcal{S}_{X,\mathfrak{m}}$ are simple, and every simple functor in D(R) is isomorphic to some $\mathcal{S}_{X,\mathfrak{m}}$.

Proof. If $\mathcal{F}$ is a subfunctor of $(X,-)$, then $J = \mathcal{F}(X)$ is a left ideal in $\mathrm{End}_R(X)$, and the inclusion $\mathcal{F} \subseteq (X,-)_J$ follows from the definition. If in addition $\mathcal{F}$ is a proper subfunctor of $(X,-)$ then J is a proper ideal, for if $1_X \in \mathcal{F}(X)$ then certainly $\mathcal{F} = (X,-)$. It follows that the maximal subfunctors of $(X,-)$ are precisely the functors $(X,-)_{\mathfrak{m}}$ with $\mathfrak{m}$ a maximal left ideal in $\mathrm{End}_R(X)$. Thus $\mathcal{S}_{X,\mathfrak{m}}$ is simple. Finally, one only has to observe that, by Yoneda's lemma, every simple functor is a quotient of a representable functor.

1.4 If M is a left R-module, the tensor product functor $-\otimes M = -\otimes_R M$ is right exact. Conversely, if $\mathcal{F}$ is any right exact functor then $\mathcal{F} \cong -\otimes\mathcal{F}(R)$ where $\mathcal{F}(R)$ has its natural structure as a left R-module. It is easy to see that $\mathrm{Hom}_{D(R)}(-\otimes M,-\otimes N) \cong \mathrm{Hom}_R(M,N)$. The next property will be needed later.

LEMMA. $\mathrm{Ext}^1_{D(R)}(\mathcal{F},-\otimes M) = 0$ for $\mathcal{F}$ coherent.

Proof. The functor $\mathcal{F}$ has a projective resolution (‡), so one can compute Ext^1 as the cohomology of the complex

$$\mathrm{Hom}((X,-),-\otimes M) \longrightarrow \mathrm{Hom}((Y,-),-\otimes M) \longrightarrow \mathrm{Hom}((Z,-),-\otimes M).$$

This is, however, isomorphic to $X\otimes M \longrightarrow Y\otimes M \longrightarrow Z\otimes M$, so it is exact.

1.5 An exact sequence $\xi : 0 \longrightarrow M \xrightarrow{\alpha} N \xrightarrow{\beta} L \longrightarrow 0$ of left R-modules is said to be *pure exact* (and α a *pure mono*, and $\mathrm{Im}(\alpha)$ a *pure submodule* of M) provided that the tensor product sequence

$$0 \longrightarrow X\otimes_R M \longrightarrow X\otimes_R N \longrightarrow X\otimes_R L \longrightarrow 0$$

is exact for every right R-module X. Of course it is always right exact. Since tensor products commute with direct limits, direct limits are exact, and every module is a direct limit of f.p. modules, it suffices for this to hold for f.p. modules X. In other words, ξ is pure exact if and only if the sequence of functors $0 \longrightarrow -\otimes M \longrightarrow -\otimes N \longrightarrow -\otimes L \longrightarrow 0$ is exact in D(R).

1.6 A left R-module M is said to be *pure-injective* if every pure exact sequence whose first term is M splits. This notion has many equivalents, for example that of algebraic compactness. For our purposes, however, we shall only need the following characterization [GJ2].

LEMMA. Up to isomorphism, the injectives in D(R) are the functors $-\otimes M$ with M pure-injective.

Proof. Let $\mathcal{F}$ be an injective functor. We show first that $\mathcal{F}$ is right exact. Namely, an exact sequence $X \longrightarrow Y \longrightarrow Z \longrightarrow 0$ of f.p. right R-modules gives an exact sequence $0 \longrightarrow (Z,-) \longrightarrow (Y,-) \longrightarrow (X,-)$ of functors. Applying the exact functor $\mathrm{Hom}(-,\mathcal{F})$ and using Yoneda's lemma one sees that the sequence $\mathcal{F}(X) \longrightarrow \mathcal{F}(Y) \longrightarrow \mathcal{F}(Z) \longrightarrow 0$ is exact, as required. Thus $\mathcal{F} \cong -\otimes M$ where $M = \mathcal{F}(R)$. Now given any pure mono $M \longrightarrow N$, the morphism $-\otimes M \longrightarrow -\otimes N$ is mono, so split, so the map $M \longrightarrow N$ is split. Thus M is pure-injective.

Conversely, suppose that M is pure-injective. Since the category D(R) has injective envelopes, the functor $-\otimes M$ can be embedded in an injective functor $-\otimes N$. This gives a pure embedding $M \longrightarrow N$. Since M is pure-injective, M is a summand of N, so $-\otimes M$ is a summand of $-\otimes N$, and hence injective.

1.7 Next we introduce another tool for studying modules, the subgroups of "finite definition" of a module. These were introduced by Gruson and Jensen, and by Zimmermann. In Azumaya's article [Az] they are called "finite matrix subgroups." An additive subgroup of a left R-module M is said to be of *finite definition* if it arises as the kernel $F_{X,x}(M)$ of a map

$$M \longrightarrow X\otimes_R M, \quad m \longmapsto x\otimes m$$

for some f.p. right R-module X, and some element $x\in X$. These subgroups are not necessarily R-submodules of M, but they are $\mathrm{End}_R(M)$-submodules.

There is an equivalent definition as follows. If $\mathcal{F}$ is a subfunctor of (R,−) and M is a left R-module, then the space $F_{\mathcal{F}}(M) = \mathrm{Hom}((R,-)/\mathcal{F},-\otimes M)$ can be regarded as the additive subgroup of $\mathrm{Hom}((R,-),-\otimes M)$ on the maps which annihilate $\mathcal{F}$. Moreover, by Yoneda's lemma, the last space can be identified with M. Thus $F_{\mathcal{F}}(M)$ is canonically an additive subgroup of M. The subgroups of M which arise in this way using f.g. subfunctors $\mathcal{F}$ of (R,−) are the subgroups of M of finite definition. To see this one only has to note that any f.g. subfunctor of (R,−) is the image of a map $(X,-)\longrightarrow(R,-)$ for some f.p. module X, and by Yoneda's lemma this map is determined by a map $R\longrightarrow X$, and hence by an element $x\in X$.

We list some basic properties of the subgroups of finite definition.

(1) The subgroups of finite definition form a lattice in M, for

$$F_{X,x}(M) \cap F_{Y,y}(M) = F_{X\oplus Y,x+y}(M), \text{ and}$$
$$F_{X,x}(M) + F_{Y,y}(M) = F_{Z,z}(M)$$

where Z is the cokernel of the map $R\longrightarrow X\oplus Y$ sending 1 to x−y, and z is the common image of x and y.

(2) If N is a pure submodule of M then $F_{X,x}(N) = N \cap F_{X,x}(M)$.

(3) If $(M_\lambda)_{\lambda\in\Lambda}$ is a family of modules then

$$F_{X,x}(\coprod_{\lambda\in\Lambda} M_\lambda) = \coprod_{\lambda\in\Lambda} F_{X,x}(M_\lambda), \text{ and}$$
$$F_{X,x}(\prod_{\lambda\in\Lambda} M_\lambda) = \prod_{\lambda\in\Lambda} F_{X,x}(M_\lambda),$$

the latter expression holding since tensor products $X\otimes_R-$ commute with products when X is a f.p. module.

2 INJECTIVE ENVELOPES OF SIMPLE FUNCTORS

Having determined the injective functors in the category D(R), it is worthwhile to characterize the injective envelopes of the simple functors. That is, to determine the indecomposable pure-injective modules M such that $-\otimes M$ has a simple subfunctor. This is not strictly necessary for our study of modules of finite endolength, but it is an important finiteness condition which will be relevant later.

2.1 In this paragraph we compute the injective envelope of the simple functor $\mathcal{S}_{X,\mathfrak{m}}$.

Lemma. Let X be a f.p. right R-module, $E = \mathrm{End}_R(X)$ and $\mathfrak{m}$ a maximal left ideal in E. If I is the injective envelope of the E-module $E/\mathfrak{m}$, and $M = \mathrm{Hom}_E(X, I)$, then the injective envelope of $\mathcal{S}_{X,\mathfrak{m}}$ is $-\otimes M$.

Proof. Let $\theta: E \longrightarrow I$ be an E-module map inducing an isomorphism from $E/\mathfrak{m}$ to $\mathrm{soc}_E(I)$, and let f be the morphism

$$(X,-) \longrightarrow \mathrm{Hom}_E(\mathrm{Hom}_R(-,X), I)$$

which when applied to a f.p. module Y sends a map $\phi \in \mathrm{Hom}_R(X,Y)$ to the map $\mathrm{Hom}_R(Y,X) \longrightarrow I$ which sends ψ to $\theta(\psi\circ\phi)$. It is clear that the kernel of f is $(X,-)_{\mathfrak{m}}$, so that $\mathcal{S}_{X,\mathfrak{m}}$ embeds in $\mathrm{Hom}_E(\mathrm{Hom}_R(-,X), I)$. Now observe that the natural map

$$-\otimes M \longrightarrow \mathrm{Hom}_E(\mathrm{Hom}_R(-,X), I)$$

is an isomorphism since both functors are right exact and they agree on R. Thus there is an embedding of $\mathcal{S}_{X,m}$ in $-\otimes M$. Now M is pure-injective since $\mathrm{Hom}_R(-,M) \cong \mathrm{Hom}_E(X\otimes -, I)$ is exact on pure exact sequences, so that if $\xi: 0 \longrightarrow M \longrightarrow Y \longrightarrow Z \longrightarrow 0$ is pure, then the map $\mathrm{Hom}_R(Y,M) \longrightarrow \mathrm{Hom}_R(M,M)$ is onto, and hence ξ splits. Finally

$$\begin{aligned}\mathrm{End}_R(M) &\cong \mathrm{Hom}_E(X\otimes_R M, I) \cong \\ &\cong \mathrm{Hom}_E(\mathrm{Hom}_E(\mathrm{Hom}_R(X,X),I),I) \cong \mathrm{End}_E(I)\end{aligned}$$

is a local ring. It follows that $-\otimes M$ is the injective envelope of the functor $\mathcal{S}_{X,m}$.

2.2 Recall that a map $\alpha: M \longrightarrow N$ of R-modules is said to be *left almost split* if it is not a split mono, and any map $M \longrightarrow X$ which is not split mono factors through α. Dually $\beta: N \longrightarrow M$ is *right almost split* if it is a not split epi, and any map $X \longrightarrow M$ which is not split epi factors through β. Taking X = M one sees that such maps can only exist if M has local endomorphism ring. Moreover, if α exists then it is not a pure mono if and only if M is pure-injective, while if β exists then it is not a pure epi (the epi in a pure exact sequence) if and only if M is f.p.. Now Auslander has shown that if M is f.p. and has local endomorphism ring then there is a right almost split map terminating at M. this paragraph we treat the case of left almost split maps.

2.3 Theorem. Let M be an indecomposable pure-injective R-module. The following statements are equivalent.

(1) M is the source of a left almost split map $\alpha: M \longrightarrow N$.

(2) $-\otimes M$ is the injective envelope of a simple functor.

(3) $M \cong \mathrm{Hom}_E(X, I)$ with X some f.p. right R-module, $E = \mathrm{End}_R(X)$ and I the injective envelope of a simple left E-module.

Proof. The equivalence (2) ⇔ (3) follows from (1.3) and (2.1). Suppose that (1) holds and let $\mathcal{S}$ be the kernel of the morphism $-\otimes M \longrightarrow -\otimes N$. If $\mathcal{S} = 0$ then the inclusion of M in N is pure, and so a split mono since M is pure- injective, a contradiction. Now if $\mathcal{S}$ is not simple, say with proper non-zero subfunctor $\mathcal{F}$ then $(-\otimes M)/\mathcal{F}$ can be embedded in an injective functor $-\otimes L$, so there is a map $\theta: M \longrightarrow L$ inducing a map $-\otimes M \longrightarrow -\otimes L$ with kernel $\mathcal{F}$. It follows that θ is not a split mono and cannot factor through α, a contradiction. Thus $\mathcal{S}$ is simple and (2) holds.

The proof of (2) ⇒ (1) is essentially contained in [A3]. In order to sketch it we need to recall several facts. If X is a f.p. right R-module and $P \xrightarrow{f} Q \longrightarrow X \longrightarrow 0$ is a projective presentation of X with P and Q f.g., then the *transpose* Tr X of X relative to this projective presentation is defined to be the f.p. left R-module which is the cokernel of the map

$$\mathrm{Hom}_R(f,R) : \mathrm{Hom}_R(Q,R) \longrightarrow \mathrm{Hom}_R(P,R).$$

If Y is another f.p. right module, then

$$\underline{\mathrm{Hom}}_R(X,Y) \cong \underline{\mathrm{Hom}}_R(\mathrm{Tr}\ Y, \mathrm{Tr}\ X)$$

where $\underline{\mathrm{Hom}}$ denotes the group of homomorphisms modulo the subgroup of those which factor through a projective. In particular there is a ring isomorphism

$$\tau : \underline{\mathrm{End}}_R(X) \longrightarrow \underline{\mathrm{End}}_R(\mathrm{Tr}\ X)^{\mathrm{op}}.$$

Suppose (2), say $-\otimes M$ is the injective envelope of a simple $\mathcal{S}$. We construct a left almost split map with source M. As a first case, suppose that $\mathcal{S}(R) \neq 0$, so by Yoneda's lemma we know that $\mathcal{S}$ is a quotient of (R,−), and hence that $\mathcal{S} \cong \mathcal{S}_{R,\mathfrak{m}}$ for some maximal left ideal $\mathfrak{m}$ in $\mathrm{End}(R_R) \cong R$. Now by (2.1) we know that $M \cong \mathrm{Hom}_R(R,I) \cong I$ is the injective envelope of a simple left R-module. In this case the projection $M \longrightarrow M/\mathrm{soc}_R(M)$ is a left almost split map. Thus we may assume that $\mathcal{S}(R) = 0$. Say $\mathcal{S} \cong \mathcal{S}_{X,\mathfrak{m}}$ where X is a f.p. right R-module and $\mathfrak{m}$ is a maximal left ideal in $E = \mathrm{End}_R(X)$, and so by (2.1) we may assume that $M = \mathrm{Hom}_E(X,I)$ where I is the injective envelope of E/m.

We now consider contravariant functors R-Mod $\longrightarrow$ $\mathbb{Z}$-Mod. By [A3] there are isomorphisms

$$\begin{aligned}\mathrm{Ext}^1_R(-,\mathrm{Hom}_E(X,I)) &\cong \mathrm{Hom}_E(\mathrm{Tor}^R_1(X,-),I)\\ &\cong \mathrm{Hom}_E(\underline{\mathrm{Hom}}_R(\mathrm{Tr}\,X,-),I), \qquad (\dagger)\end{aligned}$$

where for any left R-module Z, the group $\underline{\mathrm{Hom}}_R(\mathrm{Tr}\,X,Z)$ is considered as an E-module by means of the isomorphism τ.

We construct a map

$$h : \mathrm{Hom}_R(-,\mathrm{Tr}\,X) \longrightarrow \mathrm{Hom}_E(\underline{\mathrm{Hom}}_R(\mathrm{Tr}\,X,-),I)$$

whose image is a simple functor. Since $\mathcal{S}(R) = 0$, any endomorphism of X which factors through a projective module belongs to $\mathfrak{m}$, and so $\mathfrak{m}$ descends to a maximal left ideal $\underline{\mathfrak{m}}$ in $\underline{\mathrm{End}}_R(X)$. Let $\sigma : \underline{\mathrm{End}}_R(X) \longrightarrow I$ be an E-module map inducing an isomorphism from $\underline{\mathrm{End}}_R(X)/\underline{\mathfrak{m}}$ onto the socle of I, and let θ be the composition

$$\mathrm{End}_R(\mathrm{Tr}\,X)^{op} \twoheadrightarrow \underline{\mathrm{End}}_R(\mathrm{Tr}\,X)^{op} \xrightarrow{\tau^{-1}} \underline{\mathrm{End}}_R(X) \xrightarrow{\sigma} I$$

We now define h by sending $\psi \in \mathrm{Hom}_R(Z,\mathrm{Tr}\,X)$ to the map $\underline{\mathrm{Hom}}_R(\mathrm{Tr}\,X,Z) \longrightarrow I$ which sends ϕ to $\theta(\psi\circ\phi)$. Clearly (Ker h)(Z) is equal to

$$\{\psi \in \mathrm{Hom}_R(Z,\mathrm{Tr}\,X) \mid \psi\circ\chi \in \mathrm{Ker}(\theta)\ \forall\ \chi \in \mathrm{Hom}_R(\mathrm{Tr}\,X,Z)\}$$

and since $\mathrm{Ker}(\theta)$ is a maximal right ideal in $\mathrm{End}_R(\mathrm{Tr}\,X)$, it follows that Ker h is a maximal subfunctor of $\mathrm{Hom}_R(-,\mathrm{Tr}\,X)$, just as in the proof of (1.3). Thus the image of h is indeed simple.

Let L be a module with $(\mathrm{Im}\,h)(L) \neq 0$, for example L = Tr X suffices. We can choose an extension

$$\xi : 0 \longrightarrow \mathrm{Hom}_E(X,I) \xrightarrow{\alpha} N \xrightarrow{\beta} L \longrightarrow 0$$

whose image in $\mathrm{Ext}^1_R(L, \mathrm{Hom}_E(X, I))$ generates the simple subfunctor of $\mathrm{Ext}^1_R(-, \mathrm{Hom}_E(X, I))$ corresponding to Im(h) under the isomorphism (†). From the long exact sequence for ξ we obtain an exact sequence

$$\mathrm{Hom}_R(-, N) \xrightarrow{\mathrm{Hom}(-,\beta)} \mathrm{Hom}_R(-, L) \longrightarrow \mathrm{Ext}^1_R(-, \mathrm{Hom}_E(X, I))$$

and by definition the image of 1_L under the connecting map is ξ. Thus the cokernel of $\mathrm{Hom}(-,\beta)$ is isomorphic to Im(h) and hence is simple. The next lemma then shows that $\alpha: M \longrightarrow N$ is a left almost split map.

LEMMA [A3, Chapter II, Proposition 4.2]. Let $0 \longrightarrow M \xrightarrow{\alpha} N \xrightarrow{\beta} L \longrightarrow 0$ be an exact sequence of left R-modules. If the cokernel of $\mathrm{Hom}(-,\beta)$ is a simple contravariant functor R-Mod $\longrightarrow$ $\mathbb{Z}$-Mod, then the map α is left almost split if and only if $\mathrm{End}_R(M)$ is local.

3 Σ-PURE-INJECTIVE MODULES

A module M is said to be *Σ-pure-injective* provided that every direct sum of copies of M is pure-injective. In this section we describe a very useful characterization of the Σ-pure-injective modules, one application of which is the fact that modules of finite endolength are always direct sums of indecomposable submodules, something which is not at all obvious.

3.1 THEOREM. If M is an R-module, then the following statements are equivalent

(1) M has the dcc on subgroups of finite definition.

(2) M is Σ-pure-injective.

(3) Every product of copies of M is a direct sum of indecomposables with local endomorphism ring.

(4) Every product of copies of M is a direct sum of indecomposables of cardinality $\leq \max(\aleph_0, \mathrm{card}(R))$.

We only prove (1)⇒(2) and (1)⇒(3), the implications which are most relevant for our study of modules of finite endolength. The reader can find the other implications, and indeed a host of other equivalent statements in [GJ1], [ZH], [JL, 8.1], [P, 3.2] and in Azumaya's article [Az].

Proof of (1)⇒(2). Since $F_{X,x}(\amalg_I M) = \amalg_I F_{X,x}(M)$ it is clear that $\amalg_I M$ also has the dcc on subgroups of finite definition, so by replacing M by $\amalg_I M$, it is enough to show that M is pure-injective, that is, that $-\otimes M$ is injective. In fact it suffices to prove that $\mathrm{Ext}^1_{D(R)}((R,-)/\mathcal{F}, -\otimes M) = 0$ for all subfunctors $\mathcal{F}$ of $(R,-)$, an analogue of Baer's criterion for injectivity. To see why, let $-\otimes M \longrightarrow -\otimes L$ be the injective envelope of $-\otimes M$, so there is an exact sequence

$$0 \longrightarrow -\otimes M \longrightarrow -\otimes L \longrightarrow -\otimes (L/M) \longrightarrow 0. \qquad (\dagger)$$

If $-\otimes M$ is not injective then $L/M \neq 0$, and choosing $x \in L/M$ one obtains a map $(R,-) \longrightarrow -\otimes(L/M)$. If this has kernel $\mathcal{F}$ then by assumption the pullback of (†) via the mono $(R,-)/\mathcal{F} \longrightarrow -\otimes(L/M)$ is split. This means that $-\otimes M \oplus (R,-)/\mathcal{F}$ embeds in $-\otimes L$, which contradicts the minimality of $-\otimes L$.

Let $\xi : 0 \longrightarrow -\otimes M \longrightarrow \mathcal{E} \xrightarrow{\pi} (R,-)/\mathcal{F} \longrightarrow 0$ be an extension. We show that ξ is split. We have the equality

$$\mathrm{Hom}((R,-)/\mathcal{F}, -\otimes M) = \bigcap_{\mathcal{G}\ \text{f.g.}\ \subseteq \mathcal{F}} \mathrm{Hom}((R,-)/\mathcal{G}, -\otimes M)$$

and since the terms in the intersection are subgroups of finite definition of M, and M has the dcc on such subgroups, one can find a f.g. subfunctor $\mathcal{H}$ of $\mathcal{F}$ with

$$\mathrm{Hom}((R,-)/\mathcal{F}, -\otimes M) = \mathrm{Hom}((R,-)/\mathcal{H}, -\otimes M).$$

That is, $\mathcal{H}$ has the property that any map $(R,-) \longrightarrow -\otimes M$ which annihilates $\mathcal{H}$, also annihilates $\mathcal{F}$.

Now let $\mathcal{G}$ be a f.g. subfunctor of $\mathcal{F}$ containing $\mathcal{H}$. Since $(R,-)/\mathcal{G}$

is coherent, Lemma (1.4) shows that the pullback of ξ with respect to the map $(R,-)/\mathcal{G} \twoheadrightarrow (R,-)/\mathcal{F}$ splits. Thus there is a map $f_{\mathcal{G}} : (R,-) \longrightarrow \mathcal{E}$ which annihilates $\mathcal{G}$, and such that the composition $\pi \circ f_{\mathcal{G}}$ is equal to the natural projection of $(R,-)$ onto $(R,-)/\mathcal{F}$. Now the difference $f_{\mathcal{G}} - f_{\mathcal{H}}$ actually maps $(R,-)$ into $-\otimes M$, and it annihilates $\mathcal{H}$, so that it also annihilates $\mathcal{F}$, and hence $f_{\mathcal{H}}$ annihilates $\mathcal{G}$. This shows that $f_{\mathcal{H}}$ annihilates any f.g. subfunctor of $\mathcal{F}$ containing $\mathcal{H}$. Thus $f_{\mathcal{H}}$ annihilates $\mathcal{F}$, so it induces a map $(R,-)/\mathcal{F} \longrightarrow \mathcal{E}$ splitting ξ, as required.

Proof of (1)⇒(3). Any product of copies of M also satisfies the hypothesis, so we only need to prove that M is a direct sum of indecomposables with local endomorphism rings. If U is a pure submodule of M then $F_{X,x}(U) = U \cap F_{X,x}(M)$, so U has the dcc on subgroups of finite definition, is pure-injective, and hence is a summand of M. Given an ascending chain of pure submodules of M, their union is again a pure submodule (since tensor products commute with direct limits), so if $0 \neq x \in M$, by Zorn's lemma we can choose a pure submodule U, maximal with respect to the condition $x \notin U$. Now $M = U \oplus X$, and if X were to decompose this would contradict the maximality of U. Thus M has an indecomposable summand. Now let A be a maximal set of indecomposable submodules of M whose sum $U = \Sigma_{V \in A} V$ is direct and is pure in M. If $U \neq M$ then it is a summand, and its complement has an indecomposable summand which could be adjoined to A, a contradiction. Finally, because the category D(R) has injective envelopes, any indecomposable injective functor has local endomorphism ring, so any indecomposable pure-injective module has local endomorphism ring.

4 MODULES OF FINITE ENDOLENGTH

In this section we describe the basic properties of modules of finite endolength, most of which are deduced directly from the properties of Σ-pure-injective modules. We finish with some examples.

4.1 PROPOSITION. An R-module has finite endolength if and only if it has the acc and the dcc on subgroups of finite definition. In this case every $End_R(M)$-submodule of M has finite definition.

This shows that the endolength of a module is the same as its pp-rank in the sense of [P]. The proposition follows directly from the lattice structure of the subgroups of finite definition and the next lemma.

LEMMA. If M is a pure-injective R-module, then every cyclic $End_R(M)$-submodule of M is an intersection of subgroups of finite definition.

Proof. An element $m \in M$ determines a morphism $(R,-) \longrightarrow -\otimes M$, and hence a monomorphism $\phi: (R,-)/\mathcal{F} \longrightarrow -\otimes M$ where $\mathcal{F}$ is the kernel. Recall that the space $Hom((R,-)/\mathcal{F}, -\otimes M)$ can be regarded as a subgroup of M, and it is clearly an $End_R(M)$-submodule. If $\theta \in Hom((R,-)/\mathcal{F}, -\otimes M)$ then since $-\otimes M$ is injective there is a factorization $\theta = \alpha \circ \phi$ for some $\alpha: -\otimes M \longrightarrow -\otimes M$. Now α corresponds to an endomorphism of M, and this shows that $Hom((R,-)/\mathcal{F}, -\otimes M)$ is the $End_R(M)$-submodule of M generated by m. Finally, we have already used in the proof of (3.1) the fact that $Hom((R,-)/\mathcal{F}, -\otimes M)$ is an intersection of subgroups of finite definition.

4.2 LEMMA. The endomorphism ring of a finite endolength module has nilpotent radical.

Proof. Nakayama's Lemma.

4.3 PROPOSITION. The class of finite endolength modules is closed under finite direct sums, and arbitrary products or direct sums of copies of one module. Moreover, if L is a pure submodule of a module M of finite endolength, then L is a direct summand and $\text{endolen}(L) \leq \text{endolen}(M)$.

Proof. Most of this can be proved using elementary means, but having proved (4.1) this all follows from the equalities in (1.7). Recall that since a pure submodule has finite endolength it is pure-injective, and hence a summand.

4.4 PROPOSITION. Indecomposable R-modules of finite endolength have local endomorphism rings and cardinality $\leq \max(\aleph_0, \mathrm{card}(R))$.

Proof. This follows from (3.1).

4.5 PROPOSITION. Every module of finite endolength is a direct sum of indecomposable modules of finite endolength. Conversely, such a direct sum has finite endolength if and only if there are only finitely many isomorphism classes of indecomposables involved.

The only part not contained in (3.1) is the "only if", which follows from the lemma below.

LEMMA. If $M_1,\ldots,M_n$ are non-isomorphic indecomposable modules of finite endolength, then $\mathrm{endolen}(M_1\oplus\ldots\oplus M_n) = \sum_{i=1}^{n} \mathrm{endolen}(M_i)$.

Proof. Since the M_i have local endomorphism rings, the ring

$$\mathrm{End}_R(M_1\oplus\ldots\oplus M_n)/\mathrm{rad}\ \mathrm{End}_R(M_1\oplus\ldots\oplus M_n)$$

is isomorphic to the product $\prod_{i=1}^{n} \mathrm{End}_R(M_i)/\mathrm{rad}\ \mathrm{End}_R(M_i)$, and the assertion follows.

REMARKS. (1) Since the indecomposables have local endomorphism rings, the decomposition of a finite endolength module into indecomposable summands is essentially unique.

(2) A result of Garavaglia, see [P, Exercise 2, p200], shows that an indecomposable module M has finite endolength if and only if every product of copies of M is isomorphic to a direct sum of

copies of M. One direction is as follows. If M has finite endolength, then $\prod_I M$ is a direct sum of indecomposable modules, and it has endolength equal to that of M. Now since it has M as a summand, it cannot have any other indecomposable summands by the lemma above.

4.6 PROPOSITION. If a module of finite endolength is either artinian or noetherian, then it has finite length.

This follows from a simple generalization of a theorem of Lenagan,

THEOREM. If a bimodule is artinian on one side and noetherian on the other, then it has finite length on each side.

Proof. Let ${}_RM_S$ be such a bimodule, artinian over R and noetherian over S. Since M has both chain conditions on sub-bimodules, to prove the theorem we may assume that M is simple as a bimodule. Now $\mathrm{soc}_R(M)$ is a non-zero sub-bimodule, so equal to M. Therefore ${}_RM$ is semisimple, and hence of finite length since it is artinian. Now the usual form of Lenagan's Theorem shows that M_S has finite length, see [MR, 4.1.6] or [GW, 7.10].

4.7 EXAMPLES. (1) If R is a ring without invariant basis number [C, §0.2], for example the endomorphism ring of an infinite dimensional vector space, then there are no non-zero modules M of finite endolength, for by definition $R^n \cong R^m$ for some $n \neq m$, and then $R^n \otimes_R M \cong R^m \otimes_R M$ as $\mathrm{End}_R(M)$-modules, but they have different lengths as such.

(2) The Jacobson density theorem shows that a simple R-module S has finite endolength if and only if $R/\mathrm{Ann}_R(S)$ is simple artinian.

(3) A ring R is said to be of *finite representation type* if it is

left artinian and has only finitely many isomorphism classes of finite length left R-modules. This is a left-right symmetric condition, and for such rings, every module is a direct sum of finitely generated modules [A2,RT]. It is proved in [P, 11.38], [ZHZ, Theorem 6] and [CB3, 1.2] that a ring R has finite representation type if and only if every R-module has finite endolength.

(4) There is a 1-1 correspondence between isomorphism classes of finite endolength modules M with $\mathrm{End}_R(M)$ a division ring, and isomorphism classes of ring-theoretic epimorphisms $\theta: R \longrightarrow S$ from R to a simple artinian ring S. The correspondence is given by sending M to the homomorphism

$$R \longrightarrow \mathrm{End}({}_{\mathrm{End}_R(M)}M),$$

and sending θ to the restriction of the simple S-module [R3].

(5) If R is a commutative ring, then the indecomposable modules of endolength 1 are precisely the quotient fields of factor rings R/P with P a prime ideal in R. More generally, if R is a prime Goldie ring then the restriction of the simple module for the simple artinian quotient ring of R is the unique faithful indecomposable module of finite endolength, see [CB3, 1.3]. It follows that if R is a noetherian ring or a PI ring, then the prime ideals in R can be identified with the indecomposable modules of finite endolength whose annihilator is prime. Thus, for example, a simple noetherian ring has a unique indecomposable module of finite endolength.

(6) If R is a Dedekind domain, then the indecomposable modules of finite endolength are the quotient field of R and the modules R/m^n with m a maximal ideal. This is because, if I is a non-zero ideal in R, then R/I is an artinian principal ideal ring, so that the indecomposable R/I-modules have form R/m^n. More generally, if

R is an hereditary noetherian prime ring, then the only indecomposable module of finite endolength and infinite length is the restriction of the simple module of the simple artinian quotient ring of R. In this case, the proper factor rings of R have finite representation type.

5 CHARACTERS

Pure-injective modules have been extensively studied by model theorists using rather sophisticated ideas, see for example [P], but for a module M of finite endolength, it appears that many of these concepts can be encoded in simple numerical data, which we call the character of M. In general, by a *character* for mod-R we mean a function χ which assigns to each f.p. right R-module X a non-negative integer $\chi(X)$, and which satisfies the two conditions

(1) $\chi(X \oplus Y) = \chi(X) + \chi(Y)$ for all f.p. modules X, Y.

(2) $\chi(Z) \leq \chi(Y) \leq \chi(X) + \chi(Z)$ for any right exact sequence $X \longrightarrow Y \longrightarrow Z \longrightarrow 0$ of f.p. modules.

This notion is adapted from Schofield's definition of a Sylvester module rank function [Sc], but we like the name "character" since they have many properties in common with group characters. We stress, however, that our characters are non-negative integer valued, and in no way involve traces. It is natural to call the number $\chi(R)$ the *degree* of χ; if it is zero then condition (2) above shows that $\chi = 0$. If M is a left R-module of finite endolength, the assignment

$$\chi_M(X) = \text{length}_{\text{End}_R(M)}(X \otimes_R M)$$

defines the character χ_M of M. Its degree is the endolength of M.

5.1 THEOREM. If $M_1, \ldots, M_n$ are non-isomorphic indecomposable R-modules of finite endolength then their characters are independent over $\mathbb{Z}$.

Proof. Suppose there is a relation $\sum_{i=1}^{n} a_i\chi_{M_i} = 0$ with $a_i\in\mathbb{Z}$. Let

$$M = \coprod_{a_i>0} a_i M_i, \qquad E = \prod_{a_i>0} \mathrm{End}_R(M_i)^{op}$$

so that M is naturally an R-E-bimodule, and let

$$N = \coprod_{a_i<0} (-a_i)M_i, \qquad F = \prod_{a_i<0} \mathrm{End}_R(M_i)^{op}$$

so that N is an R-F-bimodule. Thus the indecomposable summands of M and N are non-isomorphic, and the relation implies that

$$\mathrm{length}(X\otimes_R M_E) = \mathrm{length}(X\otimes_R N_F)$$

for all f.p. modules X. In particular, assuming that the relation is non-trivial, both M and N are non-zero. Recall that if X is a f.p. right R-module and $x\in X$ then the subgroup $F_{X,x}(M)$ is defined by an exact sequence

$$0 \longrightarrow F_{X,x}(M) \longrightarrow R\otimes_R M \longrightarrow X\otimes_R M \longrightarrow (X/xR)\otimes_R M \longrightarrow 0,$$

so we deduce that

$$\mathrm{length}(F_{X,x}(M)_E) = \mathrm{length}(F_{X,x}(N)_F). \qquad (\dagger)$$

We construct a pair U, u such that $F_{U,u}(M)$ is a simple $\mathrm{End}_R(M)$-submodule of M and $F_{U,u}(N)$ is a simple $\mathrm{End}_R(N)$-submodule of N. By (4.1) one can choose a subgroup of finite definition $F_{Y,y}(M)$ which is simple as an $\mathrm{End}_R(M)$-module, and then $F_{Y,y}(N)$ is non-zero by (†). Inside $F_{Y,y}(N)$ we can find a subgroup $F_{Z,z}(N)$ which is simple as an $\mathrm{End}_R(N)$-module. Setting $U = Y\oplus Z$ and $u = y+z$ one has $F_{U,u} = F_{Y,y} \cap F_{Z,z}$ so that $F_{U,u}(N) = F_{Y,y}(N)$ is a simple $\mathrm{End}_R(N)$-submodule of N. Also $F_{U,u}(M)$ is a submodule of $F_{Y,y}(M)$ and is non-zero by (†), so is equal to $F_{Y,y}(M)$, and hence is a simple $\mathrm{End}_R(M)$-submodule.

Choose $0 \neq m \in F_{U,u}(M)$ and $0 \neq n \in F_{U,u}(N)$. We show that

$$m \in F_{X,x}(M) \Leftrightarrow n \in F_{X,x}(N)$$

for any pair X, x. Now

$$\begin{aligned} m \in F_{X,x}(M) &\Leftrightarrow F_{U,u}(M) \subseteq F_{X,x}(M) \\ &\Leftrightarrow F_{X,x}(M) = F_{U,u}(M) + F_{X,x}(M), \end{aligned}$$

and $F_{U,u} + F_{X,x} = F_{Z,z}$ where Z is the cokernel of the map $R \longrightarrow U \oplus X$ and z is the common image of u and x. Thus

$$m \in F_{X,x}(M) \Leftrightarrow \text{length}(F_{X,x}(M)_E) = \text{length}(F_{Z,z}(M)_E).$$

The same argument for N and the property (†) then prove our assertion.

The element m determines a map $(R,-) \longrightarrow -\otimes M$ whose kernel $\mathcal{K}$ is given by

$$\begin{aligned} \mathcal{K}(X) &= \{\theta \in \text{Hom}_R(R,X) \mid \theta(1)\otimes m = 0\} \\ &= \{\theta \in \text{Hom}_R(R,X) \mid m \in F_{X,\theta(1)}(M)\}. \end{aligned}$$

Similarly the element n gives a map $(R,-) \longrightarrow -\otimes N$ whose kernel is also $\mathcal{K}$ by the statement above. Thus the injective envelope $-\otimes L$ of $(R,-)/\mathcal{K}$ embeds in both $-\otimes M$ and $-\otimes N$, so that L is a summand of both M and N. Since $L \neq 0$ this is impossible by the Krull-Schmidt theorem.

5.2 THEOREM. Every character χ can be written as a sum

$$\chi = \chi_{M_1} + \dots + \chi_{M_n}$$

with the M_i indecomposable modules of finite endolength.

The proof is given in several steps.

Step 1. We define a function on the coherent functors $\mathcal{F} \in D(R)$, which we again denote by χ, as follows. If $\mathcal{F}$ has resolution

$$0 \longrightarrow (Z,-) \longrightarrow (Y,-) \longrightarrow (X,-) \longrightarrow \mathcal{F} \longrightarrow 0$$

for a right exact sequence $X \longrightarrow Y \longrightarrow Z \longrightarrow 0$, then we set

$$\chi(\mathcal{F}) = \chi(X) - \chi(Y) + \chi(Z).$$

This is a non-negative integer since χ is a character, and it is well-defined by the long form of Schanuel's Lemma. Note that $\chi((X,-)) = \chi(X)$, so this function can really be thought of as extending χ. Standard arguments with projective resolutions show that if $0 \longrightarrow \mathcal{F} \longrightarrow \mathcal{G} \longrightarrow \mathcal{H} \longrightarrow 0$ is an exact sequence of coherent functors then $\chi(\mathcal{G}) = \chi(\mathcal{F}) + \chi(\mathcal{H})$.

Step 2. We extend the function of the previous paragraph to f.g. functors $\mathcal{F} \in D(R)$ by setting

$$\chi(\mathcal{F}) = \min \{ \chi(\mathcal{H}) \mid \mathcal{H} \text{ coherent}, \mathcal{H} \twoheadrightarrow \mathcal{F} \}.$$

This agrees with the first definition in case $\mathcal{F}$ is coherent because of the additivity of χ in that case.

If $\mathcal{G}$ is a f.g. subfunctor of $\mathcal{F}$ then $\chi(\mathcal{G}) \le \chi(\mathcal{F})$. Namely, there is a map $\theta: \mathcal{H} \twoheadrightarrow \mathcal{F}$ with $\mathcal{H}$ coherent and $\chi(\mathcal{H}) = \chi(\mathcal{F})$. Since $\mathcal{G}$ is f.g. one can find a f.g. subfunctor $\mathcal{K}$ of $\theta^{-1}(\mathcal{G})$ mapping onto $\mathcal{G}$. Now $\mathcal{K}$ is a f.g. subfunctor of $\mathcal{H}$, so is coherent. Thus $\chi(\mathcal{G}) \le \chi(\mathcal{K}) \le \chi(\mathcal{H}) = \chi(\mathcal{F})$.

Step 3. We extend the function χ to all functors $\mathcal{F} \in D(R)$, with χ now taking values in $\mathbb{N} \cup \{\infty\}$. Namely, we set

$$\chi(\mathcal{F}) = \max \{ \chi(\mathcal{G}) \mid \mathcal{G} \text{ f.g.}, \mathcal{G} \subseteq \mathcal{F} \}.$$

This agrees with the definition for f.g. functors by the observation above. We show that if $0 \longrightarrow \mathcal{F} \xrightarrow{\alpha} \mathcal{G} \xrightarrow{\beta} \mathcal{H} \longrightarrow 0$ is exact, then

$$\chi(\mathcal{G}) = \chi(\mathcal{F}) + \chi(\mathcal{H}) \qquad (\dagger)$$

with the usual conventions if any term is ∞. Our proof involves a sequence of special cases. We have already observed "Case 0", that (†) holds when $\mathcal{F}$, $\mathcal{G}$ and $\mathcal{H}$ are coherent.

Case 1. Suppose that all f.g. subfunctors of $\mathcal{G}$ and $\mathcal{H}$ are coherent. If $\mathcal{L}$ is a f.g. subfunctor of $\mathcal{G}$ then the sequence

$$0 \longrightarrow \alpha^{-1}(\mathcal{L}) \longrightarrow \mathcal{L} \longrightarrow \beta(\mathcal{L}) \longrightarrow 0$$

is exact. Since $\mathcal{L}$ and $\beta(\mathcal{L})$ are finitely generated, they are coherent by the assumption, and hence so is $\alpha^{-1}(\mathcal{L})$, and so by Case 0 we have

$$\chi(\mathcal{L}) = \chi(\alpha^{-1}(\mathcal{L})) + \chi(\beta(\mathcal{L})). \qquad (‡)$$

Now $\alpha^{-1}(\mathcal{L})$, $\mathcal{L}$ and $\beta(\mathcal{L})$ are subfunctors of $\mathcal{F}$, $\mathcal{G}$ and $\mathcal{H}$ respectively, so

$$\chi(\alpha^{-1}(\mathcal{L})) \le \chi(\mathcal{F}), \quad \chi(\mathcal{L}) \le \chi(\mathcal{G}), \quad \chi(\beta(\mathcal{L})) \le \chi(\mathcal{H}).$$

Moreover, by taking $\mathcal{L}$ large enough we can ensure that $\alpha^{-1}(\mathcal{L})$, $\mathcal{L}$ and $\beta(\mathcal{L})$ contain any given f.g. subfunctors of $\mathcal{F}$, $\mathcal{G}$ and $\mathcal{H}$. Now the equality (†) follows on taking the supremum of (‡) over all $\mathcal{L}$.

Case 2. Suppose that $\mathcal{G}$ is coherent. Since $\mathcal{H}$ is f.g. so one can find a surjection $\mathcal{L} \twoheadrightarrow \mathcal{H}$ with $\mathcal{L}$ coherent and $\chi(\mathcal{L}) = \chi(\mathcal{H})$. Form the pullback

$$\begin{array}{ccccccccc}
 & & & & 0 & & 0 & & \\
 & & & & \downarrow & & \downarrow & & \\
 & & & & \mathcal{K} & = & \mathcal{K} & & \\
 & & & & \downarrow & & \downarrow & & \\
0 & \longrightarrow & \mathcal{F} & \longrightarrow & \mathcal{E} & \longrightarrow & \mathcal{L} & \longrightarrow & 0 \\
 & & \| & & \downarrow & & \downarrow & & \\
0 & \longrightarrow & \mathcal{F} & \longrightarrow & \mathcal{G} & \longrightarrow & \mathcal{H} & \longrightarrow & 0 \\
 & & & & \downarrow & & \downarrow & & \\
 & & & & 0 & & 0 & &
\end{array}$$

Note that $\mathcal{E}$ embeds in $\mathcal{G} \oplus \mathcal{L}$ which is coherent, so that f.g.

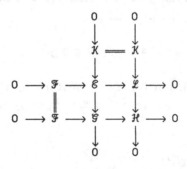

subfunctors of $\mathcal{E}$ are coherent. Thus $\chi(\mathcal{E}) = \chi(\mathcal{F}) + \chi(\mathcal{L})$ and $\chi(\mathcal{E}) = \chi(\mathcal{G}) + \chi(\mathcal{K})$, and so $\chi(\mathcal{G}) = \chi(\mathcal{F}) + \chi(\mathcal{H}) - \chi(\mathcal{K})$. Now if $\chi(\mathcal{K}) \neq 0$ then $\mathcal{K}$ has a f.g. (and hence coherent) subfunctor $\mathcal{I}$ with $\chi(\mathcal{I}) \neq 0$. But then $\mathcal{L}/\mathcal{I} \twoheadrightarrow \mathcal{H}$, so that

$$\chi(\mathcal{H}) \leq \chi(\mathcal{L}/\mathcal{I}) = \chi(\mathcal{L}) - \chi(\mathcal{I}) < \chi(\mathcal{L}),$$

a contradiction.

Case 3. Suppose that all f.g. subfunctors of $\mathcal{G}$ are coherent. The proof is the same as Case 1, using Case 2 to prove (‡).

Case 4. Suppose that $\mathcal{G}$ is finitely generated. We can find a surjection $\mathcal{L} \twoheadrightarrow \mathcal{G}$ with $\mathcal{L}$ coherent. This gives a commutative exact diagram

$$\begin{array}{ccccccccc}
 & & 0 & & 0 & & & & \\
 & & \downarrow & & \downarrow & & & & \\
 & & \mathcal{K} & = & \mathcal{K} & & & & \\
 & & \downarrow & & \downarrow & & & & \\
0 & \to & \mathcal{E} & \to & \mathcal{L} & \to & \mathcal{H} & \to & 0 \\
 & & \downarrow & & \downarrow & & \| & & \\
0 & \to & \mathcal{F} & \to & \mathcal{G} & \to & \mathcal{H} & \to & 0 \\
 & & \downarrow & & \downarrow & & & & \\
 & & 0 & & 0 & & & &
\end{array}$$

so that $\chi(\mathcal{L}) = \chi(\mathcal{E}) + \chi(\mathcal{H})$ and $\chi(\mathcal{L}) = \chi(\mathcal{K}) + \chi(\mathcal{G})$ since $\mathcal{L}$ is coherent, and $\chi(\mathcal{E}) = \chi(\mathcal{K}) + \chi(\mathcal{F})$ since $\mathcal{E}$ is a subfunctor of $\mathcal{L}$, so that f.g. subfunctors of $\mathcal{E}$ are coherent. Thus $\chi(\mathcal{G}) = \chi(\mathcal{F}) + \chi(\mathcal{H})$, as required.

General case. The proof is the same as in Case 1, now using Case 4 to prove (‡).

Step 4. If M is a module with the property that

$$\mathrm{Hom}(\mathcal{H}, -\otimes M) = 0 \text{ for all functors } \mathcal{H} \text{ with } \chi(\mathcal{H}) = 0 \qquad (*)$$

then M has endolength $\leq \chi(R)$.

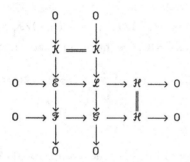

Proof. If not, then by (4.1) we can find a strictly increasing chain

$$F_{\mathcal{F}_0}(M) < F_{\mathcal{F}_1}(M) < \dots < F_{\mathcal{F}_d}(M) < F_{\mathcal{F}_{d+1}}(M)$$

of subgroups of finite definition, where $d = \chi(R)$ and the $\mathcal{F}_i$ are f.g. subfunctors of $(R,-)$. Moreover, we may assume that

$$\mathcal{F}_{d+1} \le \mathcal{F}_d \le \dots \le \mathcal{F}_1 \le \mathcal{F}_0,$$

if necessary by replacing $\mathcal{F}_i$ by $\sum_{j\ge i} \mathcal{F}_j$. Consider the coherent functors

$$(R,-)/\mathcal{F}_{d+1} \twoheadrightarrow (R,-)/\mathcal{F}_d \twoheadrightarrow \dots \twoheadrightarrow (R,-)/\mathcal{F}_1 \twoheadrightarrow (R,-)/\mathcal{F}_0.$$

Each of these functors is a quotient of $(R,-)$ so has χ bounded by $\chi(R)$, and they have χ decreasing, so at some stage there are two functors with the same χ, say $(R,-)/\mathcal{F}_{i+1} \twoheadrightarrow (R,-)/\mathcal{F}_i$. Since the kernel $\mathcal{K}$ of this map has $\chi(\mathcal{K}) = 0$, we have an exact sequence

$$\begin{array}{ccccccc} 0 \longrightarrow & \mathrm{Hom}((R,-)/\mathcal{F}_i, -\otimes M) & \longrightarrow & \mathrm{Hom}((R,-)/\mathcal{F}_{i+1}, -\otimes M) & \longrightarrow & \mathrm{Hom}(\mathcal{K}, -\otimes M) \\ & \| & & \| & & \| \\ & F_{\mathcal{F}_i}(M) & & F_{\mathcal{F}_{i+1}}(M) & & 0 \end{array}$$

which contradicts the assumption that the chain is strictly increasing.

Step 5. It follows from the considerations above and Lemma (4.5) that there are at most $\chi(R)$ non-isomorphic indecomposable modules with the property (*). Let them be $M_1,\dots,M_r$, and denote by a_i the smallest value of $\chi(\mathcal{F})$ with $\mathcal{F}$ a non-zero subfunctor of $-\otimes M_i$. Since $-\otimes M_i$ has no subfunctors with $\chi = 0$ it follows that $a_i \ne 0$. We claim that for all $\mathcal{G}$ with $\chi(\mathcal{G}) < \infty$ one has

$$\chi(\mathcal{G}) = \sum_{i=1}^{r} a_i \, \mathrm{length}_{\mathrm{End}(M_i)}(\mathrm{Hom}(\mathcal{G}, -\otimes M_i)). \qquad (\#)$$

On specializing to the case when $\mathcal{G} = (X,-)$ one deduces that

$$\chi(X) = \sum_{i=1}^{r} a_i \chi_{M_i}(X),$$

which proves the theorem. We first prove (#) in two special cases.

Case 1. Suppose that every non-zero subfunctor $\mathcal{H}$ of $\mathcal{G}$ has $\chi(\mathcal{H}) = \chi(\mathcal{G})$. Clearly we may assume that $\chi(\mathcal{G}) \neq 0$. The injective envelope $-\otimes M$ of $\mathcal{G}$ has the property (*) since if $\mathcal{H}$ is a functor with $\chi(\mathcal{H}) = 0$ and $\theta: \mathcal{H} \longrightarrow -\otimes M$ is a map, then $\mathcal{G} \cap \mathrm{Im}(\theta)$ is a subfunctor of $\mathcal{G}$ with $\chi = 0$, so is zero, and hence $\theta = 0$ since $\mathcal{G}$ is essential in $-\otimes M$. Thus M has finite endolength and is a direct sum of copies of the M_i. In particular there is a non-zero map from $\mathcal{G}$ to some $-\otimes M_i$. Now any non-zero map $\mathcal{G} \longrightarrow -\otimes M_j$ must be mono, so that $-\otimes M_j$ is the injective envelope of $\mathcal{G}$. Thus $\mathrm{Hom}(\mathcal{G}, -\otimes M_j) = 0$ for $j \neq i$, and the injective property for $-\otimes M_i$ shows that $\mathrm{Hom}(\mathcal{G}, -\otimes M_i)$ is simple as an $\mathrm{End}_R(M_i)$-module, proving (#), since clearly $a_i = \chi(\mathcal{G})$.

Case 2. Suppose that every subfunctor $\mathcal{H}$ of $\mathcal{G}$ has $\chi(\mathcal{H}) \in \{0, \chi(\mathcal{G})\}$. For any functor $\mathcal{F}$, the sum $T\mathcal{F}$ of the subfunctors $\mathcal{L} \subseteq \mathcal{F}$ with $\chi(\mathcal{L}) = 0$ is the unique largest subfunctor of $\mathcal{F}$ with $\chi = 0$. Thus $\mathcal{G}/T\mathcal{G}$ satisfies the condition of Case 1, and then (#) follows since $\chi(\mathcal{G}) = \chi(\mathcal{G}/T\mathcal{G})$ and $\mathrm{Hom}(\mathcal{G}, -\otimes M_i) \cong \mathrm{Hom}(\mathcal{G}/T\mathcal{G}, -\otimes M_i)$.

General case. We can filter $\mathcal{G}$ by subfunctors

$$0 = \mathcal{G}_0 < \mathcal{G}_1 < \dots < \mathcal{G}_h = \mathcal{G}$$

so that whenever $\mathcal{G}_i \leq \mathcal{H} \leq \mathcal{G}_{i+1}$ then $\chi(\mathcal{H}) \in \{\chi(\mathcal{G}_i), \chi(\mathcal{G}_{i+1})\}$. Now $\mathcal{G}_{i+1}/\mathcal{G}_i$ satisfies the condition of case 2, and (#) follows since both sides of (#) are additive on short exact sequences.

5.3 Let us say that a non-zero character χ is *irreducible* if it cannot be written as a sum $\chi = \chi_1 + \chi_2$ with χ_1 and χ_2 non-zero characters. The previous two theorems may now be reformulated as follows.

COROLLARY. (1) The assignment $M \longmapsto \chi_M$ induces a bijection between the isomorphism classes of indecomposable modules of finite endolength and the irreducible characters.

(2) The irreducible characters are independent over $\mathbb{Z}$.

(3) Every character is a sum of irreducible characters.

5.4 REMARK. The Sylvester module rank functions, upon which our characters are based, were used by Schofield [Sc] to classify suitable equivalence classes of ring homomorphisms from R to a simple artinian ring. Quite why these homomorphisms are related to finite endolength modules is not clear to us, except in the case of a ring epi, in which case Example (4.7)(4) applies.

6 DUALITY

In this section we use characters to define a 1-1 correspondence between the indecomposable left and right R-modules of finite endolength. This seems to be a special case of a duality studied by Herzog [H]. When he introduced Sylvester module rank functions in [Sc], Schofield observed that these were equivalent to Sylvester map rank functions, and that the latter were left-right symmetric for the ring R. In the context of characters this takes the following form. If χ is a character for mod-R we define a character $D\chi$ on f.p. left R-modules, so formally a character for mod-R^{op}, as follows. If X is a f.p. left R-module, let

$$P \xrightarrow{\alpha} Q \longrightarrow X \longrightarrow 0 \qquad (\dagger)$$

be a projective presentation of X. We define

$$(D\chi)(X) = \chi(Q^*) - \chi(P^*) + \chi(\mathrm{Coker}(\alpha^*))$$

where $(-)^* = \mathrm{Hom}_R(-,R)$ is the duality between f.g. projective left and right R-modules, so that $\mathrm{Coker}(\alpha^*)$ is the transpose of X with respect to the resolution (†). One can check that $D\chi$ is well-defined, and when this is done, that $DD\chi = \chi$, so that D defines a duality between the characters on mod-R and on mod-R^{op}. Note also that χ and $D\chi$ have the same degree.

6.1 If $\chi = \chi_M$ then there is a much simpler expression for $D\chi$. Combined with the fact that every character is a sum of characters of this form, this gives an indirect proof that $D\chi$ is well-defined.

Lemma. $D\chi_M(X) = \mathrm{length}_{\mathrm{End}_R(M)}(\mathrm{Hom}_R(X,M))$ for f.p. left R-modules X.

Proof. If (†) is a projective presentation, then the diagram

$$\begin{array}{ccccccccc} & & Q^* \otimes_R M & \longrightarrow & P^* \otimes_R M & \longrightarrow & \mathrm{Coker}(\alpha^*) \otimes_R M & \longrightarrow & 0 \\ & & \| & & \| & & & & \\ 0 \longrightarrow \mathrm{Hom}_R(X,M) & \longrightarrow & \mathrm{Hom}_R(Q,M) & \longrightarrow & \mathrm{Hom}_R(P,M) & & & & \end{array}$$

commutes and has exact rows. Now count lengths.

6.2 The duality D clearly induces a 1-1 correspondence between irreducible characters for mod-R and mod-R^{op}, and hence between the isomorphism classes of indecomposable left and right R-modules of finite endolength, say $M \longleftrightarrow DM$ with $\chi_{DM} = D\chi_M$. Note in particular that M and DM have the same endolength. The proposition below shows how to construct DM.

6.3 **Proposition.** If M is an indecomposable R-module of finite endolength, $E = \mathrm{End}_R(M)$, and I is the injective envelope of the unique simple left E-module, then $\mathrm{Hom}_E(M,I)$ is a direct sum of copies of DM.

Proof. Let $S = \mathrm{soc}_E(I)$ and $F = \mathrm{End}_E(I)$ so that $N = \mathrm{Hom}_E(M,I)$ is an F-R-bimodule. Since I is the injective envelope of S, the F-module $\mathrm{Hom}_E(S,I)$ is simple, and so $\mathrm{length}_F(\mathrm{Hom}_E(Z,I)) = \mathrm{length}_E(Z)$ for any E-module Z of finite length. In particular N has finite length over F. Now there is an isomorphism $\mathrm{Hom}_E(\mathrm{Hom}_R(-,M),I) \cong \mathrm{Hom}_E(M,I)\otimes_R-$ on f.p. left R-modules since both functors are right exact and they agree on R, and so

$$\begin{aligned}\mathrm{length}_E(\mathrm{Hom}_R(X,M)) &= \mathrm{length}_F(\mathrm{Hom}_E(\mathrm{Hom}_R(X,M),I)) \\ &= \mathrm{length}_F(N\otimes_R X)\end{aligned}$$

and hence $\mathrm{length}_F(N\otimes_R X) = D\chi_M(X) = \chi_{DM}(X)$. Now N is a direct sum of copies of DM by the lemma below.

6.4 Lemma. If N is an R-E-bimodule of finite length over E and $\chi_1,\dots,\chi_n$ are the characters of the isomorphism classes of indecomposable summands of N as an R-module, then there are positive integers $a_1,\dots,a_n$ with

$$\mathrm{length}_E(X\otimes_R N) = a_1\chi_1(X) + \dots + a_n\chi_n(X)$$

for all f.p. modules X.

Proof. Let χ_i be the character of an indecomposable module M_i, so by assumption we can write $N = \coprod_{i=1}^n \coprod_{I_i} M_i$ for non-empty index sets I_i. Since

$$\mathrm{End}_R(N)/\mathrm{rad}\ \mathrm{End}_R(N) \cong \prod_{i=1}^n \mathrm{End}_R(\coprod_{I_i} M_i)/\mathrm{rad}\ \mathrm{End}_R(\coprod_{I_i} M_i)$$

the simple $\mathrm{End}_R(N)$-modules which occur in a composition series for $X\otimes_R N$ have the form $S_i = \coprod_{I_i} T_i$ for some i, where T_i is the unique simple $\mathrm{End}_R(M_i)$-module. Since $X\otimes_R N \cong \coprod_{i=1}^n \coprod_{I_i} X\otimes M_i$, the number of times that S_i occurs in the composition series is

$$\mathrm{length}_{\mathrm{End}_R(M_i)}(X\otimes M_i) = \chi_i(X).$$

Letting $a_i = \mathrm{length}_E(S_i) > 0$, the equality follows.

6.5 Remark. If M is a f.p. module of finite endolength, then $\mathrm{Hom}_E(M,I)$ is indecomposable by (2.1), so is isomorphic to DM. It then follows from (2.3) that DM is the source of a left almost split map.

7 GENERIC MODULES

Recall that a ring R is called a *noetherian* (respectively *artin*) *algebra* if its centre Z(R) is a noetherian (respectively artinian) ring, and R is a f.g. Z(R)-module. The reason for considering noetherian algebras is that they have a good supply of modules of finite endolength, the finite length modules. Indeed, a module has finite length if and only if it has finite length as a Z(R)-module, for example by Lenagan's Theorem (4.6). The next proposition shows that amongst the finite endolength modules, there is an essentially unique finiteness condition.

7.1 PROPOSITION. For an indecomposable finite endolength module M over a noetherian algebra, the following are equivalent

(1) M has finite length.

(2) M is finitely presented.

(3) M occurs as the source for a left almost split map.

Moreover these conditions are equivalent to the same conditions for DM.

Proof. (1)⇔(2) follows from Lenagan's Theorem (4.6). If (3) holds, then $M \cong \text{Hom}_E(X, I)$ by (2.3). Now X is f.g. as a Z(R)-module so E is a noetherian Z(R)-algebra, and therefore I is artinian as a Z(R)-module. Thus M is artinian, and hence of finite length by (4.6). If M is f.p. then the same argument shows that DM has finite length, and then M is the source of a left almost split map by (6.5).

7.2 In view of the above proposition, it is natural to pick out the indecomposable modules of finite endolength which have infinite length. We call these *generic modules*. We then say that a noetherian algebra R is *generically trivial* if it has no generic modules, is *generically tame* if for all d there are only finitely many isomorphism classes of generic modules of endolength d, and is *generically wild* if there is a generic module whose endomorphism ring is not a PI ring. Note that since the endomorphism ring $\mathrm{End}_R(M)$ of a generic module has nilpotent radical, it is a PI ring if and only if the division ring $\mathrm{End}_R(M)/\mathrm{rad}\ \mathrm{End}_R(M)$ is finite dimensional over its centre. The definitions above are in terms of generic left R-modules. By (6.2) the notions of generic triviality and generic tameness are left-right symmetric. We do not know if the same is true for generic wildness.

7.3 We first consider the question of generic triviality. Recall that an artin algebra R is said to have *strongly unbounded representation type* if, for infinitely many d, there are infinitely many non-isomorphic indecomposable R-modules of length d. The important part is the existence of one such d, for then the existence of infinitely many d has been proved by Smalø [Sm]. This condition is, however, impossible if R is a finite ring. A more natural condition is to use finite length modules of bounded endolength. The following result is proved in §9.

THEOREM. If R is an artin algebra, then R has a generic module if and only if there are infinitely many non-isomorphic indecomposable finite length R-modules of some fixed endolength. If in addition the simple R-modules have infinite underlying sets, these statements are equivalent to R having strongly unbounded representation type.

The example of a discrete valuation ring shows that the equivalence fails for noetherian algebras without some extra assumptions, while an example of Ringel, see [CB3, 1.6], shows that it fails in general for artinian rings.

The Second Brauer-Thrall Conjecture, which is now proved, asserts that if a finite dimensional algebra over an algebraically closed field is of infinite representation type, then it has strongly unbounded representation type. Thus one has

COROLLARY. A finite dimensional algebra over an algebraically closed field has finite representation type if and only if it is generically trivial.

The natural extension of the Second Brauer-Thrall Conjecture is to ask whether this corollary remains true for noetherian algebras. Of course this reduces immediately to artin algebras, since in a generically trivial noetherian algebra every prime ideal must be maximal.

7.4 If k is an algebraically closed field and R and S are f.g. k-algebras, let us say that a functor F: S-Mod⟶R-Mod is a *representation embedding* if

(1) F sends indecomposable modules to indecomposable modules,

(2) F sends non-isomorphic modules to non-isomorphic modules,

(3) $F \cong M\otimes_S -$ where M is an R-S-bimodule which is f.g. projective as an S-module (and on which k acts centrally).

Equivalently, F is an exact k-linear functor which preserves products and direct sums.

Clearly a representation embedding sends f.d. modules to f.d. modules, and sends modules of finite endolength to modules of finite endolength. A f.g. k-algebra R is said to be of *wild representation type* if there is representation embedding k<x, y>-Mod⟶R-Mod, where k<x, y> is the free associative algebra on two generators. (This is the variant of Drozd's original definition of wild representation type used in [CB3].) In this case there is a representation embedding S-Mod⟶R-Mod for any f.g. k-algebra S, namely if $S = k\langle x_1,\dots,x_n\rangle/I$ then the composition

$$\text{S-Mod} \longrightarrow k\langle x_1,\dots,x_n\rangle\text{-Mod} \xrightarrow{G} k\langle x,y\rangle\text{-Mod} \xrightarrow{F} \text{R-Mod},$$

is a representation embedding. Here G is the fully faithful representation embedding used by Brenner [B]. Thus, in some sense, R is at least as bad as S, for any S.

Now suppose that R is a f.d. k-algebra, with k still an algebraically closed field. By a *one-parameter family of R-modules of dimension* d, we mean the set of modules $\{M\otimes_{k[T]}k[T]/(T-\lambda)\,|\,\lambda\in k\}$, where M is an R-k[T]-bimodule, free of rank d over k[T]. We say that R is of *tame representation type* provided that for all d > 0 there are a finite number of such one-parameter families, such that every indecomposable R-module of dimension d is isomorphic to a module in one of these families. The following theorem is fundamental.

THEOREM OF DROZD. A f.d. algebra R is either tame or wild, and not both.

This is proved in [D], see also [CB1]. With the precise definitions used here, it is discussed in [CB3]. In [CB3] we have used the method of Drozd's Theorem to study generic modules. The basic results are

THEOREM. If R is a tame f.d. algebra, then R is generically tame. In this case, if M is a generic R-module then $\mathrm{End}_R(M)/\mathrm{rad}\,\mathrm{End}_R(M)$ is a rational function field in one variable over k, the ring $\mathrm{End}_R(M)$ is split over its radical, and any two splittings are conjugate.

There are additional results which show that in this case the generic modules act in some way as "function fields" or "generic points" for the one-parameter families of f.d. modules. It is this fact which explains the terminology of "generic module", and indeed our original interest in modules of finite endolength. We shall not explain these results here, but refer the reader to [CB3]. We point out, however, the following characterization of tame and wild representation type.

Corollary. A f.d. algebra R is either generically tame or generically wild, and not both.

Proof. In view of the theorem above, one only needs to prove that if R is wild, then it is generically wild and not generically tame. Now if R is wild then there is a representation embedding $F: k\langle x,y\rangle\text{-Mod} \longrightarrow R\text{-Mod}$. To see that R is not generically tame, observe that the images of the modules $k(x)[y]/(y-\lambda)$ with $\lambda \in k$ are non-isomorphic generic R-modules of bounded endolength. For the generically wildness of R, observe that if D is the universal skew field of fractions of $k\langle x,y\rangle$ then F(D) is a generic R-module, D embeds in $\mathrm{End}_R(F(D))$, but D is not a PI ring.

We conjecture that this corollary remains true for noetherian algebras.

7.5 We finish this section with another question. Let us say that an artin algebra is *generically directed* if generic modules can never be involved in cycles $M_0 \longrightarrow M_1 \longrightarrow \ldots \longrightarrow M_n \longrightarrow M_0$ of non-zero non-isomorphisms between finite endolength indecomposables. In particular the endomorphism ring of any generic module is a division ring.

If R is a finite dimensional algebra over an algebraically closed field, R is generically directed, and R has a faithful generic module, is it true that R is either tame concealed or tubular in the sense of [R4]? Note that such an algebra must be tame, since if R is wild then one can easily construct generic modules whose endomorphism ring is not a division ring.

8 HEREDITARY ALGEBRAS

Let k be a field. In this section, by "algebra" we mean a k-algebra (not necessarily f.d.), and by "bimodule" we mean a bimodule on which k acts centrally. We prove a result which will be needed in the next section, but along the way we determine the behaviour of generic modules for f.d. hereditary algebras.

8.1 The following lemma makes an assertion of Ringel more precise, allowing the argument used in [R1, §5.4] to be extended from the category of finite dimensional modules, to the category of all modules. Let R be an algebra and let X and Y be left R-modules which are finite dimensional and finitely presented. Suppose that $\mathrm{Hom}_R(Y,X) = 0$, $\mathrm{Hom}_R(X,Y) = 0$, and that $E = \mathrm{End}_R(X)^{op}$ and $F = \mathrm{End}_R(Y)^{op}$ are semisimple algebras. Since X is f.p. and Y is f.d., the E-F-bimodule $\mathrm{Ext}^1_R(X,Y)$ is f.d.. Let M be the F-E-bimodule $\mathrm{Hom}_F(\mathrm{Ext}^1_R(X,Y),F)$ and let S be the generalized upper triangular matrix algebra $\begin{pmatrix} F & M \\ 0 & E \end{pmatrix}$.

LEMMA. There is an R-S-bimodule T, f.g. projective over S, inducing a fully faithful functor $T\otimes_S - : S\text{-Mod} \longrightarrow R\text{-Mod}$.

Proof. (1) Let P, P′ be E-modules and Q, Q′ be F-modules. Since X and Y are f.g. we have $\mathrm{Hom}_R(X\otimes_E P, X\otimes_E P') \cong \mathrm{Hom}_E(P,P')$, $\mathrm{Hom}_R(Y\otimes_F Q, Y\otimes_F Q') \cong \mathrm{Hom}_F(Q, Q')$, $\mathrm{Hom}_R(X\otimes_E P, Y\otimes_F Q') = 0$ and $\mathrm{Hom}_R(Y\otimes_F Q, X\otimes_E P') = 0$. Thus, if we are given short exact sequences of R-modules making up the rows of the diagram

$$\begin{array}{ccccccccc} 0 & \longrightarrow & Y\otimes_F Q & \longrightarrow & L & \longrightarrow & X\otimes_E P & \longrightarrow & 0 \\ & & \phi\downarrow & & \theta\downarrow & & \downarrow\psi & & \\ 0 & \longrightarrow & Y\otimes_F Q' & \longrightarrow & L' & \longrightarrow & X\otimes_E P' & \longrightarrow & 0 \end{array}$$

then for any map $\theta \in \mathrm{Hom}_R(L,L')$ there are maps ϕ, ψ making the diagram commute. Moreover θ is uniquely determined by ϕ, ψ since if θ' also makes the diagram commute then $\theta-\theta'$ induces a map $X\otimes_E P \longrightarrow Y\otimes_F Q'$.

(2) The fact that X is finitely presented implies that

$$\operatorname{Ext}^1_R(X\otimes_E P, Y\otimes_F Q) \cong \operatorname{Hom}_E(P, \operatorname{Ext}^1_R(X,Y)\otimes_F Q)$$

and in particular that

$$\operatorname{Ext}^1_R(X, Y\otimes_F M) \cong \operatorname{Ext}^1_R(X,Y)\otimes_F M \cong \operatorname{Hom}_F(M,M). \qquad (\dagger)$$

Let

$$\zeta : 0 \longrightarrow Y\otimes_F M \xrightarrow{p} N \xrightarrow{q} X \longrightarrow 0$$

be an extension corresponding to the identity endomorphism of M. Now the left and right hand terms of ζ are naturally R-E-bimodules, and since ($\dagger$) is an isomorphism of E-E-bimodules it follows that $e\zeta = \zeta e$ for all $e\in E$. Thus by (1), for each $e\in E$ there is a unique endomorphism θ of N making the diagram

$$\begin{array}{ccccccccc} 0 & \longrightarrow & Y\otimes_F M & \longrightarrow & N & \longrightarrow & X & \longrightarrow & 0 \\ & & \downarrow e & & \downarrow \theta & & \downarrow e & & \\ 0 & \longrightarrow & Y\otimes_F M & \longrightarrow & N & \longrightarrow & X & \longrightarrow & 0 \end{array}$$

commute. This gives N the structure of an R-E-bimodule in such a way that p and q are bimodule maps.

(3) Set $T = Y\oplus N$. We turn it into R-S-bimodule by defining

$$(y,n)\begin{pmatrix} f & m \\ 0 & e \end{pmatrix} = (yf, p(y\otimes m)+ne) \quad \text{for } (y,n)\in T \text{ and } \begin{pmatrix} f & m \\ 0 & e \end{pmatrix}\in S.$$

This is projective as an S-module since p is mono, and of course it is finitely generated over S since it is f.d..

(4) A left S-module U is determined by a triple (P,Q,g) where P is an E-module, Q is an F-module, and g is an F-module map $M\otimes_E P \longrightarrow Q$. Namely set $P = e_{22}U$ and $Q = e_{11}U$. Now U has a projective presentation of the form

$$0 \longrightarrow (0, M\otimes_E P, 0) \longrightarrow (P, M\otimes_E P\oplus Q, \begin{pmatrix}1\\0\end{pmatrix}) \longrightarrow U \longrightarrow 0,$$

and tensoring with T gives an exact sequence

$$0 \longrightarrow Y\otimes_F M\otimes_E P \longrightarrow N\otimes_E P \oplus Y\otimes_F Q \longrightarrow T\otimes_S U \longrightarrow 0.$$

Thus $T\otimes_S U$ fits in the pushout diagram

$$\begin{array}{ccccccccc} 0 & \longrightarrow & Y\otimes_F M\otimes_E P & \longrightarrow & N\otimes_E P & \longrightarrow & X\otimes_E P & \longrightarrow & 0 \\ & & \downarrow & & \downarrow & & \| & & \\ 0 & \longrightarrow & Y\otimes_F Q & \longrightarrow & T\otimes_S U & \longrightarrow & X\otimes_E P & \longrightarrow & 0 \end{array}$$

Clearly the lower exact sequence corresponds to the element g under the isomorphism $\mathrm{Hom}_F(M\otimes_E P, Q) \cong \mathrm{Ext}^1_R(X\otimes_E P, Y\otimes_F Q)$.

Now let $U' = (P',Q',g')$ be a second S-module. The S-module maps $\alpha: U \longrightarrow U'$ correspond to pairs (β,γ) where $\beta\in\mathrm{Hom}_E(P,P')$, $\gamma\in\mathrm{Hom}_F(Q,Q')$ and such that $g'\circ(1\otimes\beta) = \gamma\circ g$. Such a map α gives a commutative diagram

$$\begin{array}{ccccccccc} 0 & \longrightarrow & Y\otimes_F Q & \longrightarrow & T\otimes_S U & \longrightarrow & X\otimes_E P & \longrightarrow & 0 \\ & & \downarrow{\scriptstyle 1\otimes\gamma} & & \downarrow{\scriptstyle 1\otimes\alpha} & & \downarrow{\scriptstyle 1\otimes\beta} & & \\ 0 & \longrightarrow & Y\otimes_F Q' & \longrightarrow & T\otimes_S U' & \longrightarrow & X\otimes_E P' & \longrightarrow & 0 \end{array}$$

and if $1\otimes\alpha = 0$, then $1\otimes\beta = 1\otimes\gamma = 0$, and hence $\beta = \gamma = 0$, so that $T\otimes_S -$ is faithful. Conversely, if $\theta \in \mathrm{Hom}_R(T\otimes_S U, T\otimes_S U')$ then by (1) this map induces a commutative diagram with maps β, γ. Now (β,γ) is a homomorphism $\alpha: U \longrightarrow U'$ and $1\otimes\alpha = \theta$, so that $T\otimes_S -$ is full.

8.2 Let us say that an algebra R is *strictly wild* if there are f.d. left R-modules X and Y, which are finitely presented, whose endomorphism rings are division algebras, with $\mathrm{Hom}_R(X,Y) = 0$, $\mathrm{Hom}_R(Y,X) = 0$, and the product

$$p = \dim_{\mathrm{End}_R(Y)} \mathrm{Ext}^1_R(X,Y) \ . \ \dim \mathrm{Ext}^1_R(X,Y)_{\mathrm{End}_R(X)}$$

equal to at least 5. The next result is due to Ringel [R1].

Lemma. A finitely generated algebra R is strictly wild if and only if there is a finite extension field K of k and an

R-K<x,y>-bimodule T which is f.g. projective over K<x,y> and such that the tensor product functor $T\otimes_{K<x,y>}- : K<x,y>\text{-Mod} \longrightarrow R\text{-Mod}$ is fully faithful.

Proof. Suppose first that there exists such a bimodule. As in (7.4), for any K-algebra S there is a fully faithful functor $M\otimes_S - : S\text{-Mod} \longrightarrow R\text{-Mod}$ with M an R-S-bimodule, f.g. projective over S. Taking S to be strictly wild, and letting X and Y be the images of a pair of S-modules which make S strictly wild, one obtains a pair of f.d. R-modules whose Hom and Ext^1 spaces satisfy the requirements above. Because R is f.g. and X, Y are finite dimensional, they are finitely presented. Thus R is strictly wild.

To prove that a strictly wild algebra R has a fully faithful tensor product functor K<x,y>-Mod ⟶ R-Mod, it suffices to deal with algebras of the form $R = \begin{pmatrix} F & M \\ 0 & E \end{pmatrix}$ with E, F division algebras and $(\dim_F M)(\dim M_E) \geq 5$. Namely, in general, if R is strictly wild due to the existence of modules X, Y, then the algebra S constructed in (8.1) has this special form, so by assumption there is a suitable functor K<x,y>-Mod ⟶ S-Mod. The composition of this functor with the functor S-Mod ⟶ R-Mod given by Lemma (8.1) is a suitable tensor product functor K<x,y>-Mod ⟶ S-Mod.

The proof that an algebra of the form $R = \begin{pmatrix} F & M \\ 0 & E \end{pmatrix}$ has a suitable functor K<x,y>-Mod ⟶ R-Mod, follows part of the argument used in the proof of [R1, Theorem 2], working by induction on $d = \max\{\dim_K E, \dim_K F\}$ where K is the centre of the bimodule M (which is also the centre of R). If d=1, then E=F=K and $M=K^r$ with $r\geq 3$, and it is easy to find a suitable R-K<x,y>-bimodule, free of rank two over K<x,y>. Suppose, therefore that d>1. As in [R1, §5.3] one can find finite dimensional R-modules A_1, A_2 with each $\mathrm{End}_R(A_i)$ a division ring, $\dim_K \mathrm{End}_R(A_1) < d$, $\mathrm{Hom}_R(A_i,A_j) = 0$ for $i\neq j$ and $\mathrm{Ext}^1_R(A_i,A_j) \neq 0$ for all i, j. Now let $\mathfrak{U}$ be the full subcategory of R-Mod on the modules which have a finite filtration in which the quotients are isomorphic to A_1 or A_2. By [R1, §1.2 Theorem] this is an exact abelian subcategory of R-Mod, and since R is hereditary, the category $\mathfrak{U}$ is hereditary. It follows from this, and the fact that $\mathrm{Ext}^1_R(A_i, A_j) \neq 0$ for all i, j, that one can find objects X, Y of $\mathfrak{U}$ which are uniserial as objects of $\mathfrak{U}$, and with composition series in $\mathfrak{U}$ of the form

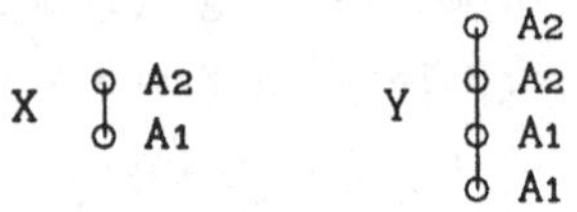

Now $\mathrm{End}_R(X)$ and $\mathrm{End}_R(Y)$ embed in $\mathrm{End}_R(A_1)$, $\mathrm{Hom}_R(X, Y) = \mathrm{Hom}_R(Y, X) = 0$, $\dim_{\mathrm{End}(Y)}\mathrm{Ext}^1_R(X, Y) \geq 3$, and $\dim \mathrm{Ext}^1_R(X, Y)_{\mathrm{End}(X)} \geq 2$, see [R1, §5.4]. Let S be the algebra constructed from X and Y in (8.1). By induction there is a functor K<x, y>-Mod $\longrightarrow$ S-Mod, and its composition with the functor S-Mod $\longrightarrow$ R-Mod gives the desired tensor product functor from K<x, y>-Mod to R-Mod.

8.3 If R is a connected f.d. hereditary algebra, there is a bilinear form defined on the Grothendieck group $K_0(R)$ of f.d. left modules modulo short exact sequences given by

$$\langle M, N\rangle = \dim_k \mathrm{Hom}_R(X, Y) - \dim_k \mathrm{Ext}^1_R(X, Y),$$

and this induces a quadratic form q_R on $K_0(R)\otimes_{\mathbb{Z}}\mathbb{Q}$. The algebra R is of finite representation type if q_R is positive definite, it is said to be *tame hereditary* if q_R is positive semidefinite but not positive definite, and *wild hereditary* if q_R is indefinite. Note that these notions are purely combinatorial, but in case the base field k is algebraically closed they coincide with the notions discussed in §7, except for the fact that a tame hereditary algebra is necessarily of infinite representation type.

8.4 THEOREM. A f.d. wild hereditary algebra is strictly wild.

Proof. If R has two simple modules this is clear. Thus suppose that R has $n > 2$ simple modules. We follows the argument of [R5, §1 Theorem] and use the terminology of that paper. Let $\Delta(R)$ be the species of R. If R^{op} is strictly wild, then clearly so is R. If S is obtained from R by reflection at a sink in $\Delta(R)$ and R is strictly wild, then so is S, for by Lemma (8.2) there is a very wide choice of modules X and Y giving the strict wildness, so we may choose X and Y to be regular, and then they correspond to S-modules. Now the argument in [R5] shows that, up to duality and reflections, one of the three cases below occurs. We verify in each one that R is strictly wild.

Case 1. R is a one-point extension of a connected hereditary algebra S of infinite representation type, so $R = \begin{pmatrix} S & M \\ 0 & D \end{pmatrix}$ where D is a division algebra and M is a non-zero S-D-bimodule which is projective as a S-module. Now if Y is an S-module regarded as an R-module, X is the simple R-module corresponding to D, and P is the projective cover of X, then X has projective resolution $0 \longrightarrow M \longrightarrow P \longrightarrow X \longrightarrow 0$, so $\mathrm{Ext}^1_R(X, Y) \cong \mathrm{Hom}_S(M, Y)$. This can be made arbitrarily large with Y an indecomposable preprojective S-module by [R5, §1 Lemma 1]. It follows that R is strictly wild.

Case 2. R has species

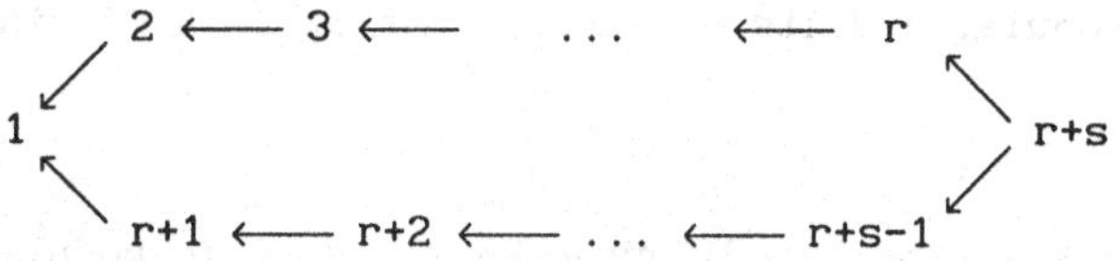

with $r \geq 2$, $s \geq 1$, and not all arrows trivially valued. Let $Y = P(1)$ and set $X = P(r+s)/\mathrm{soc}\, P(r+s)$. Note that the socle of $P(r+s)$ is the direct sum of $a'+b'$ copies of $P(1)$ where

$$a' = d'_{12}\, d'_{23} \cdots d'_{r-1,r}\, d'_{r,r+s}, \quad \text{and}$$

$$b' = d'_{1,r+1}\, d'_{r+1,r+2} \cdots d'_{r+s-1,r+s}.$$

Using the projective resolution $0 \longrightarrow (a'+b')P(1) \longrightarrow P(r+s) \longrightarrow X \longrightarrow 0$ one sees that $\dim_{\mathrm{End}(Y)} \mathrm{Ext}^1_R(X, Y) = a'+b'$. Moreover $\mathrm{End}_R(X) \cong$

$\mathrm{End}_R(P(r+s))$ so that $\dim_k \mathrm{End}_R(Y)/\dim_k \mathrm{End}_R(X) = a/a' = b/b'$, where a and b are defined in the same way as a′ and b′ but using the d_{ij} instead of the d'_{ij}. Thus the product p of (8.2) is equal to $(a'+b')(a+b)$. Now $a, a', b, b' \geq 1$, and by assumption not all are equal to 1, so we have $p \geq 6$.

Case 3. R has species

$$1 \xleftarrow{(a,b)} 2 \xleftarrow{(c,d)} 3.$$

with $abcd \geq 6$. Let $Y = P(1)$ and $X = P(3)/\mathrm{soc}\, P(3)$. Using the projective resolution $0 \longrightarrow bdP(1) \longrightarrow P(3) \longrightarrow X \longrightarrow 0$ one finds that the product p of (8.2) is equal to abcd, so R is strictly wild.

Corollary. If R is a f.d. hereditary algebra, then R is generically trivial if and only if it has finite representation type. Moreover R is either generically tame or generically wild and not both.

Proof. We may assume that R is connected. If R is wild hereditary then it is generically wild, but not generically tame as in Corollary (7.4). If R is tame hereditary then Ringel [R3, §6] has proved (with an unnecessary extra hypothesis) that R has a unique generic module, and its endomorphism ring is a PI ring by [BGL, 6.12].

Remark. In the tame hereditary case the generic module has a rather interesting endomorphism ring E. For example, if R is the generalized triangular matrix ring $\begin{pmatrix} \mathbb{H} & \mathbb{H} \\ 0 & \mathbb{R} \end{pmatrix}$ with $\mathbb{H}$ the quaternion division ring, then E is the field $\mathbb{R}(X, Y \mid X^2+Y^2+1=0)$. More generally, if R is tame hereditary and R/rad R is separable over Z(R) then the endomorphism ring of the generic module is a division ring whose centre is a function field in one variable over Z(R) of genus zero. See [CB2] for more discussion.

8.5 We also need to consider a class of non-noetherian rings which generalizes the free associative algebras. In his study of free ideal rings, Cohn has considered filtered rings

$$R_0 \subseteq R_1 \subseteq R_2 \subseteq \ldots \subseteq R,$$

defined a 'weak algorithm' for such rings, and given a construction [C, §2.5] of all filtered rings which have a weak algorithm. Supposing that R is generated by R_1 this takes the following form. Let D be a division ring, let

$$0 \longrightarrow D \xrightarrow{e} X \xrightarrow{f} Y \longrightarrow 0 \qquad (\ddagger)$$

be an exact sequence of D-D-bimodules, and set $\pi = e(1)$. Then R is the ring $X^{\otimes D}/(\pi-1)$, where $X^{\otimes D} = D \oplus X \oplus X\otimes_D X \oplus \ldots$ is the tensor ring. The terms in the filtration are the images of $X\otimes_D \ldots \otimes_D X$, so that $R_0 = D$ and $R_1 = X$. In case D is a f.d. division algebra and X and Y are f.d. bimodules we call $X^{\otimes D}/(\pi-1)$ a *skew tensor algebra*.

Lemma. If R is a skew tensor algebra given by an exact sequence (‡) with $\dim_D Y \geq 2$, then R is strictly wild.

Proof. We begin by showing that there are at least two non-isomorphic R-modules S with $\dim_D S = 1$ (so that S is simple). An R-module is determined by a left D-vector space V and a D-module map $\psi: X\otimes_D V \longrightarrow V$ satisfying $\psi(\pi\otimes v) = v$. The modules we want, correspond to the case $V = D$, say with S_ψ being the module determined by a map $\psi \in \mathrm{Hom}_D(X,D)$ with $\psi(\pi) = 1$. Now the algebraic group $D^\times = D\setminus\{0\}$ acts on the variety $W = \{\psi\in\mathrm{Hom}_D(X,D) \mid \psi(\pi)=1\}$ of such ψ via $(d.\psi)(x) = \psi(xd)d^{-1}$, and the orbits correspond to the isomorphism classes of S_ψ. Since $\dim(W) = \dim_k Y \geq 2\dim_k D = 2\dim(D^\times)$, a dimension argument, or if k is finite a counting argument, shows that there must be at least two orbits, as required.

Next we construct an exact sequence of R-R-bimodules of the form

$$\eta : 0 \longrightarrow R\otimes_D Y\otimes_D R \xrightarrow{h} R\otimes_D R \xrightarrow{m} R \longrightarrow 0.$$

For m one takes the multiplication map. Let α be the map

$$X^{\otimes D} \otimes_D X \otimes_D X^{\otimes D} \longrightarrow R \otimes_D R,$$
$$u\otimes x\otimes v \longmapsto \overline{u\otimes x} \otimes \bar{v} - \bar{u} \otimes \overline{x\otimes v}$$

where the bar denotes reduction by the ideal $(\pi-1)$. It is easily seen that α induces a map h as above. It remains to show that η is exact. Now Cohn has shown that if $\pi, x_1, \ldots, x_n$ is a left D-basis for X, then the (images in R of the) monomials in the x_i form a left D-basis for R. Thus $R\otimes_D R$ is a free left R-module with basis the elements $1\otimes q$ (q a monomial), and so the elements $q\otimes q'$ $(q, q'$ monomials) form a left D-basis for $R\otimes_D R$. It follows that Ker(m) has as left D-basis the elements of the form $qx_i\otimes q' - q\otimes x_i q'$ with q and q′ monomials and $1\le i\le n$. These elements are the images under h of the elements $q\otimes f(x_i)\otimes q'$, and a similar argument shows that these form a left D-basis for $R\otimes_D Y\otimes_D R$, which proves that η is exact.

Tensoring η with any R-module gives a projective presentation of that module, so R is hereditary, which was already clear. Moreover it follows that if S_1 and S_2 are non-isomorphic modules with $\dim_D S_i = 1$ then the S_i are finitely presented and $\mathrm{Ext}^1_R(S_i, S_j) \neq 0$ for any choice of i, j. Now the argument of [R1, §5.4] together with (8.2) shows that R is strictly wild.

8.6 Lemma. Let R be a f.d. hereditary algebra or a skew tensor algebra. If R has infinite representation type, then

(1) If k is infinite then R has infinitely many non-isomorphic indecomposable modules of some fixed dimension.

(2) R has infinitely many non-isomorphic f.d. indecomposable modules of some fixed endolength.

(3) There is an R-k(T)-bimodule, indecomposable over R, and

finite dimensional over $k(T)$.

Proof. If R is strictly wild, then this follows from Lemma (8.2). If R is a skew tensor algebra corresponding to the exact sequence (‡) of (8.5) then we may assume that $\dim_D Y = 1$ so that R is actually a skew polynomial ring $D[T;\varepsilon,\delta]$ with ε an automorphism of D/k and δ an $(\varepsilon,1)$-derivation of D/k. Now R is an hereditary noetherian domain and a PI ring since it is f.g. as a module over the (non-central) subring $k[T]$. Therefore the centre Z of R is a Dedekind domain and R is f.g. as a Z-module [MR, 13.9.16]. For (1) one takes the simple modules $R/(T-\lambda)$ with $\lambda\in k$. For (2) one takes the simple modules, since by the Nullstellensatz [MR, 13.10.3] there are infinitely many and they are f.d.. Let Q be the simple artinian quotient ring of R. By Posner's Theorem Q is f.d. over its centre which is the quotient field K of Z. Now Z is f.g. over k by the Artin-Tate lemma [MR, 13.9.10], so K is a finitely generated extension field of k of transcendence degree 1. Therefore K and Q are finite dimensional over $k(T)$ for some $T \in K$. Now for the bimodule in (3) one can take the simple Q-module.

If R is a f.d. hereditary algebra then we may assume that it is tame hereditary. Now (1) is contained in [DR, Theorem E], and (3) in [R2, Theorem 5.7] and [BGL, Corollary 6.12]. One knows, see for example [CB2], that there is a ring-theoretic epimorphism $R \longrightarrow S$ where S is a classical hereditary order, finitely generated as a k-algebra. Now S has infinitely many simple modules, and each one restricts to a f.d. indecomposable R-module of endolength bounded by the PI degree of S, proving (2).

9 LIFT CATEGORIES

In this section we prove Theorem (7.3). We use the method of matrix reductions, which enables an inductive proof, reducing to the hereditary case which we have solved in the previous section.

The proof of Drozd's Tame and Wild Theorem also uses matrix reductions, in the form of bocses, but these cannot be used with general artin algebras. Instead, we have introduced in [CB4] the notion of a "lift category" and used it to study artinian rings of finite representation type.

A *lift pair* (R,ξ) consists of a ring R and an exact sequence

$$\xi : 0 \longrightarrow M \longrightarrow E \xrightarrow{\pi} R \longrightarrow 0$$

of R-R-bimodules, and the corresponding *lift category* $\xi(R)$ has as objects the pairs (P,e) where P is a projective left R-module and e is a section for map $\pi\otimes 1 : E\otimes_R P \longrightarrow P$, and as morphisms from (P,e) to (P',e') the R-module maps $\theta : P \longrightarrow P'$ which intertwine e and e'.

$$\begin{array}{ccc} P & \xrightarrow{e} & E\otimes_R P \\ \theta\downarrow & & \downarrow 1\otimes\theta \\ P' & \xrightarrow{e'} & E\otimes_R P' \end{array}$$

Let C be a commutative artinian local ring with maximal ideal $\mathfrak{m}$ and residue field k. We now consider C-algebras, and C is supposed to act centrally on all bimodules. We say that a lift pair (R,ξ) is *C-algebraic* provided that R is an artin C-algebra and E is f.g. as a C-module.

Let (R,ξ) be a C-algebraic lift pair and let $J = \text{rad } R$. Let us say that an object $X = (P,e)$ in $\xi(R)$ is *sincere* if P/JP is a sincere R-module, so involves all simple R-modules. We define the *length* of X (over C) to be $\text{length}_C(P/JP)$. If $R_X = \text{End}_{\xi(R)}(X)^{op}$, then P is naturally an R-R_X-bimodule, and we define the *endolength*, endolen(X), of X to be the length of P/JP as an R_X-module. One might have defined the last two notions without reducing modulo JP, but the definitions given provide the useful numbers, and since R is artinian the finiteness of the length or endolength is independent of the definition. We say that X is *generic* if it is indecomposable, of finite endolength, but of infinite length.

A lift pair (R,ξ) is said to be of *finite representation type* if there are only finitely many isomorphism classes of indecomposable objects in $\xi(R)$ and they all have finite length. In addition we consider the following conditions

(C1) $\xi(R)$ has infinitely many non-isomorphic indecomposable objects of some fixed length.

(C2) $\xi(R)$ has infinitely many non-isomorphic finite length objects of some fixed endolength.

(C3) $\xi(R)$ has an indecomposable object $X = (P,e)$ with a C-algebra map $C[T]_{mC[T]} \longrightarrow R_X$, such that P/JP has finite length over $C[T]_{mC[T]}$.

(C4) $\xi(R)$ has a generic object.

Note that $C[T]_{mC[T]}$ is an artinian local ring with residue field $k(T)$ and that (C3) $\Rightarrow$ (C4). Also (C1) $\Rightarrow$ (C2), but (C1) is never possible if k is finite. We adopt the *convention* that when we talk of (C1)-(C4) below, we exclude (C1) in case k is finite.

9.1 LEMMA. If (R,ξ) is a C-algebraic lift pair, R is semisimple, and the first term M of ξ is a simple bimodule or zero, then (C1)-(C4) are equivalent to (R,ξ) being of infinite representation type.

Proof. Since R is semisimple, $mR = 0$, so that R is a k-algebra. By [CB4, 2.1] the category $\xi(R)$ is equivalent to A-Mod where $A = (E^{*})^{\otimes R}/(\pi-1)$, and this equivalence preserves endolength. Moreover by [CB4, 2.2] (or at least its proof) this algebra is either a f.d. hereditary algebra, or the product of a semisimple artinian ring and a matrix ring over a skew tensor algebra. The result thus follows from Lemma (8.6). Note that since $C[T]_{mC[T]}$ is a local ring with residue field $k(T)$, any $k(T)$-module is naturally a $C[T]_{mC[T]}$-module.

9.2 Let (R,ξ) be a C-algebraic lift pair with exact sequence

$$\xi : 0 \longrightarrow M \longrightarrow E \xrightarrow{\pi} R \longrightarrow 0.$$

Let $N \subsetneq M$ be a maximal sub-bimodule, and let $J = \operatorname{rad} R$. One can form lift pairs (R,ξ_N) and $(R/J,\xi_{NJ})$ with

$$\xi_N : 0 \longrightarrow \overline{M} \longrightarrow \overline{E} \longrightarrow R \longrightarrow 0$$

where $\overline{M} = M/N$ and $\overline{E} = E/N$, and

$$\xi_{NJ} : 0 \longrightarrow \overline{M}/[\overline{M}\cap(\overline{E}J+J\overline{E})] \longrightarrow \overline{E}/(\overline{E}J+J\overline{E}) \xrightarrow{\pi_{NJ}} R/J \longrightarrow 0$$

and there are functors

$$\xi(R) \xrightarrow{\sigma_N} \xi_N(R) \xrightarrow{\rho_J} \xi_{NJ}(R/J)$$

defined as follows. If $X = (P,e)$ belongs to $\xi(R)$, then $\sigma_N(X) = (P,\bar{e})$ where $\bar{e}$ is the composition $P \xrightarrow{e} E\otimes_R P \longrightarrow \overline{E}\otimes_R P$, and if $\theta:(P,e)\longrightarrow(P',e')$ is a morphism then $\sigma_N(\theta)$ is the same R-module map, considered now as a morphism from $(P,\bar{e})$ to $(P',\bar{e}')$. If $Z = (P,f) \in \xi_N(R)$ then $\rho_J(Z) = (R/J\otimes_R P,\bar{f})$ where $\bar{f}$ is the composition

$$R/J\otimes_R P \xrightarrow{1\otimes f} R/J\otimes_R \overline{E}\otimes_R P \xrightarrow{p} \overline{E}/(\overline{E}J+J\overline{E})\otimes_R P$$

with p the natural projection, and if $\theta:(P,f)\longrightarrow(P',f')$ is a morphism then $\rho_J(\theta) = 1\otimes\theta$. It is shown in [CB4, 3.1 and 4.1] that σ_N and ρ_J are both dense and reflect isomorphisms, and that ρ_J is full, so that it is a representation equivalence.

One can apply (9.1) to the lift pair $(R/J,\xi_{NJ})$, and in the next two paragraphs we investigate the consequences for (R,ξ).

9.3 In this paragraph we treat the case when $(R/J,\xi_{NJ})$ is of infinite representation type. We begin with some lemmas.

LEMMA a. Let R be an artin C-algebra and S a C-algebra such that S/mS is a separably generated extension field of k. If N is an $R\otimes_C S$-module which is projective as an R-module, then it is projective as an $R\otimes_C S$-module.

Proof. Let J = rad R and let L be the image of $J\otimes_C S$ in $R\otimes_C S$. Since R is artinian, m annihilates R/J, so that

$$(R\otimes_C S)/L \cong (R/J)\otimes_C S \cong (R/J)\otimes_k (S/mS),$$

and by the assumption on S/mS, this is semisimple. Since also L is nilpotent, it is the radical of $R\otimes_C S$. Now $R\otimes_C S$ is a semiprimary ring, so the module N has a projective cover, a map $\alpha: P \twoheadrightarrow N$ with P projective and Ker(α) superfluous in P, and moreover a submodule of any $R\otimes_C S$-module M is superfluous if and only if it is contained in LM. Thus Ker(α) $\subseteq$ LP = JP, and since R is semiprimary it follows that Ker(α) is superfluous as an R-submodule of P. However α splits as an R-module map, so Ker(α) = 0, and hence N is a projective $R\otimes_C S$-module.

LEMMA b. Let R be an artin C-algebra and let S be a C-algebra which is projective over C, and with S/mS a separably generated extension field of k. If N is an $R\otimes_C S$-module, then its projective cover as an $R\otimes_C S$-module is a projective cover as an R-module. In particular this assertion holds for $S = C[T]_{mC[T]}$.

Proof. As in Lemma a, the ring $R\otimes_C S$ is semiprimary with rad($R\otimes_C S$) the image of $J\otimes_C S$, and if $\alpha: P \twoheadrightarrow N$ is a projective cover then Ker(α) is superfluous as an R-submodule of P. Since S is projective as a C-module, P is a projective R-module, and hence α is an R-module projective cover. For $S = C[T]_{mC[T]}$ note that S/mS $\cong$ k(T) is separably generated over k, and that S is flat over C, and hence projective since C is artinian.

LEMMA c. Let A be an artin C-algebra. If k is a perfect field then A has a C-subalgebra S with S + rad A = A, S $\cap$ rad A = rad S and rad S = mS.

Proof. Since S = A satisfies the first two inequalities, one can choose S to be a C-subalgebra of A, with $\text{length}_C(S)$ minimal, amongst those subalgebras satisfying the first two equalities. Since mS is nilpotent it is contained in rad S, and for a contradiction we may suppose that S/mS is not semisimple. Now S/mS is a finite dimensional k-algebra, so split over its radical by the Wedderburn-Malcev Theorem. Taking a splitting T/mS, one obtains a strictly smaller subalgebra T with T + rad S = S, T ∩ rad S = mS and rad T ⊆ mS, so that T also satisfies the first two inequalities. A contradiction.

PROPOSITION. Under the hypotheses of (9.2), if $(R/J, \xi_{NJ})$ is of infinite representation type, then (C1)-(C4) hold for (R, ξ).

Proof. By (9.1) the conditions (C1)-(C4) hold for $(R/J, \xi_{NJ})$. We deduce them for (R, ξ). Note first that (C1) holds because any object in $\xi_{NJ}(R/J)$ lifts to one in $\xi(R)$ of the same length, and (C4) follows from (C3), so we only need to prove (C2) and (C3). The problem is to lift the objects in $\xi_{NJ}(R/J)$ to objects in $\xi(R)$ *of the same endolength*. The approach for both of these is similar, but differs because in the case of (C3) we also need a homomorphism from $C[T]_{mC[T]}$ to the endomorphism ring of the lifted object. This is compensated for by the fact that $C[T]_{mC[T]}$ is projective over C.

(C2) In view of the implication (C1)⇒(C2) we only need to prove this when k is finite, and hence a perfect field. Let X = (Q, g) be an indecomposable object in $\xi_{NJ}(R/J)$ of finite length and with endomorphism ring F. Let $\alpha: P \twoheadrightarrow Q$ be the projective cover of Q as an R-module and let

$$A = \{\theta \in \text{End}_R(P) \mid \theta(\text{Ker}(\alpha)) \subseteq \text{Ker}(\alpha) \text{ and } \bar{\theta} \in F\}$$

where $\bar{\theta}$ is the endomorphism of Q induced by θ under the assumption that $\theta(\mathrm{Ker}(\alpha)) \subseteq \mathrm{Ker}(\alpha)$. The assignment $\theta \longmapsto \bar{\theta}$ is a homomorphism $A \longrightarrow F$ and the induced map $A/\mathrm{rad}\,A \longrightarrow F/\mathrm{rad}\,F$ is an isomorphism since α is a projective cover. Let S be the subalgebra of A chosen by Lemma c. Now P can be regarded as an $R\otimes_C S$-module and it is projective over $R\otimes_C S$ by Lemma a, since S has the property that $S/mS \cong F/\mathrm{rad}\,F$ is a finite extension field of k, hence separable. Now in the diagram

$$\begin{array}{ccc} P & \xrightarrow{\alpha} & Q \\ {\scriptstyle f}\downarrow & & \downarrow{\scriptstyle g} \\ E\otimes_R P & \xrightarrow{p\otimes\alpha} & \bar{E}/(\bar{E}J+J\bar{E})\otimes_R Q \end{array}$$

with p the projection, the modules all have natural structures as $R\otimes_C S$-modules, and the maps are all $R\otimes_C S$-module maps. Since $p\otimes\alpha$ is epi, there is a map f making the diagram commute. Also, the map $\pi\otimes 1: E\otimes_R P \longrightarrow P$ has a section e_0 as an $R\otimes_C S$-module map, and if $e: P \longrightarrow E\otimes_R P$ is defined by

$$e = f + e_0\circ(1 - (\pi\otimes 1)\circ f)$$

then $Y = (P,e)$ is an object in $\xi(R)$, and it has image X in $\xi_{NJ}(R/J)$. Moreover the fact that e is an $R\otimes_C S$-module map means that S is contained in $\mathrm{End}_{\xi(R)}(Y)$. Now the isomorphism $S/\mathrm{rad}\,S \longrightarrow A/\mathrm{rad}\,A \longrightarrow F/\mathrm{rad}\,F$ shows that X and Y have the same endolength. Finally (C2) follows by using the infinite family of objects X of the same endolength.

(C3) Since $\xi_{NJ}(R/J)$ satisfies (C3) it has an indecomposable object $X = (Q,g)$, a map $C[T]_{mC[T]} \longrightarrow \mathrm{End}(X)$, and with Q of finite length over $C[T]_{mC[T]}$. By Lemma b the projective cover P of Q as an $R\otimes_C C[T]_{mC[T]}$-module is a projective cover as an R-module. Now, as in the verification of (C2) one can lift the $R\otimes_C C[T]_{mC[T]}$-module map $P \longrightarrow Q \longrightarrow \bar{E}/(\bar{E}J+J\bar{E})\otimes Q$ to a map $f: P \longrightarrow E\otimes P$, and this can then be adjusted to give an object $Y = (P,e)$ in $\xi(R)$. Since e is an $R\otimes_C C[T]_{mC[T]}$-module map, there is a natural map $C[T]_{mC[T]} \longrightarrow \mathrm{End}_{\xi(R)}(Y)$. Now $P/JP \cong Q$ has finite length over

$C[T]_{mC[T]}$, and Y has image X in $\xi_{NJ}(R/J)$ so it is indecomposable. Thus (C3) holds.

9.4 Continuing with the hypotheses of (9.2) suppose now that the lift pair $(R/J,\xi_{NJ})$ has finite representation type. Let $X = (P,\bar{e})$ be an object in $\xi_N(R)$ of finite length, and which is the direct sum of exactly n non-isomorphic indecomposable objects. Let

$$R_X = \mathrm{End}_{\xi_N(R)}(X)^{op}$$

and let ξ_X be defined via the pullback of R_X-R_X-bimodules

$$\begin{array}{ccccccccc} \xi_X : 0 & \longrightarrow & M_X & \longrightarrow & E_X & \longrightarrow & R_X & \longrightarrow & 0 \\ & & \| & & \downarrow{\scriptstyle\alpha} & & \downarrow{\scriptstyle\beta} & & \\ 0 & \longrightarrow & \mathrm{Hom}_R(P,N\otimes_R P) & \longrightarrow & \mathrm{Hom}_R(P,E\otimes_R P) & \longrightarrow & \mathrm{Hom}_R(P,E/N\otimes_R P) & \longrightarrow & 0 \end{array}$$

where β is the map sending 1 to $\bar{e}$. Thus (R_X,ξ_X) is a lift pair. Let $\tau_X:\xi_X(R_X)\longrightarrow\xi(R)$ be the functor defined as follows. If $Y = (Q,g)$ belongs to $\xi_X(R_X)$ then $\tau_X(Y) = (P\otimes_{R_X}Q,h)$, where h is the composition

$$P\otimes_{R_X}Q \xrightarrow{1\otimes g} P\otimes_{R_X}E_X\otimes_{R_X}Q \xrightarrow{1\otimes\alpha\otimes 1} P\otimes_{R_X}\mathrm{Hom}_R(P,E\otimes_R P)\otimes_{R_X}Q$$
$$\xrightarrow{ev\otimes 1} E\otimes_R P\otimes_{R_X}Q,$$

and ev is the evaluation map, and if $\theta:(Q,g)\longrightarrow(Q',g')$ is a morphism then $\tau_X(\theta) = 1\otimes\theta$. It is shown in [CB4, 4.2] that τ_X is fully faithful, and that it induces an equivalence from $\xi_X(R_X)$ to the full subcategory of $\xi(R)$ on those objects whose image under σ_N is a summand of a direct sum of copies of X. Now the representation equivalence ρ_J shows that $\xi_N(R)$ has only finitely many non-isomorphic indecomposable objects, and if X is the direct sum of all of them, then τ_X is an equivalence (since there is a direct sum preserving representation equivalence from $\xi_N(R)$ to the category of modules for an algebra of finite representation type, so that every object is a direct sum of indecomposables).

Lemma. If $Y = (Q, g)$ is an object in $\xi_X(R_X)$ then

$$\mathrm{endolen}(Y) \le \mathrm{endolen}(\tau_X(Y)) \le \mathrm{length}_C(X).\mathrm{endolen}(Y).$$

If in addition Y is sincere, then either endolen(Y) is less than $\mathrm{endolen}(\tau_X(Y))$, or $\mathrm{length}_C(M_X) < \mathrm{length}_C(M)$.

Proof. Since τ_X is fully faithful,

$$\mathrm{endolen}(\tau_X(Y)) = \mathrm{length}(P/JP \otimes_{R_X} Q_F) \quad \text{where}$$
$$F = \mathrm{End}_{\xi_X(R_X)}(Y)^{\mathrm{op}}.$$

If S is a simple right R_X-module, then S is a summand of R_X/J_X, where $J_X = \mathrm{rad}\, R_X$, so that

$$\mathrm{length}(S \otimes_{R_X} Q_F) \le \mathrm{length}(R_X/J_X \otimes_{R_X} Q_F) = \mathrm{endolen}(Y).$$

Taking a composition series of P/JP as a right R_X-module one obtains

$$\begin{aligned} \mathrm{endolen}(\tau_X(Y)) &\le \mathrm{length}(P/JP_{R_X}).\mathrm{endolen}(Y) \\ &\le \mathrm{length}_C(X).\mathrm{endolen}(Y), \end{aligned}$$

which is the second inequality. Now P/JP is a sincere right R_X-module, since P is faithful, so sincere as a right R_X-module, but for each r, $J^rP/J^{r+1}P$ is a quotient of $(J^r/J^{r+1}) \otimes_{R/J} P/JP$, which is a summand of a direct sum of copies of P/JP. On the other hand, since the indecomposable summands of X are non-isomorphic, R_X is basic, that is, R_X/J_X is isomorphic as a right R_X-module to the direct sum of one copy of each simple. It follows that

$$\mathrm{length}(R_X/J_X \otimes_{R_X} Q_F) \le \mathrm{length}(P/JP \otimes_{R_X} Q_F). \qquad (\dagger)$$

which is the first inequality.

Now suppose that Y is sincere, so that Q/J_XQ is a sincere R_X-module, and hence $S \otimes_{R_X} Q \neq 0$ for any simple right R_X-module S. If the inequality (†) is not strict, then P/JP must have length exactly n as an R_X-module, that is, endolength(X) = n. Now the functor $\rho_J: \xi_N(R) \longrightarrow \xi_{NJ}(R/J)$ is a representation equivalence, and by [CB4, 2.1] there is an equivalence $\xi_{NJ}(R/J) \longrightarrow$A-Mod, where

$$A = ([\bar{E}/(\bar{E}J+J\bar{E})]^*)^{\otimes R/J}/(\pi_{NJ}-1)$$

is a f.d. hereditary k-algebra of finite representation type. Both of these functors preserve endolength, so the image X′ of X in A-Mod has endolength n. Recall that X is a direct sum of exactly n non-isomorphic indecomposable summands, so there is a decomposition $X' = U_1 \oplus \ldots \oplus U_n$ into non-isomorphic indecomposable summands. Now the endolength of the direct sum is the sum of the endolengths by Lemma (4.5), so all U_i must have endolength 1, and therefore be simple. This means that P/JP is a direct sum of distinct simples, and hence P is a summand of R. Since $M_X = \mathrm{Hom}_R(P, N \otimes_R P)$ this implies that

$$\mathrm{length}_C(M_X) \leq \mathrm{length}_C(N) < \mathrm{length}_C(M),$$

as required.

9.5 Theorem. (C1)-(C4) are equivalent for C-algebraic lift pairs.

Proof. Let (R,ξ) be such a lift pair. We assume that (C2) or (C4) holds, and wish to prove that the rest hold. Thus for some d there is either a generic object G of endolength d, or an infinite family $(N_\lambda)_{\lambda\in\Lambda}$ of finite length indecomposable objects of endolength d. We use induction on d and $\mathrm{length}_C(M)$.

If M = 0, then $\xi(R) \cong$ R-Proj has finite representation type, which is impossible under our assumption. Thus one can pick a maximal sub-bimodule $N \subseteq M$ and make the constructions of (9.2).

If $\xi_{NJ}(R/J)$ has infinite type the conclusions are given by Proposition (9.3), so suppose that $\xi_{NJ}(R/J)$ has finite type.

For X as in (9.4) the functor $\tau_X: \xi_X(R_X) \longrightarrow \xi(R)$ is fully faithful, and in case X is the direct sum of all indecomposables this is an equivalence. By choosing X carefully we can ensure that either G or infinitely many of the N_λ are the images under τ_X of *sincere* objects in $\xi_X(R_X)$. Now by Lemma (9.4) the lift pair (R_X, ξ_X) either satisfies the hypotheses for some $d' < d$, or for d but with $\text{length}_C(M_X) < \text{length}_C(M)$. By the induction, the lift pair (R_X, ξ_X) satisfies (C1)-(C4), and then the second inequality in Lemma (9.4) ensures that (R, ξ) satisfies (C1)-(C4).

9.6 Let A be an artin C-algebra with radical L, and let (R, ξ) be the lift pair with $R = \begin{pmatrix} A & 0 \\ 0 & A \end{pmatrix}$, and

$$\xi : 0 \longrightarrow \begin{pmatrix} 0 & L \\ 0 & 0 \end{pmatrix} \longrightarrow \begin{pmatrix} A & L \\ 0 & A \end{pmatrix} \longrightarrow \begin{pmatrix} A & 0 \\ 0 & A \end{pmatrix} \longrightarrow 0.$$

Clearly (R, ξ) is C-algebraic. By [CB4, 1.7] there is an equivalence between $\xi(R)$ and the category $P^1(A)$ of triples (P', P'', α) with P' and P'' projective A-modules and $\alpha: P' \longrightarrow P''$ a map with $\text{Im}(\alpha) \subseteq LP''$. If $P^2(A)$ denotes the subcategory of $P^1(A)$ on the triples with $\text{Ker}(\alpha) \subseteq LP'$, then every object in $P^1(A)$ is the direct sum of an object in $P^2(A)$ and a triple of the form $(P', 0, 0)$, and the functor $P^2(A) \longrightarrow A\text{-Mod}$ sending (P', P'', α) to $\text{Coker}(\alpha)$ is a representation equivalence.

Lemma. Let $X = (P, e) \in \xi(R)$ correspond to an object in $P^2(A)$ with image $N \in A\text{-Mod}$ under the cokernel functor. If S is a C-subalgebra of $\text{End}_{\xi(R)}(X)^{op}$, and $J = \text{rad } R$, then

$$\text{length}(N_S) \le \text{length}(A_A).\text{length}(P/JP_S) \quad \text{and}$$

$$\text{length}(P/JP_S) \le (\text{length}(A_A) + 1).\text{length}(N_S).$$

Moreover $\text{endolen}(X) \le (\text{length}(A_A) + 1).\text{endolen}(N)$.

Proof. Let X correspond to the triple (P', P'', α) in $P^2(A)$, so there is an exact sequence

$$P' \xrightarrow{\alpha} P'' \longrightarrow N \longrightarrow 0 \qquad (\ddagger)$$

of A-S-bimodules. Now $R = \begin{pmatrix} A & 0 \\ 0 & A \end{pmatrix}$ and $P = \begin{pmatrix} P' \\ P'' \end{pmatrix}$, so that

$$\text{length}(P/JP_S) = \text{length}(P'/LP'_S) + \text{length}(P''/LP''_S).$$

If T is a simple right A-module, then $\text{length}(T\otimes_A P''_S) \le \text{length}(P''/LP''_S)$, so a composition series of A_A gives

$$\text{length}(P''_S) \le \text{length}(A_A).\text{length}(P''/LP''_S) \qquad (*)$$

and the first inequality follows. Now X corresponds to an object in $P^2(A)$, so the exact sequence $(\ddagger)$ is a minimal projective presentation, and thus $P''/LP'' \cong N/LN$ and $P'/LP' \cong \text{Im}(\alpha)/L\,\text{Im}(\alpha)$, a subquotient of P''. Therefore

$$\text{length}(P''/LP''_S) \le \text{length}(N_S)$$

and then $(*)$ implies that

$$\text{length}(P'/LP'_S) \le \text{length}(A_A).\text{length}(N_S),$$

giving the second inequality. Finally, taking $S = \text{End}_{\xi(R)}(X)^{op}$, the fact that the cokernel functor is full means that $\text{endolen}(N) = \text{length}(N_S)$.

Theorem. Let A be an artin C-algebra. Consider the following statements.

(1) For some $d\in\mathbb{N}$ there are infinitely many non-isomorphic indecomposable A-modules of length d over C.

(2) For some $d\in\mathbb{N}$ there are infinitely many non-isomorphic indecomposable A-modules which are of endolength d, and have finite length over C.

(3) There is an $A\text{-}C[T]_{mC[T]}$-bimodule, indecomposable over A, and of finite length over $C[T]_{mC[T]}$.

(4) A has a generic module.
Then (2)-(4) are equivalent. If in additiion the field k is infinite, then they are also equivalent to (1).

Proof. If A satisfies one of (1)-(4), then it satisfies (2) or (4), and by the lemma, the lift pair (R,ξ) satisfies (C2) or (C4). By Theorem (9.5) the lift pair satisfies (C1)-(C4). Now there are only finitely many indecomposable objects in $P^1(A)$ which do not belong to $P^2(A)$, and they all correspond to objects in ξ(R) of finite length over C. Thus the lemma enables one to deduce (1)-(4).

REMARK. Theorem (7.3) follows on reducing to the case when the artin algebra R is connected, and then setting C = Z(R). If the simple R-modules have infinite underlying sets, then k is infinite. Note that a family of R-modules has bounded length if and only if it has bounded length over C.

REFERENCES

[A1] M. Auslander, Coherent functors, Proceedings of the conference on categorical algebra (Springer, Berlin, 1966).

[A2] M. Auslander, Representation theory of artin algebras II, Commun. Algebra, 2 (1974), 269-310.

[A3] M. Auslander, Functors and morphisms determined by objects, Representation theory of algebras, Proc. conf. Philadelphia 1976, ed R. Gordon (Dekker, New York 1978), 1-244.

[Az] G. Azumaya, Countable generatedness version of rings of pure global dimension zero, these proceedings.

[BGL] D. Baer, W. Geigle and H. Lenzing, The preprojective algebra of a tame hereditary Artin algebra, Commun. Algebra, 15 (1987), 425-457.

[B] S. Brenner, Decomposition properties of some small diagrams of modules, Symp. Math. Ist. Naz. Alta. Mat., 13 (1974), 127-141.

[C] P. M. Cohn, 'Free rings and their relations' (Academic Press, London, 1971).

[CB1] W. W. Crawley-Boevey, On tame algebras and bocses, Proc. London Math. Soc. (3), 56 (1988), 451-483.

[CB2] W. W. Crawley-Boevey, Regular modules for tame hereditary algebras, to appear in Proc. London Math. Soc.

[CB3] W. Crawley-Boevey, Tame algebras and generic modules, to appear in Proc. London Math. Soc.

[CB4] W. Crawley-Boevey, Matrix reductions for artinian rings, and an application to rings of finite representation type, to appear in J. Algebra.

[DR] V. Dlab and C. M. Ringel, On algebras of finite representation type, J. Algebra, 33 (1975), 306-394.

[D] Yu. A. Drozd, Tame and wild matrix problems, Representations and quadratic forms (Institute of Mathematics, Academy of Sciences, Ukrainian SSR, Kiev, 1979), 39-74. Amer. Math. Soc. Transl., 128 (1986), 31-55.

[G] P. Gabriel, Des catégories abéliennes, Bull. Soc. Math. France, 90 (1962), 323-448.

[GW] K. R. Goodearl and R. B. Warfield, Jr, 'An introduction to noncommutative noetherian rings' (Cambridge University Press, 1989).

[GJ1] L. Gruson and C. U. Jensen, Deux applications de la notion de L-dimension, C. R. Acad. Sci. Paris, 282 (1976), 23-24.

[GJ2] L. Gruson and C. U. Jensen, Dimensions cohomologiques relieés aux foncteurs $\varprojlim^{(i)}$, Séminaire d'Algèbre Paul Dubreil et Marie-Paule Malliavin (Springer Lec. Notes 867, Berlin, 1981), 234-294.

[H] I. Herzog, Elementary duality of modules, preprint.

[JL] C. U. Jensen and H. Lenzing, 'Model theoretic algebra' (Gordon and Breach, New York, 1989).

[MR] J. C. McConnell and J. C. Robson, 'Noncommutative Noetherian rings' (Wiley, Chichester, 1987).

[P] M. Prest, 'Model Theory and Modules' (London Math. Soc. Lec. Note Series 130, Cambridge, 1988).

[R1] C. M. Ringel, Representations of K-species and bimodules, J. Algebra, 41 (1976), 269-302.

[R2] C. M. Ringel, Infinite dimensional representations of finite dimensional hereditary algebras, Ist. Naz. Alta Mat., Symp. Math. 23 (1979) 321-412.

[R3] C. M. Ringel, The spectrum of a finite dimensional algebra, Proc. Conf. on Ring Theory, Antwerp 1978 (Dekker, New York, 1979).

[R4] C. M. Ringel, 'Tame algebras and integral quadratic forms' (Springer Lec. Notes 1099, Berlin, 1984).

[R5] C. M. Ringel, The regular components of the Auslander-Reiten quiver of a tilted algebra, Chin. Ann. of Math., 9B (1988), 1-18.

[RT] C. M. Ringel and H. Tachikawa, QF-3 rings, J. für die Reine und Angew., 272 (1975), 49-72.

[Ro] A. V. Roiter, Matrix problems and representations of BOCS's, Representation Theory I (Springer Lec. Notes 831, Berlin, 1980), 288-324.

[Sc] A. H. Schofield, 'Representations of rings over skew fields' (London Math. Soc. Lec. Note Series 92, Cambridge, 1985).

[Sm] S. O. Smalø, The inductive step of the second Brauer-Thrall conjecture, Canad. J. Math., 32 (1980), 342-349.

[ZH] B. Zimmermann-Huisgen, Rings whose right modules are direct sums of indecomposable modules, Proc. Amer. Math. Soc., 77 (1979), 191-197.

[ZHZ] B. Zimmermann-Huisgen and W. Zimmermann, On the sparsity of representations of rings of pure global dimension zero, Trans. Amer. Math. Soc., 320 (1990), 695-711.

Pairs of semi-simple algebras
(Hereditary algebras with radical-square zero)

Vlastimil Dlab, Carleton University, Ottawa

This little note represents a modest contribution illustrating an application of recently developed methods of the representation theory of algebras to questions raised in functional analysis.It is a summary of our joint results with C.M. Ringel describing an algebraic approach to the theory of pairs of semi-simple algebras, the Jones fundamental construction of a tower and its index (see [**DR4**] and [**DR5**]). Our interest in these questions was inspired by the monograph [**GHJ**] of F.M. Goodman, P. de la Harpe and V.F.R. Jones, where the classical situation of the so-called multi-matrix algebras is considered.

The main points of this note may be summarized as follows. The basic invariant of a general pair of semi-simple algebras is a bimodule together with a vector space which determine the corresponding weighted valued graph. Whereas, in the classical situation, this graph (known then as the Bratteli diagram) determines a pair of multi-matrix algebras completely, in a general situation, the bimodule plays an essential role. This can easily be seen on the following simple example of a central and non-central embedding of the real algebra of the complex numbers $\mathbb{C}$ into the real algebra of the 2×2 complex matrices. Clearly, the two embeddings

$$c \mapsto \begin{pmatrix} c & 0 \\ 0 & c \end{pmatrix} \quad \text{and} \quad c = a + bi \mapsto \begin{pmatrix} a & -b \\ b & a \end{pmatrix}$$

are not equivalent; however, their weighted graphs are equal: $(1) \xrightarrow{(2,2)} (2)$. The first embedding is characterized by the $\mathbb{C}$–$\mathbb{C}$-bimodule ${}_{\mathbb{C}}\mathbb{C}_{\mathbb{C}} \oplus {}_{\mathbb{C}}\mathbb{C}_{\mathbb{C}}$ (with the canonical operations), while the second embedding is described by the $\mathbb{C}$–$\mathbb{C}$-bimodule ${}_{\mathbb{C}}\mathbb{C}_{\mathbb{C}} \oplus {}_{\mathbb{C}}\mathbb{C}_{\overline{\mathbb{C}}}$ (with the bimodule operation on ${}_{\mathbb{C}}\mathbb{C}_{\overline{\mathbb{C}}}$ given by $c_1 \cdot c \cdot c_2 = c_1 c \bar{c}_2$).

In fact, the entire theory of pairs of finite dimensional semi-simple algebras is equivalent to the theory of finite dimensional hereditary algebras $\mathcal{A}$ whose radical $\operatorname{Rad}\mathcal{A}$ satisfies $(\operatorname{Rad}\mathcal{A})^2 = 0$. The equivalence is achieved by the natural bijection

$$A \subseteq B \quad \longleftrightarrow \quad \begin{pmatrix} A & B \\ 0 & B \end{pmatrix}.$$

The Jones fundamental construction of a tower of semi-simple algebras is, in this equivalence, related to the weighted preprojective component of the corresponding algebra and the index is expressed in a simple manner in terms of the largest real part of the eigenvalues of the associated Coxeter transformation. It turns out that this relationship explains the "discrete" nature of the set of all possible values of the index. These are intimately connected to the representation type of the respective algebra.

1. A few remarks on finite dimensional associative algebras

This brief section is included for the convenience of a reader not fully familiar with some of the recent developments in the theory of finite dimensional associative algebras and, in particular, of the hereditary algebras.

Throughout, K is a fixed field and $\mathcal{A}$ is a finite dimensional algebra. Recall that the (Jacobson) radical $\operatorname{Rad}\mathcal{A}$ of $\mathcal{A}$ is the intersection of the maximal ideals of $\mathcal{A}$. Thus, $\overline{\mathcal{A}} = \mathcal{A}/\operatorname{Rad}\mathcal{A}$ is a semi-simple finite dimensional algebra; by the Wedderburn–Artin theorem, $\overline{\mathcal{A}}$ is a finite direct product of the full matrix algebras over division K-algebras. We shall call $\mathcal{A}$ basic if $\overline{\mathcal{A}}$ is a product of division K-algebras.

Let $\{S(1), S(2), \ldots, S(n)\}$ be the (ordered) set of all simple (right) $\mathcal{A}$-modules and $\{P(1), P(2), \ldots, P(n)\}$ the set of the respective indecomposable projective $\mathcal{A}$-modules; in other words, $P(t)$ is the projective cover of $S(t)$, $1 \leq t \leq n$.

(i) Morita equivalence

Consider a decomposition of the (right) regular representation $\mathcal{A}_{\mathcal{A}}$ of $\mathcal{A}$ into indecomposable (projective) direct summands:

$$\mathcal{A}_{\mathcal{A}} = \bigoplus_{t=1}^{n} \bigoplus_{i=1}^{r_t} P_i(t) \quad \text{with } r_t \geq 1 \text{ for all } 1 \leq t \leq n.$$

Observe that the algebra $\mathcal{A}$ is basic if and only if $r_t = 1$ for all $1 \leq t \leq n$. Furthermore, with appropriately defined multiplication of the endomorphisms,

$$\mathcal{A} \cong \operatorname{End} \mathcal{A}_{\mathcal{A}}.$$

Consider the (right) $\mathcal{A}$-modules

$$M = M(k_1, k_2, \ldots, k_n) = \bigoplus_{t=1}^{n} \bigoplus_{i=1}^{k_t} P_i(t) \quad \text{with } k_t \geq 1 \text{ for all } 1 \leq t \leq n;$$

the respective K-algebras $\operatorname{End} M_{\mathcal{A}}$ form a Morita equivalence class of K-algebras. Thus, each of the equivalence classes contains a (unique) basic K-algebra $\operatorname{End}\big(\bigoplus_{t=1}^{n} P(t)\big)$.

The fundamental result concerning Morita equivalence of algebras is the following theorem.

Theorem. *Two finite dimensional K-algebras $\mathcal{A}$ and $\mathcal{B}$ are Morita equivalent if and only if their categories of (right finite dimensional) modules* mod-$\mathcal{A}$ *and* mod-$\mathcal{B}$ *are equivalent.*

Thus, given a division K-algebra F, the matrix algebras $\operatorname{Mat}(k, F)$, $k \in \mathbb{N}$, are all Morita equivalent to F: their module categories are equivalent to the category of vector spaces over F. Every finite dimensional semi-simple K-algebra is Morita equivalent to a product of division K-algebras. Let us point out that whereas the concepts of a simple module, an indecomposable module, a projective module etc. are Morita invariant, the K-dimension of a module is not !

Given an $\mathcal{A}$-module $X_{\mathcal{A}}$, define its dimension type $\mathbf{dim}\, X$ by $\mathbf{dim}\, X = (x_1, x_2, \ldots, x_n)$, where x_t is the multiplicity of the simple module $S(t)$ in a composition series of X. In other words, x_t is the length of the E_t-module $\operatorname{Hom}(P(t), X)$, where $E_t = \operatorname{End} P(t)$. We shall see in (v) that some indecomposable modules are determined by their dimension types.

(ii) K-species and (valued) graphs

Given a basic K-algebra $\mathcal{A}$, the semi-simple K-algebra $\Lambda = \mathcal{A}/\operatorname{Rad}\mathcal{A}$ (uniquely) decomposes into the product

$$\Lambda = F_1 \times F_2 \times \cdots \times F_n$$

of division K-algebras F_i, $1 \le i \le n$. Furthermore, the Λ–Λ-bimodule ${}_\Lambda M_\Lambda = \operatorname{Rad}\mathcal{A}/(\operatorname{Rad}\mathcal{A})^2$ has a unique decomposition

$$M = \bigoplus_{1\le i,j\le n} {}_iM_j \text{ into the } F_i\text{–}F_j\text{-spaces } {}_iM_j = F_iMF_j.$$

The data $\mathcal{S}(\mathcal{A}) = (F_i, {}_iM_j\ ;\ 1 \le i,j \le n)$ is said to be the K-species of $\mathcal{A}$. Recall the homological meaning of the ingredients of $\mathcal{S}(\mathcal{A})$:

$$F_i = \operatorname{End}\big(S(i)\big) \quad \text{and} \quad {}_jM_i^* = \operatorname{Ext}^{(1)}_{\mathcal{A}}\big(S(j), S(i)\big).$$

Moreover, we associate an oriented valued graph $\Gamma(\mathcal{A})$ to the K-algebra $\mathcal{A}$. By definition, an (oriented, symmetrizable) valued graph is a pair of non-negative integral $n \times n$ matrices

$$\big(U = (u_{ij}), V = (v_{ij})\big)$$

such that there is a positive invertible diagonal $n \times n$ matrix $D = (d_{ij})$ satisfying $UD = DV$. The index set $\{1, 2, \dots, n\}$ is the set of the vertices of the graph and if $u_{ij} \ne 0$ (and thus $v_{ij} \ne 0$), we say that there is an arrow from i to j with valuation (u_{ij}, v_{ij}). Now, the graph $\Gamma(\mathcal{A}) = \Gamma\big(\mathcal{S}(\mathcal{A})\big)$ is defined by

$$u_{ij} = \dim({}_iM_j)_{F_j}, \quad v_{ij} = \dim_{F_i}({}_iM_j) \quad \text{and} \quad d_{ii} = \dim_K F_i.$$

Thus, $\mathcal{A}$ is a semi-simple algebra if and only if its graph $\Gamma(\mathcal{A})$ consists of a number of discrete vertices, with no arrows.

(iii) Tensor algebras

Let Λ be a semi-simple K-algebra and ${}_\Lambda M_\Lambda$ a Λ–Λ-bimodule with K operating centrally. Then the tensor algebra $\mathcal{T}(M)$ of M is the graded algebra

$$\mathcal{T}(M) = \bigoplus_{t\ge 0} N^{(t)},$$

where $N^{(0)} = \Lambda$, $N^{(1)} = M$ and $N^{(t+1)} = N^{(t)} \otimes_\Lambda M$ for $t \ge 1$, with componentwise addition and multiplication induced by taking tensor products.

Given an (abstract) K-species $\mathcal{S} = (F_i, {}_iM_j\,;\ 1 \le i,j \le n)$ with division K-algebras F_i and F_i–F_j-bimodules ${}_iM_j$ (K operating centrally), the tensor algebra $\mathcal{T}(\mathcal{S})$ of $\mathcal{S}$ is the tensor algebra of ${}_\Lambda M_\Lambda = \bigoplus_{1\le i,j\le n} {}_iM_j$ with the canonical operations by $\Lambda = F_1 \times F_2 \times \cdots F_n$.

In general, $\mathcal{T}(\mathcal{S})$ is an infinite dimensional K-algebra. In fact, $\mathcal{T}(\mathcal{S})$ is finite dimensional if and only if there is no oriented cycle in the graph $\Gamma(\mathcal{S})$; in that case, $\mathcal{T}(\mathcal{S})$ is hereditary in the sense that $\operatorname{Rad}\mathcal{T}(\mathcal{S})$ is a projective (right) $\mathcal{T}(\mathcal{S})$-module (and thus every submodule of a projective module is projective).

(iv) Hereditary algebras

The basic results on the representation type of hereditary algebras can be summarized as follows (see [**DR1**] and [**DR3**]); we formulate the statements for the connected algebras, i.e. those having only trivial central idempotents, or equivalently, those whose graphs are connected.

Theorem. *(i) A finite dimensional hereditary connected K-algebra is of finite representation type if and only if it is Morita equivalent to a tensor K-algebra over a Dynkin graph.*

(ii) A finite dimensional hereditary connected K-algebra $\mathcal{A}$ is of tame representation type (i.e. there is no exact full embedding of the category of finite dimensional modules over the polynomial algebra in two non-commuting variables into mod-$\mathcal{A}$*) if and only if $\mathcal{A}$ is of type $\tilde{A}_n(\varepsilon,\delta)$ or is Morita equivalent to a tensor K-algebra over a Euclidean graph.*

Let us remark that we shall deal in this note with hereditary algebras $\mathcal{A}$ such that $(\operatorname{Rad}\mathcal{A})^2 = 0$, and thus the algebras of type $\tilde{A}_n(\varepsilon,\delta)$, which cannot have this property, will not appear.

(v) Coxeter functors, Coxeter transformations

Given a K-species $\mathcal{S} = (F_i, {}_iM_j\ ;\ 1 \le i,j \le n)$ whose graph $\Gamma(\mathcal{S})$ has no oriented cycles, the tensor algebra $\mathcal{A} = \mathcal{T}(\mathcal{S})$ is a finite dimensional hereditary K-algebra. There is a pair of endofunctors C^-, C^+ of the category mod-$\mathcal{A}$ called Coxeter functors having the following properties: If $X \in$ mod-$\mathcal{A}$ is indecomposable, then

(a) $C^-X = 0$ if and only if X is injective; otherwise C^-X is indecomposable and $C^+C^-X \cong X$;

(b) $C^+X = 0$ if and only if X is projective; otherwise C^+X is indecomposable and $C^-C^+X \cong X$.

The elements of the family

$$\mathcal{P} = \{C^{-t}P(i) \mid t \geq 0,\ \mathcal{P}(i) \text{ indecomposable projective, } 1 \leq i \leq n\}$$

are the (indecomposable) preprojective $\mathcal{A}$-modules. This family is finite if and only if the graph $\Gamma(\mathcal{S})$ is Dynkin; in this case, $\mathcal{P}$ is the family of all indecomposable $\mathcal{A}$-modules.

The modules $X \in \mathcal{P}$ are uniquely defined by their dimension type $\mathbf{dim}\,X \in \mathbb{R}^n$. In fact, there is a linear transformation $c = c_\Gamma : \mathbb{R}^n \longrightarrow \mathbb{R}^n$ defined in terms of the graph $\Gamma(\mathcal{S})$ above, called the Coxeter transformation, such that

$$c^{-1}(\mathbf{dim}\,X) = \mathbf{dim}\,(C^-X) \text{ for every } X \in \mathcal{P};$$

c can be expressed explicitly as a composition of the n reflections (corresponding to the vertices of the graph) defining the Weyl group of $\Gamma(\mathcal{S})$; see e.g. [**DR1**].

The matrix M_c of the Coxeter transformation c with respect to the standard basis can be written as

$$M_c = -H(H^{-1})^{tr},$$

where $H = PD^{-1}$ with P being the (Cartan) matrix whose columns are the vectors $\mathbf{dim}\,P(1)$, $\mathbf{dim}\,P(2)$, ..., $\mathbf{dim}\,P(n)$. It follows that the characteristic polynomial χ_c of c is reciprocal; thus, if r is a root of χ_c, then $1/r$ is a root with the same multiplicity as r. In fact, we have the following description of the eigenvalues of c_Γ (see e.g. [**A**] and [**DR2**]).

Theorem. *If Γ is Dynkin, then all eigenvalues λ of c_Γ satisfy $|\lambda| = 1$ and $\lambda \neq 1$. If Γ is Euclidean, then all eigenvalues λ of c_Γ satisfy $|\lambda| = 1$*

and $\lambda = 1$. *Otherwise, the spectral radius* ρ *of* c_Γ *is greater than 1 and is an eigenvalue of* c_Γ.

2. Pairs $A \subseteq B$ and the algebra $\mathcal{A}(A, B)$

Given a K-pair, i.e. a pair $A \subseteq B$ of finite dimensional semi-simple K-algebras with unital embedding, consider the associated hereditary K-algebra $\mathcal{A}(A, B)$ of the upper triangular 2×2 matrices

$$\text{(i)} \qquad \mathcal{A}(A,B) = \begin{pmatrix} A & B \\ 0 & B \end{pmatrix} = \left\{ \begin{pmatrix} a & c \\ 0 & b \end{pmatrix} \mid a \in A;\, b, c \in B \right\}.$$

Thus, $\mathcal{A}(A, B)$ is the tensor algebra of the bimodule ${}_AB_B$ and is Morita equivalent to the basic tensor algebra

$$\text{(ii)} \qquad \mathcal{A} = \begin{pmatrix} F & {}_FM_G \\ 0 & G \end{pmatrix},$$

where $F = F_1 \times F_2 \times \cdots \times F_m$ and $G = G_1 \times G_2 \times \cdots \times G_n$ are basic semi-simple K-algebras Morita equivalent to A and B, respectively and ${}_FM_G$ is the appropriate F–G-bimodule. In other words,

$$(F_1, F_2, \ldots, F_m, G_1, G_2, \ldots, G_n,\ {}_iM_j = F_iMG_j,\ 1 \le i \le m,\ 1 \le j \le n)$$

is the K-species of $\mathcal{A}(A, B)$ and

$$\text{(iii)} \qquad \left(\begin{pmatrix} 0_{m\times m} & U \\ 0_{n\times m} & 0_{n\times n} \end{pmatrix}, \begin{pmatrix} 0_{m\times m} & V \\ 0_{n\times m} & 0_{n\times n} \end{pmatrix} \right),$$

where $U = (u_{ij})$ with $u_{ij} = \dim_{G_j}(F_iMG_j)$ and $V = (v_{ij})$ with $v_{ij} = \dim_{F_i}(F_iMG_j)$ is the (bipartite) valued graph $\Gamma(\mathcal{A}(A, B))$.

It turns out that all algebras of the form (i) which are Morita equivalent to (ii) have the form

$$\begin{pmatrix} \operatorname{End} X_F & \operatorname{End}(X_F \otimes {}_FM_G) \\ 0 & \operatorname{End}(X_F \otimes {}_FM_G) \end{pmatrix},$$

where X_F is the direct sum of x_i-dimensional F_i-spaces X_i for some $x_i \ge 1$, $1 \le i \le m$, with the canonical action by F. Indeed, if $\mathcal{A}_\mathcal{A} = \bigoplus_{t=1}^{m+n} P_t$ is a decomposition of the (right) regular representation into the indecomposable projective (right) $\mathcal{A}$-modules, then

$$P_i \approx F_i \oplus (\overset{n}{\underset{j=1}{\oplus}} {}_iM_j) \quad \text{and} \quad P_{m+j} \approx 0 \oplus G_j \quad \text{for} \quad 1 \leq i \leq m, \;\; 1 \leq j \leq n.$$

Choose $\mathbf{x} = (x_1, x_2, \ldots, x_m) \in \mathbb{N}^{(m)}$ and define $\mathbf{y} = (y_1, y_2, \ldots, y_n) \in \mathbb{N}^{(n)}$ by

$$y_j = \sum_{i=1}^{m} x_i u_{ij}, \quad 1 \leq j \leq n.$$

Observing that $0 \oplus {}_iM_j \approx P_{m+j}^{(u_{ij})}$, we deduce that

$$\mathcal{A}(A, B) \cong \operatorname{End}(\overset{m}{\underset{i=1}{\oplus}} P_i^{(x_i)} \oplus \overset{n}{\underset{j=1}{\oplus}} P_{m+j}^{(y_j)}),$$

i.e. that

$$A \cong \operatorname{End} X_F \quad \text{and} \quad B \cong \operatorname{End}(X_F \otimes {}_FM_G), \quad \text{where} \quad X = \overset{m}{\underset{i=1}{\oplus}} F_i^{(x_i)}.$$

Explicitly,

$$A \cong \prod_{i=1}^{m} \operatorname{Mat}(x_i, F_i), \qquad B \cong \prod_{j=1}^{n} \operatorname{Mat}(y_j, G_j),$$

and the weighted graph $\Gamma(A \subseteq B)$ if the K-pair $A \subseteq B$ is the graph (iii) together with the positive integers x_i, y_j attached to the appropriate vertices; let us point out that y_j's are determined by x_i's. Let us also mention that in the case of multi-matrix algebras (i.e. when $K = F_i = G_j$ for all $1 \leq i \leq m$ and $1 \leq j \leq n$), this notion coincides with that of Bratteli diagram [**Br**].

Theorem. *There is a one-to-one correspondence between the pairs of semi-simple algebras and the pairs consisting of a positive integral vector and a bimodule over products of division algebras.*

3. Jones fundamental construction of a tower and its index

Given a K-pair $A \subseteq B$, consider the semi-simple K-algebra $C = \operatorname{End} B_A$ and the K-pair $B \subseteq C$. If $(\mathbf{x}, {}_FM_G)$ corresponds to $A \subseteq B$, then $(\mathbf{y}, {}_GM_F^*)$ with $\mathbf{y}$ as in Section 2 and ${}_GM_F^* = \operatorname{Hom}_G({}_FM_G, {}_GG_G)$ corresponds to $B \subseteq C$. Consequently,

$$C \cong \bigoplus_{i=1}^{m} \mathrm{Mat}\,(z_i, F_i), \text{ where } z_i = \sum_{j=1}^{n} y_j v_{ij}.$$

Hence, keeping the order of the vertices of the graph fixed by their correspondence to $F_1, F_2, \ldots, F_m, G_1, G_2, \ldots, G_n$,

$$\Gamma(B \subseteq C) = \left(\begin{pmatrix} 0_{m\times m} & 0_{m\times n} \\ V^{tr} & 0_{n\times n} \end{pmatrix}, \begin{pmatrix} 0_{m\times m} & 0_{m\times n} \\ U^{tr} & 0_{n\times n} \end{pmatrix} \right)$$

with the weighting given by $\mathbf{y}$ and $\mathbf{z} = (z_1, z_2, \ldots, z_m)$.

Now, starting with $A_0 = A \subseteq B = A_1$ and repeating successively the above procedure, we arrive at a countable tower of semi-simple K-algebras

$$A_0 \subseteq A_1 \subseteq A_2 \subseteq \ldots \subseteq A_r \subseteq A_{r+1} \subseteq \ldots$$

together with its infinite graph; denoting by $\mathbf{x}_r$, $r \geq 0$ the weighting of the respective inductively defined sections of this graph, we have $\mathbf{x}_0 = \mathbf{x}$ and

$$\mathbf{x_r} = \mathbf{x}(UV^{tr})^s \in \mathbb{N}^{(m)} \quad \text{for } r = 2s \text{ even}$$

and

$$\mathbf{x_r} = \mathbf{x}(UV^{tr})^s U \in \mathbb{N}^{(n)} \quad \text{for } r = 2s+1 \text{ odd.}$$

Let us remark that we may write explicitly

$$A_r \cong \mathrm{End}\,(X_F \otimes {}_F M_G \otimes {}_G M_F^* \otimes \cdots \otimes {}_F M_G \otimes {}_G M_F^*), \text{ or}$$

$$A_r \cong \mathrm{End}\,(X_F \otimes {}_F M_G \otimes {}_G M_F^* \otimes \cdots \otimes {}_F M_G \otimes {}_G M_F^* \otimes {}_F M_G)$$

for r even or odd, respectively, and that the graph of the tower is, in fact, the preprojective component of the so-called Auslander–Reiten graph of $\mathcal{A}(A, B)$.

Define the tower transformation $t = t_{AB} : \mathbb{R}^{(m)} \longrightarrow \mathbb{R}^{(m)}$ by

$$t_{AB}(\mathbf{x}) = \mathbf{x}UV^{tr}$$

and recall **[GHJ]** that the index $[B : A]$ of $A \subseteq B$ is defined by

$$[B : A] = \limsup_{r\to\infty} |t_{AB}^r(\mathbf{x})|^{1/r},$$

where the norm $|\ldots|$ denotes the sum of the (positive) coordinates of the respective vector. By Perron–Frobenius theory, $[B : A]$ is the largest real eigenvalue of t_{AB} (and is, in fact, equal to $\lim_{r\to\infty} |t_{AB}^{r+1}(\mathbf{x})| \cdot |t_{AB}^r(\mathbf{x})|^{-1}$, independently of the choice of $\mathbf{x}$).

Let us remark that in **[GHJ]**, the values of $[B : A]$ are shown to be the squares of the norms of integral matrices and, using Kronecker's result **[K]** that the only values of such norms which are smaller than 2 are $2\cos(\pi/q)$, $q \geq 2$, the authors obtain all possible values of the index in the interval $[0,4]$.

Now, the Coxeter transformation $c_{AB} : \mathbb{R}^{(m+n)} \longrightarrow \mathbb{R}^{(m+n)}$ defined by the graph $\Gamma(A_0 \subseteq A_1)$ with respect to the order of the index set can be expressed as follows. For $\mathbf{x} \in \mathbb{R}^{(m)}$, $\mathbf{y} \in \mathbb{R}^{(n)}$,

$$c_{AB}(\mathbf{x},\mathbf{y}) = (\mathbf{x},\mathbf{y}) \begin{pmatrix} -I_{m\times m} & -U \\ V^{tr} & V^{tr}U - I_{n\times n} \end{pmatrix} =$$

$$(\mathbf{x},\mathbf{y}) \begin{pmatrix} -I_{m\times m} & 0_{m\times n} \\ V^{tr} & I_{n\times n} \end{pmatrix} \begin{pmatrix} I_{m\times m} & U \\ 0_{n\times m} & -I_{n\times n} \end{pmatrix};$$

this decomposition can be used to derive the following lemma.

Lemma. *If $\lambda \neq -1$ is an eigenvalue of c_{AB}, then $\lambda + 2 + \lambda^{-1}$ is an eigenvalue of t_{AB}, and all eigenvalues of t_{AB} are obtained in this way. In paricular, the largest real part of the eigenvalue of c_{AB} corresponds to the largest (real) eigenvalue of t_{AB}.*

As a consequence, we get the following theorems.

Theorem. *Let $A \subseteq B$ be a (connected) K-pair and c the Coxeter transformation defined by its (connected) valued graph. If the graph is Dynkin, then*

$$[B : A] = 2(r+1),$$

where r is the largest value of the real parts of the eigenvalues of c; thus, $[B : A] = 4\cos^2(\pi/q)$ for some $q \geq 3$.

Otherwise, $[B : A] = \lambda+2+\lambda^{-1}$, where λ is the largest real eigenvalue of c. Thus, $[B : A] = 4$ if and only if the graph of $A \subseteq B$ is Euclidean.

Theorem. *If the graph of the K-pair $A \subseteq B$ is neither Dynkin nor Euclidean, then the least possible value of $[B : A]$ is $\rho_0 = 4.026417949\ldots$*

and the least accumulation point of the set of values of the index is $\rho_* = 4.079595623\ldots$

The values ρ_0 and ρ_* can be easily deduced from the results of [**X**] and [**Z**]. In fact, $\rho_0 = \lambda_0 + 2 + \lambda_0^{-1}$, where $\lambda_0 = 1.176280818\ldots$ is the largest real root of the irreducible polynomial

$$x^{10} + x^9 - x^7 - x^6 - x^5 - x^4 - x^3 + x + 1,$$

and $\rho_* = \lambda_* + 2 + \lambda_*^{-1}$, where

$$\lambda_* = \sqrt[3]{\tfrac{1}{2} + \sqrt{\tfrac{23}{108}}} + \sqrt[3]{\tfrac{1}{2} - \sqrt{\tfrac{23}{108}}}.$$

Answering a question of V. F. R. Jones in [**J2**] about possible connections between the values of $[B : A]$ and the representation type of $\mathcal{A}(A, B)$, we conclude with the following theorem.

Theorem. *Let $A \subseteq B$ be a (connected) K-pair. Then*

(i) $[B : A] < 4$ if and only if $\mathcal{A}(A, B)$ is of finite representation type;

(ii) $[B : A] = 4$ if and only if $\mathcal{A}(A, B)$ is of tame representation type;

(iii) $[B : A] > 4$ if and only if $\mathcal{A}(A, B)$ is of wild representation type.

Let us append some additional information on the set of values of $[B : A]$ resulting from the studies of maximum eigenvalues of graphs by A.J. Hoffman ([**H**]) and J.B. Shearer ([**S**]).

Theorem. *Let $A \subseteq B$ be a (connected) K-pair. Then there is a sequence of accumulation points $\rho_n = \beta_n + 2 + {\beta_n}^{-1}$ of the set of values of the index $[B : A]$ given by the largest real solution β_n of the equation*

$$x^{n+1} = \sum_{t=0}^{n-1} x^t.$$

All real numbers $\rho \geq \overline{\rho} = 2 + \sqrt{5} = 4.236067977\ldots$ are accumulation points of the set of values of $[B : A]$.

Observe that $\rho_1 = 4$, $\rho_2 = \rho_*$ and that $\overline{\rho} = \tau + 2 + \tau^{-1}$ with $\tau = \frac{1}{2}(1 + \sqrt{5})$ satisfies $\lim_n \rho_n = \overline{\rho}$.

4. Some examples

In this final section, we wish to present three simple illustrations of the concepts and results.

(i) Consider the K-pair of the group algebras $KH \subseteq KG$ for an algebraically closed field K and a subgroup H of a finite group G. Let, in paricular, $H = \mathcal{S}_5$ be the group of permutations of $\{1,2,3,4,5\}$, $G = \mathcal{S}_6$ the group of permutations of $\{1,2,3,4,5,6\}$ and $K = \mathbb{C}$. The irreducible representations ρ_r, $1 \le r \le 7$, of $\mathcal{S}_5$ are in one-to-one correspondence with the Young frames

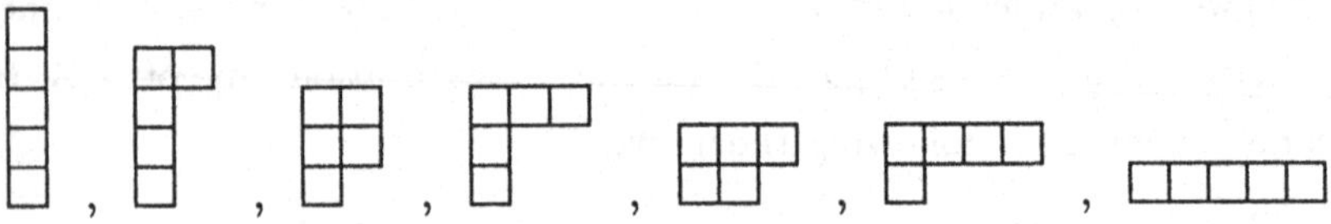

and have dimensions 1, 4, 5, 6, 5, 4, 1, respectively. The group $\mathcal{S}_6$ has 11 irreducible representations σ_s, $1 \le s \le 11$, of dimensions 1, 5, 9, 5, 10, 16, 10, 5, 9, 5, 1 whose restrictions to $\mathcal{S}_5$ are respectively, ρ_1, $\rho_1 \oplus \rho_2$, $\rho_2 \oplus \rho_3$, ρ_3, $\rho_2 \oplus \rho_4$, $\rho_3 \oplus \rho_4 \oplus \rho_5$, $\rho_4 \oplus \rho_6$, ρ_5, $\rho_5 \oplus \rho_6$, $\rho_6 \oplus \rho_7$, ρ_7. The graph is given by

$$U = V = \begin{pmatrix} 1 & 1 & 0 & 0 & 0 & & & & & & \\ & 1 & 1 & 0 & 1 & 0 & & & & 0 & \\ & & 1 & 1 & 0 & 1 & 0 & & & & \\ & & & 0 & 1 & 1 & 1 & 0 & & & \\ & & & & 0 & 1 & 0 & 1 & 1 & & \\ & 0 & & & & 0 & 1 & 0 & 1 & 1 & \\ & & & & & & 0 & 0 & 0 & 1 & 1 \end{pmatrix},$$

the largest value of the Coxeter transformation is $2+\sqrt{3}$ and $[\mathbb{C}\mathcal{S}_6 : \mathbb{C}\mathcal{S}_5] = 2+\sqrt{3}+2+2-\sqrt{3} = 6$. Notice that $[\mathbb{C}\mathcal{S}_6 : \mathbb{C}\mathcal{S}_5] = [\mathcal{S}_6 : \mathcal{S}_5]$; in fact, always

$$[KG : KH] = [G : H].$$

(ii) Let $\mathrm{Mat}\,(3,\mathbb{R}) = A \subseteq B = \mathrm{Mat}\,(3,\mathbb{C}) \times \mathrm{Mat}\,(6,\mathbb{R})$. The algebra $\mathcal{A}(A,B)$ is Morita equivalent to

$$\mathcal{A} = \begin{pmatrix} \mathbb{R} & \mathbb{C} & \mathbb{R}^2 \\ 0 & \mathbb{C} & 0 \\ 0 & 0 & \mathbb{R} \end{pmatrix},$$

i.e. $F = \mathbb{R}$, $G = \mathbb{C} \times \mathbb{R}$, ${}_F M_G = {}_1M_1 \oplus {}_1M_2$ with ${}_1M_1 = {}_{\mathbb{R}}\mathbb{C}_{\mathbb{C}}$, ${}_1M_2 = {}_{\mathbb{R}}\mathbb{R}_{\mathbb{R}} \oplus {}_{\mathbb{R}}\mathbb{R}_{\mathbb{R}}$. The graph $\Gamma(A \subseteq B)$ is

$$\left(\begin{pmatrix} 0 & 1 & 2 \\ 0 & 0 & 0 \\ 0 & 0 & 0 \end{pmatrix}, \begin{pmatrix} 0 & 2 & 2 \\ 0 & 0 & 0 \\ 0 & 0 & 0 \end{pmatrix} \right) \quad \text{with weighting } (3,3,6).$$

The matrix of the Coxeter transformation c_{AB} is

$$\begin{pmatrix} -1 & -1 & -2 \\ 2 & 1 & 4 \\ 2 & 2 & 3 \end{pmatrix};$$

hence, the eigenvalues of c_{AB} are -1, $2-\sqrt{3}$, $2+\sqrt{3}$:

$$[B : A] = 6.$$

Of course, this result is trivial: $t_{AB}(r) = 6r$, $r \in \mathbb{R}$.

(iii) Let F_1 be a field with subfields F_2 and G_1 of index 2 and 3, respectively (take e.g. $F_1 = \mathbb{Q}(\sqrt[3]{2}, i\sqrt{3})$, $F_2 = \mathbb{Q}(\sqrt[3]{2})$, $G_1 = \mathbb{Q}(i\sqrt{3})$). Consider the pair $A \subseteq B$ with $A = F_1 \times F_2$, $B = \mathrm{Mat}\,(3,G_1) \times G_2 \times \mathrm{Mat}\,(2,G_3)$, where $G_2 = G_3 = F_1$, and such that $\mathcal{A}(A,B)$ is Morita equivalent to

$$\mathcal{A} = \begin{pmatrix} F_1 & 0 & F_1 & F_1 & F_1 \\ 0 & F_2 & 0 & 0 & F_1 \\ 0 & 0 & G_1 & 0 & 0 \\ 0 & 0 & 0 & F_1 & 0 \\ 0 & 0 & 0 & 0 & F_1 \end{pmatrix}.$$

The graph $\Gamma(A \subseteq B)$ is given by $U = \begin{pmatrix} 3 & 1 & 1 \\ 0 & 0 & 1 \end{pmatrix}$, $V = \begin{pmatrix} 1 & 1 & 1 \\ 0 & 0 & 2 \end{pmatrix}$; thus

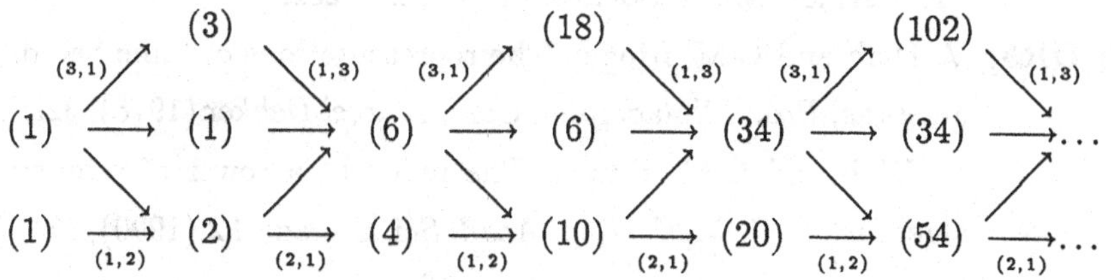

is the weighted graph of the corresponding tower. The matrix of the Coxeter transformation c_{AB} is

$$\begin{pmatrix} -1 & 0 & -3 & -1 & -1 \\ 0 & -1 & 0 & 0 & -1 \\ 1 & 0 & 2 & 1 & 1 \\ 1 & 0 & 3 & 0 & 1 \\ 1 & 2 & 3 & 1 & 2 \end{pmatrix}$$

and the eigenvelues of c_{AB} are

$$-1, \qquad \frac{3+\sqrt{17} \pm \sqrt{6\sqrt{17}+10}}{4}, \qquad \frac{3-\sqrt{17} \pm i\sqrt{6\sqrt{17}-10}}{4};$$

hence, $[B:A] = \dfrac{7+\sqrt{17}}{2} = 5.561552815\ldots$

REFERENCES

[A] N. A'Campo, Sur les valuers propres de la transformation de Coxeter, *Invent. Math.* **33** (1976), 61–67.

[B] N. Bourbaki, *Groupes et algèbres de Lie*, Chap. 4, 5 et 6, Hermann 1968.

[Br] O. Bratteli, Inductive limits of finite dimensional C^*-algebras, *Trans. Amer. Math. Soc.* **171** (1972), 195–234.

[C] A. Connes, Indice des sous-facteurs, algèbres de Hecke et théorie des noeuds, *Astérisque* **133–134** *Soc. Math. France* (1986), 289–308.

[DR1] V. Dlab and C.M. Ringel, Indecomposable representations of graphs and algebras, *Mem. Amer. Math. Soc.* No. **173** (1976).

[DR2] V. Dlab and C.M. Ringel, Eigenvalues of Coxeter transformations and the Gelfand–Kirillov dimension of the preprojective algebras, *Proc. Amer. Math. Soc.* **83** (1981), 228-232.

[DR3] V. Dlab and C.M. Ringel, The representations of tame hereditary algebras, *Proc. Philadelphia Conf. Marcel Dekker* (1978), 329–353.

[DR4] V. Dlab and C.M. Ringel, The index of a tower of semi-simple algebras, *C. R. Math. Rep. Acad. Sci. Canada* **12** (1990), 171–175.

[DR5] V. Dlab and C.M. Ringel, Towers of semi-simple algebras, *J. Funct. Analysis* (to appear)

[GHJ] F.M. Goodman, P.de la Harpe and V.F.R. Jones, *Coxeter graphs and towers of algebras*, Springer–Verlag 1989.

[H] A.J. Hoffman, On limit points of spectral radii of non-negative symmetric integral matrices, *Springer LNM* **303** (1972), 165–172.

[J1] V.F.R. Jones, Index for subfactors, *Invent. Math.* **72** (1983), 1–25.

[J2] V.F.R. Jones, Index for subrings of rings, *Contemp. Math.* **43** (Amer. Math. Soc. 1985), 181–190.

[K] L. Kronecker, Zwei Sätze über Gleichungen mit ganzzahligen Koeffizienten, *Crelle* (1857), Oeuvres I, 105–108.

[S] J.B. Shearer, On the distribution of the maximum eigenvalue of graphs, *Linear Alg. and Appl.* **114** (1989), 17–20.

[X] Changchang Xi, On wild hereditary algebras with small growth numbers, *Comm. Alg.* (to appear)

[Z] Yingbo Zhang, Eigenvalues of Coxeter transformations and the structure of regular components of an Auslander–Reiten quiver, *Comm. Alg.* **17** (1989), 2347–2362.

The Module Theoretical Approach to Quasi-hereditary Algebras

Vlastimil Dlab and Claus Michael Ringel

Quasi-hereditary algebras were introduced by L.Scott [S] in order to deal with highest weight categories as they arise in the representation theory of semi–simple complex Lie algebras and algebraic groups. Since then, also many other algebras arising naturally, such as the Auslander algebras, have been shown to be quasi-hereditary. It seems to be rather surprising that the class of quasi-hereditary algebras, defined in purely ring–theoretical terms, has not been studied before in ring theory.

The central concept of the theory of quasi-hereditary algebras are the notions of a standard and a costandard module; these modules depend in an essential way, on a (partial) ordering of the set of all simple modules. So we start with a finite dimensional algebra A, and a partial ordering of the simple A–modules, in order to define the standard modules $\Delta(i)$ and the costandard modules $\nabla(i)$; we have to impose some additional conditions on their endomorphism rings, and on the existence of some filtrations, in order to deal with quasi-hereditary algebras. This is the content of the first chapter. The second chapter collects some properties of quasi-hereditary algebras, in particular those needed in later parts of the paper. The third chapter presents the process of standardization: here, we give a characterization of the categories of Δ–filtered modules over quasi-hereditary algebras. In fact, we show that given indecomposable A–modules $\Theta(1),\dots,\Theta(n)$ over a finite–dimensional algebra such that $\operatorname{rad}(\Theta(i),\Theta(j)) = 0$, and $\operatorname{Ext}^1(\Theta(i),\Theta(j)) = 0$ for all $i \geq j$, the category $\mathcal{F}(\Theta)$ of all A–modules with a Θ–filtration is equivalent (as an exact category) to the category of all Δ–filtered modules over a quasi-hereditary algebra.

The forth and the fifth chapter consider cases when the standard modules over a quasi-hereditary algebra have special homological properties: first, we assume that any $\Delta(i)$ has projective dimension at most 1, then we deal with the case that the dominant dimension of any $\Delta(i)$ is at least 1. Both these properties, as well as their duals, are satisfied for the Auslander algebras of a uniserial algebra, and we are going to present the Auslander–Reiten quivers of the category of Δ–filtered modules for some examples.

These notes give a unified treatment of some basic results of Cline–Parshall–Scott [PS,CPS], Dlab–Ringel [DR2], Donkin [D], Ringel [R2] and Soergel [So]; they are intended as a guideline for understanding further investigations in [PS] and [R4]. We will not cover those developments of the theory of quasi-hereditary algebras which are formulated in terms of the internal ring structure (these results are rather easily available in the literature, some of the references are listed in the bibliography at the end of the paper).

1. Definition of a quasi-hereditary algebra

Let k be a field, and A a finite–dimensional k–algebra. We denote by A–mod the category of all (finite–dimensional left) A–modules. If Θ is a class of A–modules (closed under isomorphisms), $\mathcal{F}(\Theta)$ denotes the class of all A–modules M which have a Θ–filtration, i.e. a filtration $M = M_0 \supseteq M_1 \supseteq \cdots \supseteq M_{t-1} \supseteq M_t \supseteq \cdots \supseteq M_m = 0$ such that all factors M_{t-1}/M_t, $1 \leq t \leq m$, belong to Θ.

Let $E(\lambda)$, $\lambda \in \Lambda$, be the simple A–modules (one from each isomorphism class), and we assume that the index set Λ is endowed with a partial ordering. If M is an A–module, we denote the Jordan–Hölder multiplicity of $E(\lambda)$ in M by $[M : E(\lambda)]$. For each $\lambda \in \Lambda$, let $P(\lambda)$ be the projective cover, and $Q(\lambda)$ the injective hull of $E(\lambda)$. Denote by $\Delta(\lambda) = \Delta_A(\lambda) = \Delta_\Lambda(\lambda)$ the maximal factor module of $P(\lambda)$ with composition factors of the form $E(\mu)$ where $\mu \leq \lambda$; these modules $\Delta(\lambda)$ are called the *standard* modules, and we set $\Delta = \{\Delta(\lambda) | \lambda \in \Lambda\}$. Similarly, denote by $\nabla(\lambda) = \nabla_A(\lambda) = \nabla_\Lambda(\lambda)$ the maximal submodule of $Q(\lambda)$ with composition factors of the form $E(\mu)$ where $\mu \leq \lambda$; in this way, we obtain the set $\nabla = \{\nabla(\lambda) | \lambda \in \Lambda\}$ of *costandard* modules. Let us point out that $\nabla(\lambda)$ is the dual of a corresponding standard module: Let $D = \mathrm{Hom}_k(-, k)$ be the duality with respect to the base field. Let A^o be the opposite algebra of A, with simple modules $E_{A^o}(\lambda) = DE_A(\lambda)$ (note that we use the same index set!). Then $\nabla_A(\lambda) = D\Delta_{A^o}(\lambda)$. It follows that any statement on standard modules yields a corresponding statement for costandard modules, we often will refrain from stating the dual results explicitly.

Note that the only module M with $\mathrm{Hom}(\Delta(\lambda), M) = 0$ for all $\lambda \in \Lambda$ is the zero module $M = 0$. [For $M \neq 0$, let $E(\lambda)$ be a submodule, then $\mathrm{Hom}(\Delta(\lambda), M) \neq 0$.] Dually, the only module M with $\mathrm{Hom}(M, \nabla(\lambda)) = 0$ is the zero module.

Given a set $\mathcal{X}$ of A–modules, then for any A–module M, we denote by $\eta_{\mathcal{X}} M$ the *trace* of $\mathcal{X}$ in M, it is the maximal submodule of M generated by $\mathcal{X}$.

The standard modules may be characterized as follows:

Lemma 1.1. *For any A–module M, and $\lambda \in \Lambda$ the following assertions are equivalent:*
(i) $M \cong \Delta(\lambda)$,
(ii) $\operatorname{top} M \cong E(\lambda)$, *all composition factors of M are of the form $E(\mu)$, with $\mu \leq \lambda$, and* $\operatorname{Ext}^1(M, E(\mu)) = 0$ *for all $\mu \leq \lambda$,*
(iii) $M \cong P(\lambda)/\eta_{\{P(\mu) \mid \mu \not\leq \lambda\}} P(\lambda)$.

Lemma 1.2. *Let M be an A–module, and $\lambda, \mu \in \Lambda$. Then:*
(a) $\operatorname{Hom}(\Delta(\lambda), M) \neq 0$ *implies* $[M : E(\lambda)] \neq 0$.
(b) $\operatorname{Hom}(\Delta(\lambda), \Delta(\mu)) \neq 0$ *implies* $\lambda \leq \mu$.
(c) $\operatorname{Hom}(\Delta(\lambda), \nabla(\mu)) \neq 0$ *implies* $\lambda = \mu$.

For the proof of (c), we use both (a) and its dual statement.

The sets Δ and ∇ depend, in an essential way, on the given partial ordering of Λ. We will say that two partially ordered sets Λ, Λ' used as index sets for the simple A–modules are *equivalent* provided the sets $\{\Delta_\Lambda(\lambda) \mid \lambda \in \Lambda\}$ and $\{\Delta_{\Lambda'}(\lambda) \mid \lambda \in \Lambda'\}$ coincide, and $\{\nabla_\Lambda(\lambda) \mid \lambda \in \Lambda\}$ and $\{\nabla_{\Lambda'}(\lambda) \mid \lambda \in \Lambda'\}$ coincide.

In general, the standard, and the costandard modules will change when we refine the ordering. In order to avoid this to happen, we usually will consider only adapted orderings in the sense of the following definition: A partial ordering Λ of the set of simple A–modules $\{E(\lambda) \mid \lambda \in \Lambda\}$ is said to be ***adapted***, provided the following condition holds: for every A–module M with $\operatorname{top} M \cong E(\lambda_1)$ and $\operatorname{soc} M \cong E(\lambda_2)$, where λ_1 and λ_2 are incomparable, there is some $\mu \in \Lambda$ with $\mu > \lambda_1$ and $\mu > \lambda_2$ such that $[M : E(\mu)] \neq 0$. [Observe that we may weaken the condition as follows: we only have to require the existence of some $\mu \in \Lambda$ with $\mu > \lambda_1$ *or* $\mu > \lambda_2$ such that $[M : E(\mu)] \neq 0$. Indeed, in case the weaker condition is satisfied, assume that there exists a module M with top $E(\lambda_1)$, socle $E(\lambda_2)$, where λ_1 and λ_2 are incomparable, and such that there is no $\mu \in \Lambda$ with $\mu > \lambda_1, \mu > \lambda_2$ and $[M : E(\mu)] \neq 0$. We may assume that M is of smallest possible length, and we know that there is at least a μ with $[M : E(\mu)] \neq 0$ and either $\mu > \lambda_1$ or $\mu > \lambda_2$. Assume we have $\mu > \lambda_1$. Now M has a submodule M' with top $E(\mu)$. Note that μ cannot be comparable with λ_2. The minimality of M implies that there is ν such that $\nu > \mu, \nu > \lambda_2$, and $[M' : E(\nu)] \neq 0$. But $\nu > \mu > \lambda_1, \nu > \lambda_2$, and $[M : E(\nu)] \neq 0$, contrary to our assumption.] As an example of non–adapted partial orderings, the reader should have in mind any non–semisimple algebra with the discrete ordering of the simple modules.

If Λ' is a refinement of Λ, and Λ is adapted, then clearly $\Delta_{\Lambda'}(\lambda) = \Delta_\Lambda(\lambda)$ and $\nabla_{\Lambda'}(\lambda) = \nabla_\Lambda(\lambda)$ for all $\lambda \in \Lambda$, thus Λ and Λ' are equivalent, and Λ' also is adapted. Thus, for Λ adapted, we always may assume that we deal with a total ordering. In such a case, we may replace Λ by the equivalent index set $\{1, 2, \ldots, n\}$ with its natural ordering.

In case we deal with an adapted partial ordering, we may reformulate Lemma 1.1 as follows:

Lemma 1.1′. *For any A–module M, and $\lambda \in \Lambda$, where Λ is adapted, the following assertions are equivalent:*
(i) $M \cong \Delta(\lambda)$,
(ii) $\operatorname{top} M \cong E(\lambda)$, *all composition factors of M are of the form $E(\mu)$, with $\mu \leq \lambda$, and $\operatorname{Ext}^1(M, E(\mu)) \neq 0$ implies $\mu > \lambda$,*
(iii) $M \cong P(\lambda)/\eta_{\{P(\mu) \mid \mu > \lambda\}} P(\lambda)$.

As an immediate consequence, we see:

Lemma 1.3. *Assume Λ is adapted. Let M be an A–module, and $\lambda, \mu \in \Lambda$. Then*
(a) If $\operatorname{Ext}^1(\Delta(\lambda), M) \neq 0$, *then* $[M : E(\mu)] \neq 0$, *for some* $\mu > \lambda$.
(b) If $\operatorname{Ext}^1(\Delta(\lambda), \Delta(\mu)) \neq 0$, *then* $\lambda < \mu$.
(c) $\operatorname{Ext}^1(\Delta(\lambda), \nabla(\mu)) = 0$.

For (a), note that $\operatorname{Ext}^1(\Delta(\lambda), M) \neq 0$ implies that $\operatorname{Ext}^1(\Delta(\lambda), E) \neq 0$ for some composition factor E of M. Let $E = E(\mu)$, then $\mu > \lambda$, according to Lemma 1.1′(ii). For the proof of (b), assume $\operatorname{Ext}^1(\Delta(\lambda), \Delta(\mu)) \neq 0$. Then $[\Delta(\mu) : E(\nu)] \neq 0$, for some $\nu > \lambda$, according to (a). However $[\Delta(\mu) : E(\nu)] \neq 0$ implies $\nu \leq \mu$, therefore $\lambda < \nu \leq \mu$. Similarly, for the proof of (c), we use (a) in order to see that $\operatorname{Ext}^1(\Delta(\lambda), \nabla(\mu)) \neq 0$ implies $\lambda < \mu$. But in this case, the duality also yields the dual statement $\lambda > \mu$, so we obtain a contradiction.

The main interest will lie on the subcategory $\mathcal{F}(\Delta)$ of all A–modules with a Δ–filtration. First of all, let us point out that usually $\mathcal{F}(\Delta)$ is closed under direct summands:

Lemma 1.4. *Let $\Lambda = \{1, 2, \ldots, n\}$, with the canonical ordering. For any A–module M, and $0 \leq i \leq n$, let $\eta_i M = \eta_{\{P(j) \mid j > i\}} M$. Then M belongs to $\mathcal{F}(\Delta)$ if and only if for all $1 \leq i \leq n$, the factor $\eta_{i-1} M / \eta_i M$ is a direct sum of copies of $\Delta(i)$.*

Proof: Assume M belongs to $\mathcal{F}(\Delta)$. According to Lemma 1.3 (b), there is a filtration $M = M_0 \supseteq M_1 \supseteq \cdots \supseteq M_n = 0$ such that for all $1 \leq i \leq n$, the factor M_{i-1}/M_i is isomorphic to a direct sum of copies of $\Delta(i)$. It follows that $M_i = \eta_i M$.

Lemma 1.5. *Assume Λ is adapted. Then $\mathcal{F}(\Delta)$ is closed under kernels of epimorphisms.*

Proof: We may assume that $\Lambda = \{1, 2, \ldots, n\}$ with its canonical ordering. Let X, Y belong to $\mathcal{F}(\Delta)$, and let $f : X \to Y$ be an epimorphism, say with kernel K. Clearly $f(\eta_i X) = \eta_i Y$ for all i, since any map $P(j) \to Y$ lifts to X. For any $1 \leq i \leq n$, we obtain the following commutative diagram

$$\begin{array}{ccccccccc}
0 & \longrightarrow & \eta_i X \cap K & \xrightarrow{\iota} & \eta_i X & \xrightarrow{f} & \eta_i Y & \longrightarrow & 0 \\
 & & \downarrow & & \downarrow{\scriptstyle \iota_X} & & \downarrow{\scriptstyle \iota_Y} & & \\
0 & \longrightarrow & \eta_{i-1} X \cap K & \xrightarrow{\iota} & \eta_{i-1} X & \xrightarrow{f} & \eta_{i-1} Y & \longrightarrow & 0
\end{array}$$

with exact rows (the maps being the canonical inclusions or induced by f.) As cokernels of the vertical maps, we obtain the exact sequence $0 \to K_i \to X\langle i\rangle \to Y\langle i\rangle \to 0$ with $K_i = \eta_{i-1}X \cap K/\eta_i X \cap K$, $X\langle i\rangle = \eta_{i-1}X/\eta_i X$, and $Y\langle i\rangle = \eta_{i-1}Y/\eta_i Y$. Now, both $X\langle i\rangle$, and $Y\langle i\rangle$ are direct sums of copies of $\Delta(i)$, thus K_i as a submodule of $X\langle i\rangle$ has only composition factors of the form $E(j)$, with $j \leq i$. Since we know that $\mathrm{Ext}^1(\Delta(i), E(j)) = 0$, for $j \leq i$, it follows that the cokernel sequence splits, thus K_i is a direct sum of copies of $\Delta(i)$. In this way, we see that K has the filtration $K = \eta_0 X \cap K \supseteq \eta_1 X \cap K \supseteq \cdots \supseteq \eta_n X \cap K = 0$, with factors in $\mathcal{F}(\Delta)$, thus K belongs to $\mathcal{F}(\Delta)$.

We consider now the case that the endomorphism rings of standard modules and of costandard modules are division rings. Note that modules with a division ring as endomorphism ring are called *Schurian.*

Lemma 1.6. *The following statements are equivalent, for any* $\lambda \in \Lambda$:
(i) $\Delta(\lambda)$ *is a Schurian module.*
(ii) $[\Delta(\lambda) : E(\lambda)] = 1$.
(iii) If M *is an* A*-module with top and socle isomorphic to* $E(\lambda)$*, and* $[M : E(\mu)] \neq 0$ *only for* $\mu \leq \lambda$*, then* $M \cong E(\lambda)$.
*(ii)** $[\nabla(\lambda) : E(\lambda)] = 1$.
*(i)** $\nabla(\lambda)$ *is a Schurian module.*

With these preparations, we are able to present the definition of a quasi-hereditary algebra.

Theorem 1. *Assume that* Λ *is adapted, and that all standard modules are Schurian. Then the following conditions are equivalent:*
(i) $\mathcal{F}(\Delta)$ *contains* ${}_A A$.
(ii) $\mathcal{F}(\Delta) = \{X \mid \mathrm{Ext}^1(X, \nabla) = 0\}$.
(iii) $\mathcal{F}(\Delta) = \{X \mid \mathrm{Ext}^i(X, \nabla) = 0$ *for all* $i \geq 1\}$.
(iv) $\mathrm{Ext}^2(\Delta, \nabla) = 0$

An algebra A with an adapted partial ordering Λ, whose standard modules are Schurian and such that the equivalent conditions of Theorem 1 are satisfied, is said to be *quasi-hereditary*. The usual definition is (i), or an equivalent form in terms of "heredity chains", see [S, PS, CPS, DR2], and Soergel [So] presented the last condition (iv). In fact, the decisive implication (iv) $\Longrightarrow$ (ii) can be traced back to Donkin [D]. Condition (iv) obviously is self–dual, thus we may add the dual form of the remaining conditions:

*(i)** $\mathcal{F}(\nabla)$ *contains* $D(A_A)$.
*(ii)** $\mathcal{F}(\nabla) = \{Y \mid \operatorname{Ext}^1(\Delta, Y) = 0\}$.
*(iii)** $\mathcal{F}(\nabla) = \{Y \mid \operatorname{Ext}^i(\Delta, Y) = 0$ *for all* $i \geq 1\}$.

Since under the assumption that Λ is adapted, $\mathcal{F}(\Delta)$ is closed under direct summands, and under kernels of surjective maps, the condition (i) may be reformulated that $\mathcal{F}(\Delta)$ contains all projective modules, or also that $\mathcal{F}(\Delta)$ is resolving. Recall that a full subcategory of A–mod is said to be *resolving* provided it is closed under extensions, kernels of surjective maps, and contains all projective modules. Of course, there is the dual concept of a *coresolving* subcategory (closed under extensions, cokernels of injective maps, and containing all injective modules). So, condition (i)* may be reformulated that $\mathcal{F}(\nabla)$ contains all injective modules, or also that $\mathcal{F}(\nabla)$ is coresolving.

Proof of Theorem 1: (iii) implies (iv): this is trivial.

(iv) implies (ii): According to Lemma 1.3 (c), we know that any module X in $\mathcal{F}(\Delta)$ satisfies $\operatorname{Ext}^1(X, \nabla) = 0$. We are going to prove the converse. We may assume that $\Lambda = \{1, 2, \ldots, n\}$. Let X be a module with $\operatorname{Ext}^1(X, \nabla) = 0$. Let i be minimal with $\eta_i X = 0$. By induction on i, we are going to show that X belongs to $\mathcal{F}(\Delta)$. For $i = 0$, we deal with the zero module, so nothing has to be shown. So assume $i \geq 1$. Let $X' = \eta_{i-1}X$, and $X'' = X/X'$.

First, let us show that X'' belongs to $\mathcal{F}(\Delta)$. For $s < i$, we have $\operatorname{Hom}(X', \nabla(s)) = 0$, since X' is generated by $P(i)$, and $[\nabla(s) : E(i)] = 0$. For $s > i$, we have $\operatorname{Hom}(X', \nabla(s)) = 0$, since $\eta_{s-1}X' = 0$, whereas $\nabla(s)$ is cogenerated by $Q(s)$. The exact sequence $0 \to X' \to X \to X'' \to 0$ induces for any s an exact sequence

$$\operatorname{Hom}(X', \nabla(s)) \to \operatorname{Ext}^1(X'', \nabla(s)) \to \operatorname{Ext}^1(X, \nabla(s)).$$

We have seen that the first term is zero for $s \neq i$, and by assumption, the last term is always zero, thus $\operatorname{Ext}^1(X'', \nabla(s)) = 0$ for $s \neq i$. The same is true for $s = i$, according to the dual of Lemma 1.1', since the composition factors of X'' are of the form $E(j)$, with $j < i$. Since $\operatorname{Ext}^1(X'', \nabla) = 0$, and $\eta_{i-1}X'' = 0$, we know by induction that X'' belongs to $\mathcal{F}(\Delta)$.

Next, let us note that $\operatorname{Ext}^1(X', \nabla) = 0$. Namely, the exact sequence $0 \to X' \to X \to X'' \to 0$ yields for any s an exact sequence

$$\operatorname{Ext}^1(X, \nabla(s)) \to \operatorname{Ext}^1(X', \nabla(s)) \to \operatorname{Ext}^2(X'', \nabla(s)),$$

again, the first term is zero by assumption, the last term is zero according to condition (iv), and the fact that X'' belongs to $\mathcal{F}(\Delta)$.

Since X' is generated by $P(i)$, and $\eta_i X' = 0$, it follows that there exists an exact sequence $0 \to K \to Z \to X' \to 0$, where Z is a direct sum of copies of $\Delta(i)$, and K is contained in the radical of Z. In particular, $\eta_{i-1}K = 0$. The exact sequence $0 \to K \to Z \to X' \to 0$ yields an exact sequence

$$\mathrm{Hom}(Z, \nabla(s)) \to \mathrm{Hom}(K, \nabla(s)) \to \mathrm{Ext}^1(X', \nabla(s)),$$

the last term is always zero, as we have shown, the first term is zero at least for $s \neq i$, according to Lemma 1.2 (c), thus we see that $\mathrm{Hom}(K, \nabla(s)) = 0$, for $s \neq i$. The same is true for $s = i$, since the composition factors of K are of the form $E(j)$ with $j < i$, and $\nabla(i)$ is cogenerated by $Q(i)$. However, $\mathrm{Hom}(K, \nabla) = 0$ implies $K = 0$, as we have remarked before Lemma 1.1. This shows that X' is isomorphic to a direct sum of copies of $\Delta(i)$, thus both X' and X'' belong to $\mathcal{F}(\Delta)$, and therefore also X.

(ii) implies (i): this again is trivial.

(i) implies (iii): We assume that ${}_AA$ belongs to $\mathcal{F}(\Delta)$. We show that any module X in $\mathcal{F}(\Delta)$ satisfies $\mathrm{Ext}^i(X, \nabla) = 0$ for all $i \geq 1$. According to Lemma 1.3 (c), we know it for $i = 1$, thus consider some $i \geq 2$. Take any exact sequence $0 \to X' \to P \to X \to 0$, where P is a free A-module. Since both P and X belong to $\mathcal{F}(\Delta)$, also X' belongs to $\mathcal{F}(\Delta)$, according to Lemma 1.5. On the other hand, $\mathrm{Ext}^i(X, \nabla(\mu)) = \mathrm{Ext}^{i-1}(X', \nabla(\mu))$, and, by induction, the latter group is zero. Thus we see

$$\mathcal{F}(\Delta) \subseteq \{X \mid \mathrm{Ext}^i(X, \nabla) = 0 \text{ for all } i \geq 1\} \subseteq \{X \mid \mathrm{Ext}^1(X, \nabla) = 0\},$$

in particular, the first inclusion shows that (iv) is satisfied. Since we know already that (iv) implies (ii), we see that all inclusions are equalities. This finishes the proof.

2. Some properties of quasi-hereditary algebras

Let A be a quasi-hereditary algebra with respect to Λ. We are going to give bounds for the Loewy length, and the projective or injective dimension of modules in terms of Λ. Also, we will consider the Jordan–Hölder multiplicities of selected modules.

If Λ' is a subset of Λ, let $\mathrm{h}(\Lambda')$ be the maximal number m such that there exists a chain $\lambda_0 < \lambda_1 < \cdots < \lambda_m$ in Λ, with all $\lambda_i \in \Lambda'$.

Given an A-module M, its *support* $\operatorname{supp} M$ is the set of all $\lambda \in \Lambda$, such that $[M : E(\lambda)] \neq 0$.

Lemma 2.1. *Let M be an A-module, and $h = \mathrm{h}(\operatorname{supp} M)$. Then the Loewy length of M is at most $2^{h+1} - 1$.*

Remark: The algebra A does not have to be quasi-hereditary in order that the assertion holds. It is sufficient to know that Λ is adapted, and that the standard modules are Schurian.

Proof: We use induction on h. For $h = -1$, we have $M = 0$, and the zero module has Loewy length zero. Assume now that $h \geq 0$, and let $\mu_1, \ldots, \mu_t$ be the maximal elements of $\operatorname{supp} M$. Let M' be the smallest submodule of M such that none of the elements μ_i, with $1 \leq i \leq t$, belongs to the support of M/M'. Let M'' be the largest submodule of M' such that none of the elements μ_i, with $1 \leq i \leq t$ belongs to its support. By induction, the Loewy length of both M/M' and of M'' is at most $2^h - 1$. We claim that $N = M'/M''$ is semisimple. Otherwise, there are submodules $N'' \subset N' \subseteq N$ such that N'/N'' has length at least two, and simple top $E(\mu_r)$ and simple socle $E(\mu_s)$, with $r, s \in \{1, \ldots, t\}$. Note that N'/N'' is a factor module of $\Delta(\mu_r)$, since μ_r is maximal in $\operatorname{supp} N'/N''$, and Λ is adapted (see Lemma 1.1'). However, $[N'/N'' : E(\mu_s)] \neq 0$ shows that $\mu_s \leq \mu_r$. Thus $\mu_s = \mu_r$, since μ_s is maximal. But since $\Delta(\mu_r)$ is standard, we cannot have $[N'/N'' : E(\mu_r)] \geq 2$. This contradiction shows that N has to be semisimple, and therefore the Loewy length of M is at most $2 \cdot (2^h - 1) + 1 = 2^{h+1} - 1$.

Given $\lambda \leq \mu$ in Λ, we denote by $[\lambda, \mu]$ the subset $\{\nu \mid \lambda \leq \nu \leq \mu\}$, the interval between λ and μ. Similarly, $[\lambda, -] = \{\nu \mid \lambda \leq \nu\}$ is the principal filter generated by λ, and $[-, \mu] = \{\nu \mid \nu \leq \mu\}$ is the principal ideal generated by μ.

Lemma 2.2. *Let $\lambda \in \Lambda$. Then*

$$\operatorname{proj.dim.} \Delta(\lambda) \leq \mathrm{h}([\lambda, -]), \quad \operatorname{proj.dim.} E(\lambda) \leq \mathrm{h}(\Lambda) + \mathrm{h}([-, \lambda]).$$

Consequently, the projective dimension of a module in $\mathcal{F}(\Delta)$ is bounded by $\mathrm{h}(\Lambda)$, *and the global dimension of A is bounded by* $2\,\mathrm{h}(\Lambda)$.

Proof: If λ is maximal, then $\Delta(\lambda)$ is projective. So assume that λ is not maximal. There is a submodule U of $P(\lambda)$ such that $\Delta(\lambda) = P(\lambda)/U$, and U is filtered with factors some standard modules $\Delta(\mu)$ with $\mu > \lambda$. Note that for $\mu > \lambda$, we have $\mathrm{h}([\mu, -]) < \mathrm{h}([\lambda, -])$, thus by induction we have

$$\operatorname{proj.dim.} U \leq \max_{\mu > \lambda} \operatorname{proj.dim.} \Delta(\mu) \leq \max_{\mu > \lambda} \mathrm{h}([\mu, -]) < \mathrm{h}([\lambda, -]).$$

It follows that $\operatorname{proj.dim.} \Delta(\lambda) \leq 1 + \operatorname{proj.dim.} U \leq \mathrm{h}([\lambda, -])$.

If λ is minimal, then $E(\lambda) = \Delta(\lambda)$, thus by the previous considerations, $\operatorname{proj.dim.} E(\lambda) \leq \mathrm{h}(\Lambda)$. Assume that λ is not minimal. The composition factors of $\operatorname{rad} \Delta(\lambda)$ are of the form $E(\nu)$ with $\nu < \lambda$, thus by induction proj. dim. $E(\nu) \leq \text{h}(\Lambda) + h([-,\nu]) < \text{h}(\Lambda) + h([-,\lambda])$, and proj. dim. $\Delta(\lambda) \leq \text{h}(\Lambda)$. This shows that proj. dim. $E(\lambda) \leq \text{h}(\Lambda) + \text{h}([-,\lambda])$.

In order to consider the various Jordan–Hölder multiplicities of a module as a vector with integer coefficients, it seems to be convenient to replace Λ by a totally ordered refinement. Thus, let $\Lambda = \{1, 2, \ldots, n\}$, with the canonical ordering. For any A–module M, we may consider the n-tuple $\mathbf{dim}M$ with coordinates $(\mathbf{dim}M)_j = [M : E(j)]$, called the *dimension vector* of M, and we consider $\mathbf{dim}M$ as an element of the Grothendieck group $K_0(A) = \mathbb{Z}^n$. The sets Δ and ∇ yield $n \times n$–matrices $\mathbf{dim}\Delta, \mathbf{dim}\nabla$, (the rows with index i being $\mathbf{dim}\Delta(i), \mathbf{dim}\nabla(i)$, respectively).

Lemma 2.3. *Let $\Lambda = \{1, 2, \ldots, n\}$, and assume the standard modules are Schurian. Then both matrices $\mathbf{dim}\Delta, \mathbf{dim}\nabla$ are unipotent lower triangular matrices.*

In particular, we see that under the assumptions of Lemma 2.3, both the dimension vectors $\mathbf{dim}\Delta(\lambda)$, as well as the dimension vectors $\mathbf{dim}\nabla(\lambda)$, form a $\mathbb{Z}$-basis of $K_0(A)$. The basis $\mathbf{dim}\Delta(\lambda)$ of $K_0(A)$ will be called the *standard* basis, the basis $\mathbf{dim}\nabla(\lambda)$ the *costandard* basis of $K_0(A)$. If M is an A–module, the coefficients of $\mathbf{dim}M$ expressed in the standard basis will be denoted by $[M : \Delta(i)]$, its coefficients in terms of the costandard basis will be $[M : \nabla(i)]$, thus

$$\mathbf{dim}M = \sum_{i=1}^{n} [M : \Delta(i)] \mathbf{dim}\Delta(i) = \sum_{i=1}^{n} [M : \nabla(i)] \mathbf{dim}\nabla(i).$$

For M in $\mathcal{F}(\Delta)$, the number of copies of $\Delta(i)$ in any Δ–filtration of M is $[M : \Delta(i)]$. Namely, if d_i is the number of copies of $\Delta(i)$ in some Δ–filtration of M, then $\mathbf{dim}M = \sum d_i \mathbf{dim}\Delta(i)$, therefore $d_i = [M : \Delta(i)]$.

For any $\lambda \in \Lambda$, let $d_\lambda = \dim_k \operatorname{End}(\Delta(\lambda))$.

Lemma 2.4. *The restriction of the functor* $\operatorname{Hom}(-, \nabla(\lambda))$ *to* $\mathcal{F}(\Delta)$ *is exact, and for* $M \in \mathcal{F}(\Delta)$, *we have*

$$\dim_k \operatorname{Hom}(M, \nabla(\lambda)) = d_\lambda \cdot [M : \Delta(\lambda)].$$

Proof: We know from 1.3 (c) that $\operatorname{Ext}^1(\mathcal{F}(\Delta), \nabla(\lambda)) = 0$, thus the restriction of the functor $\operatorname{Hom}(-, \nabla(\lambda))$ to $\mathcal{F}(\Delta)$ is exact. Also, by 1.2 (c), we know that $\operatorname{Hom}(\Delta(\mu), \nabla(\lambda)) = 0$ for $\lambda \neq \mu$, thus given a Δ–filtration of M, the functor $\operatorname{Hom}(-, \nabla(\lambda))$ will pick out those factors which are of the form $\Delta(\lambda)$.

The following equality is sometimes called Bernstein–Gelfand–Gelfand reciprocity law:

Lemma 2.5. *For all λ, μ in Λ*

$$[P(\mu) : \Delta(\lambda)] \cdot d_\lambda = [\nabla(\lambda) : E(\mu)] \cdot d_\mu.$$

Proof: Lemma 2.4 with $M = P(\mu)$ shows that the left hand side is equal to $\dim_k \operatorname{Hom}(P(\mu), \nabla(\lambda))$. However, for any A–module N, we have $\dim_k \operatorname{Hom}(P(\mu), N) = [N : E(\mu)] \cdot d_\mu$.

Consider the case where $d_\lambda = 1$ for all $\lambda \in \Lambda$ (for example, if the base field k is algebraically closed), then we can reformulate the reciprocity law as follows: recall that the Cartan matrix $C(A)$ of A is, by definition, the matrix whose columns are just the transpose of the vectors $\mathbf{dim}P(\lambda)$. Then

$$C(A) = (\mathbf{dim}\Delta)^t \cdot (\mathbf{dim}\nabla);$$

in particular, the determinant of the Cartan matrix is equal to 1.

3. Standardization

Let $\mathcal{C}$ be an abelian k–category and $\Theta = \{\Theta(\lambda) | \lambda \in \Lambda\}$ a finite set of objects of $\mathcal{C}$. The set Θ is said to be *standardizable* provided the following conditions are satisfied:

(F) $\dim_k \operatorname{Hom}(\Theta(\lambda), \Theta(\mu)) < \infty$, and $\dim_k \operatorname{Ext}^1(\Theta(\lambda), \Theta(\mu)) < \infty$, for all $\lambda, \mu \in \Lambda$.

(D) The quiver with vertex set Λ which has an arrow $\lambda \to \mu$ (and just one) provided $\operatorname{rad}(\Theta(\lambda), \Theta(\mu)) \neq 0$ or $\operatorname{Ext}^1(\Theta(\lambda), \Theta(\mu)) \neq 0$, has no (oriented) cycles.

(Here, $\operatorname{rad}(\Theta(\lambda), \Theta(\mu))$ denotes the set of non–invertible maps $\Theta(\lambda) \to \Theta(\mu)$.) Let us point out that condition (D) asserts, in particular, that all the objects $\Theta(\lambda)$ are Schurian, and do not have self–extensions.

Given a standardizable set Θ indexed by Λ, then condition (D) defines a partial ordering on Λ, with $\lambda \leq \mu$ provided there exists a chain of arrows $\lambda = \lambda_0 \to \lambda_1 \to \cdots \to \lambda_m = \mu$. The set Λ with this ordering may be called the *weight set* for Θ.

As before, we denote by $\mathcal{F}(\Theta)$ the full subcategory of all objects in $\mathcal{C}$ having a Θ–filtration.

Theorem 2. *Let Θ be a standardizable set of objects of an abelian category $\mathcal{C}$. Then there exists a quasi-hereditary algebra A, unique up to Morita equivalence, such that the subcategory $\mathcal{F}(\Theta)$ of $\mathcal{C}$ and the category $\mathcal{F}(\Delta_A)$ of all Δ_A–filtered A–modules are equivalent.*

Proof: Without loss of generality, we may refine Λ to a total ordering, thus we may assume that $\Lambda = \{1, 2, \ldots, n\}$, with its natural ordering.

First, let us observe that for any $1 \leq i \leq n$, there exists an indecomposable Ext–projective object $P_\Theta(i)$ of $\mathcal{F}(\Theta)$ with an epimorphisms $P_\Theta(i) \to \Theta(i)$ with kernel in $\mathcal{F}(\Theta)$. In order to show the existence, we fix some i. We want to construct inductively indecomposable objects $P(i,m)$, with $i \leq m \leq n$, such that there is an exact sequence $0 \to K(i,m) \to P(i,m) \to \Theta(i) \to 0$, with $K(i,m)$ in $\mathcal{F}(\Theta(i+1), \ldots, \Theta(m))$ and such that $\mathrm{Ext}^1(P(i,m), \Theta(j)) = 0$, for $1 \leq j \leq m$. Let $P(i,i) = \Theta(i)$, and therefore $K(i,i) = 0$; the condition (D) shows that the Ext–condition is satisfied. Now assume $i < m$, and that $P(i, m-1)$ and $K(i, m-1)$ are already defined. Let $d(i,m) = \dim \mathrm{Ext}^1(P(i,m-1), \Theta(m))_{\mathrm{End}(\Theta(m))}$, then there is a "universal extension"

$$0 \to d(i,m)\Theta(m) \to P(i,m) \to P(i,m-1) \to 0$$

(the induced map $\mathrm{Hom}(d(i,m)\Theta(m), \Theta(m)) \to \mathrm{Ext}^1(P(i,m-1), \Theta(m))$ being surjective). It is easy to see that $\mathrm{Ext}^1(P(i,m), \Theta(j)) = 0$, for all $j \leq m$. Also, since $\mathrm{Hom}(\Theta(m), P(i,m-1)) = 0$, it follows that $P(i,m)$ is indecomposable. We define $K(i,m)$ as the kernel of the composition of the given maps $P(i,m) \to P(i,m-1)$ and $P(i,m-1) \to \Theta(i)$, thus $K(i,m)$ is an extension of $d(i,m)\Theta(m)$ by $K(i,m-1)$. This finishes the induction step. We define $P_\Theta(i) = P(i,n)$.

Given any object $X \in \mathcal{F}(\Theta)$, we claim that there exists an exact sequence $0 \to X' \to P_0(X) \to X \to 0$, with $P_0(X) \in \operatorname{add} P_\Theta$, and X' again in $\mathcal{F}(\Theta)$. For $X = \Theta(i)$, we take $P_0(\Theta(i)) = P(i)$, and we proceed by induction: assume there is given an object X in $\mathcal{F}(\Theta)$ with a non–zero proper submodule U such that both U and $Y = X/U$ belong to $\mathcal{F}(\Theta)$. By induction, there are epimorphisms $\epsilon_U : P_0(U) \to U$ and $\epsilon_Y : P_0(Y) \to Y$ such that $P_0(U), P_0(Y)$ belong to $\operatorname{add} P_\Theta$, whereas the kernels U' of ϵ_U, and Y' of ϵ_Y belong to $\mathcal{F}(\Theta)$. Let $\iota : U \to X$ be the inclusion map, and $\pi : X \to Y$ the projection. Since $\mathrm{Ext}^1(P_0(Y), U) = 0$, there is some $\alpha : P_0(Y) \to X$ such that $\alpha\pi = \epsilon_Y$. Then $[\epsilon_U\iota, \alpha] : P_0(U) \oplus P_0(Y) \to X$ is surjective, and its kernel is an extension of U' by Y'.

Let $P = \bigoplus_{i=1}^n P_\Theta(i)$, and A its endomorphism ring. We consider the functor $F = \mathrm{Hom}(P, -) : \mathcal{C} \to A$–Mod. Condition (F) asserts that A is a finite–dimensional algebra, and that the images $F(X)$, where X belongs to $\mathcal{F}(\Theta)$, are finite dimensional A–modules. Finally, since $\mathrm{Ext}^1(P, X) = 0$, for $X \in \mathcal{F}(\Theta)$, we see that F is exact on exact sequences $0 \to X \to Y \to Z \to 0$ in $\mathcal{C}$, with $X \in \mathcal{F}(\Theta)$.

Let $P_A(i) = F(P_\Theta(i))$, and consider for $1 \leq i \leq n$, the modules $\Delta(i) = F(\Theta(i))$. Since F is exact on exact sequences of $\mathcal{C}$ whose objects lie inside $\mathcal{F}(\Theta)$, it follows that F maps $\mathcal{F}(\Theta)$ into $\mathcal{F}(\Delta)$. We claim that the restriction of F to $\mathcal{F}(\Theta)$ is fully faithful. Of course, this is true for the restriction of F to $\operatorname{add} P_\Theta$. Let X be in $\mathcal{F}(\Theta)$. As we have seen above, there is a map $\delta_X : P_1(X) \to P_0(X)$ in $\mathcal{C}$ such that its cokernel is X, and such that the kernel X'' and the image X' of δ_X both belong to $\mathcal{F}(\Theta)$. We denote the projection map by $\epsilon_X : P_0(X) \to X$. Note that under F, the exact sequences $P_1(X) \to P_0(X) \to X \to 0$ goes to a projective presentation of $F(X)$. Now assume X, Y in $\mathcal{F}(\Theta)$ are given. Let $f : X \to Y$ be a map with $F(f) = 0$. Since $\operatorname{Ext}^1(P_0(X), Y') = 0$, and $\operatorname{Ext}^1(P_1(X), Y'') = 0$, it follows that there are maps $f_0 : P_0(X) \to P_0(Y)$, and $f_1 : P_1(X) \to P_1(Y)$ such that $f_0\epsilon_Y = \epsilon_X f$, and $f_1\delta_Y = \delta_X f_0$. Since $F(f) = 0$, there is a map g' : $F(P_0(X)) \to F(P_1(Y))$ such that $gF(\delta_Y) = F(f_0)$. However, $g' = F(g)$ for some $g : P_0(X) \to P_1(Y)$, and $f_0 = g\delta_Y$, using the fact that the restriction of F to $\operatorname{add} P_\Theta$ is fully faithful. Hence $\epsilon_X f = f_0\epsilon_Y = g\delta_Y\epsilon_Y = 0$, thus $f = 0$. This shows that the restriction of F to all of $\mathcal{F}(\Theta)$ is faithful. In order to show that it is full, let $f' : F(X) \to F(Y)$ be a map. Since there are given projective presentations, we obtain maps $f_i' : F(P_i(X)) \to F(P_i(Y))$, for $i = 0, 1$ such that $f_0'F(\epsilon_Y) = F(\epsilon_X)f'$, and $f_1'F(\delta_Y) = F(\delta_X)f_0'$. Since the restriction of F to $\operatorname{add} P_\Theta$ is fully faithful, we can write $f_i' = F(f_i)$, with maps $f_i : P_i(X) \to P_i(Y)$, and we have $f_1\delta_Y = \delta_X f_0$. Since $\delta_X f_0 \epsilon_Y = 0$, there is $f : X \to Y$ in $\mathcal{C}$ such that $f_0\epsilon_Y = \epsilon_X f$. Under F we obtain $F(\epsilon_X)F(f) = F(f_0\epsilon_Y) = F(\epsilon_X)f'$, therefore $F(f) = f'$, since $F(\epsilon_X)$ is an epimorphism. Thus, the restriction of F to $\mathcal{F}(\Theta)$ is also full.

As a consequence, we see that A is quasi-hereditary relative to the ordering $\{1, 2, \dots, n\}$, and that the modules $\Delta(i)$ are the standard modules. For, $P_A(i) = F(P_\Theta(i))$ has a Δ–filtration, the upper factor being $\Delta(i)$, the remaining ones being of the form $\Delta(j)$, with $j > i$. In particular, $\Delta(i)$ has simple top $E(i)$. Since $\operatorname{Hom}(P_A(j), \Delta(i)) \neq 0$ only for $j \le i$, it follows that $\Delta(i)$ is the maximal factor module of $P_A(i)$ with composition factors of the form $E(j)$, where $j \le i$.

It remains to be seen that any A–module in $\mathcal{F}(\Delta)$ is of the form $F(X)$ with X in $\mathcal{F}(\Theta)$. Let M be a non–zero module in $\mathcal{F}(\Delta)$, let U be a submodule isomorphic to some $\Delta(i)$, such that also M/U belongs to $\mathcal{F}(\Delta)$. Denote by $\iota : \Delta(i) \to M$ a monomorphism with image U, and by $\pi : M \to M/U$ the projection map. By induction, there is an object Y in $\mathcal{F}(\Theta)$ such that $F(Y) = M/U$. As we know, there is an epimorphism $\epsilon_Y : P_0(Y) \to Y$ with $P_0(Y) \in \operatorname{add} P_\Theta$, such that its kernel Y' also belongs to $\mathcal{F}(\Theta)$. Let $u : Y' \to P_0(Y)$ be the inclusion map. Since $F(P_0(Y))$ is projective, there is a map $\alpha : F(P_0(Y)) \to M$ such that $\alpha\pi = F(\epsilon_Y)$. Then $[\iota, \alpha] : \Delta(i) \oplus F(P_0(Y)) \to M$ is surjective, and its kernel is easily seen to be isomorphic to $F(Y')$, with kernel map of the form $[\phi, F(u)] : F(Y') \to \Delta(i) \oplus F(P_0(Y))$ where $\phi : F(Y') \to \Delta(i)$ is some map. Since the objects $F(Y')$, and $\Delta(i) = F(\Theta(i))$ are images under F, and F is full, there is a map $h : Y' \to \Theta(i)$ with $F(h) = \phi$. With u also $[h, u] : Y' \to \Theta(i) \oplus P_0(Y)$ is a monomorphism, let X be its cokernel. Since $u = [h,u]\begin{bmatrix}0\\1\end{bmatrix}$, the cokernel X of $[h,u]$ maps onto the cokernel Y of u, say by $e : X \to Y$, and the kernel of e is the same as the kernel of $\begin{bmatrix}0\\1\end{bmatrix}$, thus just $\Theta(i)$. In this way, we see that X as an extension of $Y \in \mathcal{F}(\Theta)$ and $\Theta(i)$ belongs to $\mathcal{F}(\Theta)$. The exact sequence

$$0 \longrightarrow Y' \xrightarrow{[h,u]} \Theta(i) \oplus P_0(Y) \longrightarrow X \longrightarrow 0$$

goes under F to an exact sequence, since Y' belongs to $\mathcal{F}(\Theta)$, thus $F(X)$ is isomorphic to the cokernel of $F([h,u]) = [\phi, F(u)]$, thus to M. This finishes the proof.

Note that if Θ is a standardizable set of an abelian k–category $\mathcal{C}$, this set is also standardizable when considered in the opposite category $\mathcal{C}^o$, of course then its weight set will be changed to the opposite partially ordered set. It follows that for any statement dealing with standardizable sets, there also is a corresponding dual statement.

In particular, we see that given a standardizable set Θ of modules, the category $\mathcal{F}(\Theta)$ always has sufficiently many Ext–projective modules. (This is a consequence of Theorem 2 as stated, but actually, it was the first step in its proof, and one may refer to the proof in order to get further properties of the Ext–projective objects from the construction presented there.) By duality, the category $\mathcal{F}(\Theta)$ also has sufficiently many Ext–injective modules.

Of course, given a quasi-hereditary algebra A, the set of standard modules is a standardizable set, its weight set will be called the *weight set of* A. Here, the Ext–projective objects of $\mathcal{F}(\Delta)$ are just the projective modules. The Ext–injective objects of $\mathcal{F}(\Delta)$ have been considered in [R2]. The indecomposable Ext–injective A–modules have been denoted by $T(1), \ldots, T(n)$, where $\Delta(\lambda)$ is embedded into $T(\lambda)$, with $T(\lambda)/\Delta(\lambda) \in \mathcal{F}(\{\Delta(\mu) | \mu < \lambda\})$, and $T = \bigoplus_\lambda T(\lambda)$ has been called the *characteristic module* of A.

If A is quasi-hereditary, then also the set of costandard modules is a standardizable set, its weight set is the opposite of the weight set of A. Note that the Ext–projective modules in $\mathcal{F}(\nabla)$ will belong to $\mathcal{F}(\Delta)$, according to Theorem 1,(ii), and they are Ext–injective in $\mathcal{F}(\Delta)$. Thus, the indecomposable Ext–projective modules in $\mathcal{F}(\nabla)$ are just the modules $T(1), \ldots, T(n)$. Note that the construction of $P_\nabla(\lambda)$ shows that $[P_\nabla(\lambda) : E(\lambda)] = 1$, and that $[P_\nabla(\lambda) : E(\mu)] \neq 0$ only in case $\mu \leq \lambda$ (here, $\leq$ is the given ordering of the weight set of A.) This implies that $P_\nabla(\lambda) = T(\lambda)$. Altogether, we see:

Proposition 3.1. *Let A be a quasi-hereditary algebra. Then there are indecomposable modules $T(\lambda), \lambda \in \Lambda$, and exact sequences*

$$0 \to \Delta(\lambda) \to T(\lambda) \to X(\lambda) \to 0,$$

and

$$0 \to Y(\lambda) \to T(\lambda) \to \nabla(\lambda) \to 0,$$

where $X(\lambda)$ *is filtered with factors* $\Delta(\mu), \mu < \lambda$, *and* $Y(\lambda)$ *is filtered with factors* $\nabla(\mu), \mu < \lambda$, *such that the module* $T = \bigoplus_{\lambda\in\Lambda} T(\lambda)$ *satisfies*

$$\operatorname{add} T = \mathcal{F}(\Delta) \cap \mathcal{F}(\nabla).$$

The characteristic module $T = \bigoplus_{\lambda\in\Lambda} T(\lambda)$ is a generalised tilting and cotilting module; a general context for the existence of such a module has been exhibited by Auslander and Reiten [AR]; they also show:

Proposition 3.2. *Let* T *be the characteristic module*

$$\mathcal{F}(\Delta) = \{X \in A\text{–mod} |\ \operatorname{Ext}^i(X,T) = 0 \text{ for all } i \geq 1\},$$

and

$$\mathcal{F}(\nabla) = \{Y \in A\text{–mod} |\ \operatorname{Ext}^i(T,Y) = 0 \text{ for all } i \geq 1\}.$$

A short version of the proof of Auslander and Reiten may be found in [R2].

Note that any subset of a standardizable set again is standardizable. In particular, given the set Δ of standard modules over a quasi-hereditary algebra, any subset will be standardizable and we obtain corresponding quasi-hereditary algebras. Of course, starting with an ideal of Λ, we will obtain just one of the factor algebras A/I, where I is an ideal belonging to a heredity chain of A. Similarly, starting with a filter of Λ, we will obtain a quasi-hereditary algebra of the form eAe, where e is an idempotent such that the ideal generated by e belongs to a heredity chain. These two extreme cases have been considered in [PS] and in [DR2].

4. Standard modules of small projective dimension

In this section, we are going to consider quasi-hereditary algebras with the property that all the standard modules have projective dimension at most 1, and also those satisfying the dual property that all costandard modules have injective dimension at most 1. In fact, in case both properties are satisfied, we will see that a category rather similar to $\mathcal{F}(\Delta)$ can be described very nicely.

Lemma 4.1. *Let A be a quasi-hereditary algebra. Then the following conditions are equivalent:*
(i) The projective dimension of any standard module is at most 1.
(ii) The projective dimension of the characteristic module T is at most 1.
(iii) The subcategory $\mathcal{F}(\nabla)$ is closed under factor modules.
(iv) All divisible modules belong to $\mathcal{F}(\nabla)$.

Recall that a module is said to be *divisible,* provided it is generated by an injective module. Dually, the *torsionless* modules are those which are cogenerated by projective modules. There is the following dual statement:

Lemma 4.1*. *Let A be a quasi-hereditary algebra. Then the following conditions are equivalent:*
(i) The injective dimension of any costandard module is at most 1.
(ii) The injective dimension of the characteristic module T is at most 1.
(iii) The subcategory $\mathcal{F}(\Delta)$ is closed under submodules.
(iv) All torsionless modules belong to $\mathcal{F}(\Delta)$.

Let us remark that quasi-hereditary algebras satisfying the latter conditions have been studied rather carefully in [DR5]; there, one may find additional equivalent properties.

Proof of Lemma 4.1. (i) implies (ii): This is trivial, since T belongs to $\mathcal{F}(\Delta)$.

(ii) implies (iii): Let $M \in \mathcal{F}(\nabla)$, and let N be a submodule of M. We apply $\mathrm{Ext}^1(T,-)$ to the exact sequence $0 \to N \to M \to M/N \to 0$ and obtain a surjective map $\mathrm{Ext}^1(T,M) \to \mathrm{Ext}^1(T,M/N)$, since $\mathrm{proj.dim.}\,T \leq 1$. The first group is zero, since $M \in \mathcal{F}(\nabla)$, thus $\mathrm{Ext}^1(T,M/N) = 0$. We use Proposition 3.2 in order to conclude that $M/N \in \mathcal{F}(\nabla)$, again taking into account that $\mathrm{proj.dim.}\,T \leq 1$.

(iii) implies (iv): This is trivial, since the injective modules belong to $\mathcal{F}(\nabla)$.

(iv) implies (i): Let Y be an arbitrary A–module, we want to show that $\mathrm{Ext}^2(\Delta(\lambda),Y) = 0$. Let $0 \to Y \to Q(Y) \to Y' \to 0$ be exact, with $Q(Y)$ injective. Then $\mathrm{Ext}^2(\Delta(\lambda),Y) \cong \mathrm{Ext}^1(\Delta(\lambda),Y')$. Now, Y' is divisible, thus by assumption Y' belongs to $\mathcal{F}(\nabla)$, therefore $\mathrm{Ext}^1(\Delta(\lambda),Y') = 0$. This completes the proof.

Note that under the equivalent conditions exhibited in Lemma 4.1, the characteristic module T is a tilting module in the sense of [HR1], thus it defines a torsion pair $(\mathcal{G}(T),\mathcal{H}(T))$, where

$$\mathcal{G}(T) = \{Y \in A\text{–mod} \mid \mathrm{Ext}^1(T,Y) = 0\},$$

and

$$\mathcal{H}(T) = \{Y \in A\text{–mod} \mid \mathrm{Hom}(T,Y) = 0\}.$$

The following lemma is an immediate consequence of Proposition 3.2 and Lemma 4.1.

Lemma 4.2. *Assume that the projective dimension of any standard module is at most 1. Then $\mathcal{G}(T) = \mathcal{F}(\nabla)$.*

Theorem 3. *Assume that the projective dimension of any standard module and the injective dimension of any costandard module is at most 1. Let ϕ be the endofunctor of A–mod defined by $\phi(M) = M/\eta_T M$. This functor ϕ induces an equivalence between $\mathcal{F}(\Delta)/\langle T\rangle$ and $\mathcal{H}(T)$.*

By definition, the category $\mathcal{F}(\Delta)/\langle T\rangle$ has the same objects as $\mathcal{F}(\Delta)$, the morphisms being residue classes of maps in $\mathcal{F}(\Delta)$, two maps $f, g : X \to Y$ belong to the same residue class if and only if $f - g$ factors through a direct sum of copies of T. Similarly, we may consider A–mod$/\langle T\rangle$, and $\mathcal{F}(\Delta)/\langle T\rangle$ is a full subcategory. Note that the isomorphism classes of indecomposable objects in A–mod$/\langle T\rangle$ are just the isomorphism classes of the indecomposable A–modules which do not belong to add T.

Proof: We know that $(\mathcal{G}(T), \mathcal{H}(T))$ is a torsion pair. Now $\eta_T M$ is the torsion submodule of M, thus $M/\eta_T M$ belongs to $\mathcal{H}(T)$. Of course, $\eta_T T = T$, thus $\phi(T) = 0$, therefore ϕ induces a functor A–mod$/\langle T\rangle \to \mathcal{H}(T)$, which we also denote by ϕ. We want to show that the restriction of ϕ to $\mathcal{F}(\Delta)/\langle T\rangle$ is fully faithful and dense.

First of all, let Y belong to $\mathcal{H}(T)$. Take a universal extension $0 \to mT \to \tilde{Y} \to Y \to 0$ of Y by copies of T. In the corresponding long exact sequence

$$\mathrm{Hom}(mT, T) \to \mathrm{Ext}^1(Y, T) \to \mathrm{Ext}^1(\tilde{Y}, T) \to \mathrm{Ext}^1(mT, T)$$

the connecting homomorphism is surjective. Since $\mathrm{Ext}^1(T, T) = 0$, it follows that $\mathrm{Ext}^1(\tilde{Y}, T) = 0$. Our assumption that the injective dimension of T is at most 1 and Proposition 3.2 imply that $\tilde{Y}$ belongs to $\mathcal{F}(\Delta)$. Of course, the image of mT in $\tilde{Y}$ is just $\eta_T\tilde{Y}$, thus $\phi(\tilde{Y}) = Y$. This shows that our functor is dense.

Given M in $\mathcal{F}(\Delta)$, we claim that $\eta_T M$ always belongs to add T. As a submodule of $M \in \mathcal{F}(\Delta)$, it also belongs to $\mathcal{F}(\Delta)$, since inj. dim. $T \leq 1$; as a module in $\mathcal{G}(T)$, it belongs to $\mathcal{F}(\nabla)$, thus to $\mathcal{F}(\Delta) \cap \mathcal{F}(\nabla) = \mathrm{add}\, T$.

Let $M_1, M_2 \in \mathcal{F}(\Delta)$, and let $f : M_1 \to M_2$ be a map. Assume that $\phi(f) : M_1/\eta_T M_1 \to M_2/\eta_T M_2$ is the zero map, thus f maps into $\eta_T M_2$, thus f factors through a module in add T. This shows that $\phi : \mathcal{F}(\Delta)/\langle T\rangle \to \mathcal{H}(T)$ is faithful.

In order to show that $\phi : \mathcal{F}(\Delta)/\langle T\rangle \to \mathcal{H}(T)$ is full, let us consider again $M_1, M_2 \in \mathcal{F}(\Delta)$, and let $g : M_1/\eta_T M_1 \to M_2/\eta_T M_2$ be a map. Denote by $\pi_i : M_i \to M_i/\eta_T M_i$ the canonical projections. Since $\mathrm{Ext}^1(M_1, \eta_T M_2) = 0$, the map $\pi_1 g : M_1 \to M_2/\eta_T M_2$ can be lifted to M_2, thus there is $g' : M_1 \to$

M_2 such that $g'\pi_2 = \pi_1 g$. But this means that $\phi(g') = g$, thus our functor is also full. This completes the proof.

5. Quasi-hereditary algebras with many projective–injective modules

In this section we consider quasi-hereditary algebras such that the projective cover of any costandard module is injective.

Lemma 5.1. *Let A be a quasi-hereditary algebra. The following conditions are equivalent:*
(i) The projective cover of any costandard module is injective.
(ii) The projective cover of T is injective.
(iii) The projective cover of $D(A_A)$ is injective, and $\operatorname{top}\nabla(\lambda)$ *belongs to* $\operatorname{add}\operatorname{top}D(A_A)$, *for all* $\lambda \in \Lambda$.
(iv) Every module in $\mathcal{F}(\nabla)$ is divisible, and $\mathcal{F}(\nabla)$ is closed under projective covers.

Proof:(i) implies (iv): If the projective cover of any costandard module is injective, the same is true for the projective cover $P(M)$ of any module M in $\mathcal{F}(\nabla)$. Thus any module M in $\mathcal{F}(\nabla)$ is generated by an injective module. Also, $P(M)$ as an injective module again belongs to $\mathcal{F}(\nabla)$.

(iv) implies (iii): The module $D(A_A)$ belongs to $\mathcal{F}(\nabla)$, thus also its projective cover $P(D(A_A))$. Also, as a module in $\mathcal{F}(\nabla)$, we know that $P(D(A_A))$ is divisible. But a projective divisible module is injective. Also, since $\nabla(\lambda)$ is divisible, we know that $\operatorname{top}\nabla(\lambda)$ is in $\operatorname{add}\operatorname{top}D(A_A)$.

(iii) implies (ii): Since $T \in \mathcal{F}(\nabla)$, we know that every composition factor of $\operatorname{top}T$ belongs to some $\operatorname{top}\nabla(\lambda)$, thus to $\operatorname{add}\operatorname{top}D(A_A)$. Thus, the projective cover of T belongs to $\operatorname{add}P(D(A_A))$, and therefore is injective.

(ii) implies (i): For every $\lambda \in \Lambda$, we know that $T(\lambda)$ maps onto $\nabla(\lambda)$, thus the projective cover $P(\nabla(\lambda))$ is a direct summand of the projective cover $P(T)$. This completes the proof.

Dually, we may consider the case where the injective envelope of any standard module is projective. (Recall that the *dominant dimension* $\operatorname{dd}M$ of a module M (as introduced by Tachikawa) is greater or equal to 1 if and only if its injective hull is projective.)

Theorem 4. *Let A be a quasi-hereditary algebra, and assume the projective cover of any costandard module is injective. Let* $B = A/\eta_T A$. *Then* $\mathcal{H}(T) = B$–mod.

Proof: By Lemma 5.1, we know that the projective cover of T is injective, thus belongs to $\operatorname{add}T$. It follows that $\eta_T M = \eta_{P(T)}M$, for any module M, and therefore $\operatorname{Hom}(T, M) = 0$ if and only if $\operatorname{Hom}(P(T), M) = 0$.

However, the modules M with $\mathrm{Hom}(P(T), M) = 0$ are just the $A/\eta_{P(T)}A$–modules, thus the B–modules.

6. The preprojective algebra of type A_n

The preprojective algebra of a finite graph has been introduced by Gelfand and Ponomarev [GP] in order to study the preprojective representations of a finite quiver without oriented cycles. A general account which covers the more general situation of a valued graph (thus dealing with the preprojective representations of a finite species) is [DR1].

It seems to be convenient to start with the following rather fancy definition of a graph (possibly with loops and multiple edges): a *graph* G is a quiver with a fixpointfree involution σ on the set of arrows such that for any arrow $\alpha : x \to y$, the arrow $\sigma(\alpha)$ points from y to x. (The usual definition will replace the two arrows α and $\sigma(\alpha)$ by a single edge between x and y.) Important graphs for representation theory are the Dynkin diagrams $\mathsf{A}_n, \mathsf{D}_n, \mathsf{E}_6, \mathsf{E}_7, \mathsf{E}_8$, and the Euclidean diagrams $\tilde{\mathsf{A}}_n, \tilde{\mathsf{D}}_n, \tilde{\mathsf{E}}_6, \tilde{\mathsf{E}}_7, \tilde{\mathsf{E}}_8$; note that according to our convention we have to draw the graph D_7 as follows:

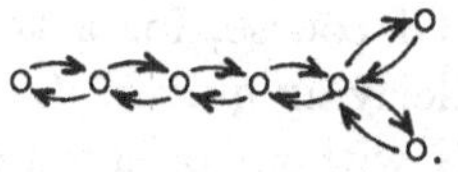

Given a Dynkin diagram of the form $\mathsf{A}_n, \mathsf{D}_m$, or E_6, where $n \geq 2$, and $m \geq 5$ is odd, we denote by ν the unique automorphism of order precisely 2. For the remaining Dynkin diagrams, we denote by ν the identity automorphism.

The *preprojective algebra* $\mathcal{P}(G)$ of the graph G is the factor ring of of the path algebra kG (here, G is considered as a quiver) modulo the ideal $\langle \rho_x | \, x \in G_0 \rangle$ generated by the elements $\rho_x = \sum_{t(\alpha)=x} \sigma(\alpha) \cdot \alpha$ (where $t(\alpha)$ denotes the terminal vertex of the arrow α, and G_0 is the set of vertices of G). [Note that we may consider any graph as a (stable) polarized translation quiver, as defined in [R1], using the identity map on G_0 as translation, and then $\mathcal{P}(G)$ is just the mesh algebra.]

The following result is due to Riedtmann [Rm] and Gelfand–Ponomarev [GP], see also [Ro]. Several proofs are available, we may refer also to [G], [DR1], and [R1].

Proposition 6.1. *The preprojective algebra $\mathcal{P}(G)$ is finite dimensional if and only if G is the disjoint union of Dynkin diagrams $\mathsf{A}_n, \mathsf{D}_n, \mathsf{E}_6, \mathsf{E}_7, \mathsf{E}_8$.*

Proposition 6.2. *Let G be a disjoint union of Dynkin diagrams $\mathsf{A}_n, \mathsf{D}_n, \mathsf{E}_6, \mathsf{E}_7, \mathsf{E}_8$. Then $\mathcal{P}(G)$ is a self-injective algebra with Nakayama functor given by ν.*

Proof: Let e_x be the primitive idempotent of $\mathcal{P}(G)$ corresponding to the vertex x of G. We consider the indecomposable projective $\mathcal{P}(G)$–module $\mathcal{P}(G)e_x$. It follows from the hammock considerations in [RV] that the socle of $\mathcal{P}(G)e_x$ is simple and not annihilated by $e_{\nu(x)}$. Consequently, $\mathcal{P}(G)$ is self–injective, and ν is its Nakayama permutation.

We denote by Ω the Heller functor: $\Omega(M)$ is the kernel of a projective cover $P(M) \to M$, and we recall from [G] that for any self–injective algebra B with Nakayama functor ν, we have $\tau = \Omega^2\nu = \nu\Omega^2$.

We are going to consider the case $G = \mathbb{A}_n$ in more detail.

Proposition 6.3. *Let G be a Dynkin diagram of type $\mathbb{A}_n$. The algebra $B = \mathcal{P}(G)$ is representation–finite in case $n \leq 4$. For $n = 5$, the algebra is of tubular type $\mathbb{E}_8$.*

The proof will use the universal cover $\tilde{B}$ of B, as introduced by Bongartz and Gabriel [BG]. Note that $\tilde{B}$ is an infinite dimensional algebra without 1, but with sufficiently many idempotents. We may construct $\tilde{B}$ as the mesh algebra of the translation quiver $\mathbb{Z}\mathbb{A}_n$. We recall that a graph G is said to be a *tree* provided it is connected, but is no longer connected when a pair $\alpha, \sigma(\alpha)$ is deleted. Of course, for a tree, there are no loops, and σ is determined by the underlying quiver. Given a tree G, there exists a stable translation quiver $\mathbb{Z}G$ without oriented cycles such that $\mathbb{Z}G/\tau \cong G$, and $\mathbb{Z}G$ is unique up to isomorphism. The translation quivers $\mathbb{Z}G$ have been introduced by Riedtmann [Rm], those of the form $\mathbb{Z}\mathbb{A}_n$ already have appeared in [GR].

Proof of Proposition 6.3. The consideration of representation–finite algebras is by now standard, so we only deal with the case $n = 5$ (and here, the arguments are similar to those used in [HR2]). There exists a convex subquiver of $\mathbb{Z}\mathbb{A}_5$ of the form

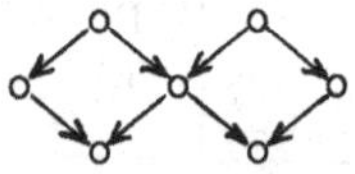

and the corresponding factor algebra C_0 of $\tilde{B}$ is tame concealed of type $\tilde{\mathbb{D}}_6$, thus of tubular type (4,2,2). The convex subquiver

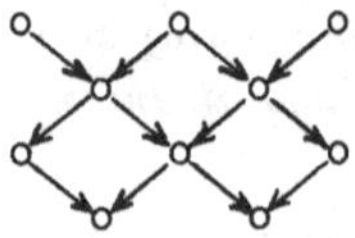

of $\mathbb{Z}\mathbb{A}_5$ (with the induced relations) yields a tubular extension C of C_0 of extension type (6,3,2), thus C is a tubular algebra of tubular type $\mathbb{E}_8$.

Note that the algebra C_∞ obtained from C by removing all sinks is a tame concealed algebra of type $\tilde{\mathsf{E}}_7$, and the opposite algebra C^o of C is a tubular extension of C_∞. This completes the proof.

Let us add the following remark:

Lemma 6.4. *Let G be a Dynkin diagram of type A_n. For $B = \mathcal{P}(G)$, we have*

$$\Omega^3 E(x) \cong E(\nu(x)) \cong \tau^3 E(x).$$

Proof: It is easy to verify that $\Omega \operatorname{rad} P(x) \cong P(x)/\operatorname{soc} P(x)$, and this implies that $\Omega^3 E(x) \cong E(\nu(x))$. The second isomorphism is a direct consequence: $\tau^3 E(x) \cong \nu^3 \Omega^6 E(x) \cong E(\nu(x))$.

In case $n \leq 5$, any indecomposable non–projective $\mathcal{P}(\mathsf{A}_n)$–module M satisfies

$$\Omega^3 M \cong \nu(M) \cong \tau^3 M,$$

and it seems that this is true for any n.

7. The Auslander algebra of $k[T]/\langle T^n \rangle$

Let $R_n = k[T]/\langle T^n \rangle$, this is a representation finite algebra, with indecomposable modules $M(i)$, $1 \leq i \leq n$, where $M(i)$ is of length $n - i + 1$. Let $A_n = \operatorname{End}(\bigoplus_i M(i))$ be its Auslander–algebra with the corresponding indexing of the simple A_n–modules, thus the indecomposable projective A_n–modules embed as follows into each other

$$P(1) \supset P(2) \supset \cdots \supset P(n-1) \supset P(n).$$

Note that A_n is quasi-hereditary in a unique way, with weight set the canonically ordered set $\{1, 2, \ldots, n\}$.

Lemma 7.1. *The standard modules are the serial modules with socle $E(1)$. The class $\mathcal{F}(\Delta)$ of all Δ-filtered modules is just the set of all torsionless modules, and also the set of all modules with socle generated by $P(1)$.*

Of course, dually, the costandard modules are the serial ones with top $E(1)$, and $\mathcal{F}(\nabla)$ is the set of all divisible modules, and also the set of all modules generated by $P(1)$. The modules $T(i)$ are the indecomposable modules with top and socle isomorphic to $E(1)$.

If we set $P(n+1) = 0$, then we have exact sequences

$$0 \to P(i+1) \to P(i) \to \Delta(i) \to 0$$

for all $1 \le i \le n$, therefore all standard modules have projective dimension at most 1. Dually, the costandard modules have injective dimension at most 1. This shows that we can apply Theorem 3, thus the categories $\mathcal{F}(\Delta)/\langle T\rangle$ and $\mathcal{H}(T)$ are equivalent. Also, since all costandard modules are generated by $P(1)$, and $P(1)$ is projective–injective, we see that the projective cover of any costandard module is injective. As a consequence, Theorem 4 can be applied, and it yields an equivalence of $\mathcal{H}(T)$ and B_n–mod, where $B_n = A_n/\eta_T A_n$. It remains to calculate B_n. It is easy to see that B_n is just the preprojective algebra of type $\mathbb{A}_{n-1}$, thus it follows:

Proposition 7.2. *Let $A = A_n$. The category $\mathcal{F}(\Delta)$ is finite for $n \le 5$, and of tubular type* $\textcircled{\scriptsize E}_8$ *for $n = 6$.*

Remark: It is known that the module category A_n–mod itself is finite only for $n \le 3$, and that it is of tubular type $\textcircled{\scriptsize E}_7$ for $n = 4$.

The following pages exhibit the Auslander–Reiten quivers of $\mathcal{F}(\Delta)$ for $3 \le n \le 5$, the vertical dashed lines have to be identified in order to form some kind of cylinder (for $n = 3, 5$) or Möbius strip (for $n = 4$). We use the following conventions: Let $\tilde{A}_n$ be the universal cover of the algebra A_n, note that the Galois group is just $\mathbb{Z}$. Let $3 \le n \le 6$. In this case, for any indecomposable module M in $\mathcal{F}(\Delta)$, there exists an $\tilde{A}_n$–module $\tilde{M}$ (unique up to shift by the Galois group) with push–down M. Always, the tables present the support and the Jordan–Hölder multiplicities of the modules $\tilde{M}$. For a better identification of the support of different modules $\tilde{M}$ inside the quiver of $\tilde{A}_n$, one vertex is encircled. As our presentation has shown, for the indecomposable modules M in $\mathcal{F}(\Delta)$ which are not relative injective, it is sufficient to know the factor module $M/\eta_T M = M/\eta_{P(1)} M$. Let $\mathcal{P}(1)$ be the set of indecomposable projective $\tilde{A}_n$–modules with push–down of the form $P(1)$. In our tables, the composition factors belonging to $\eta_{\mathcal{P}(1)}\tilde{M}$ are given by crosses, the remaining ones by a digit.

Acknowledgement

The authors are endebted to I. Agoston for helpful comments concerning the final presentation of the paper.

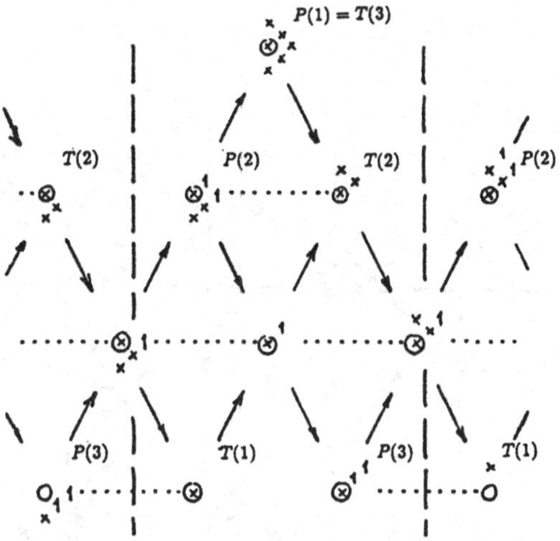

Table 1: The Auslander–Reiten quiver of $\mathcal{F}(\Delta)$ for A_3

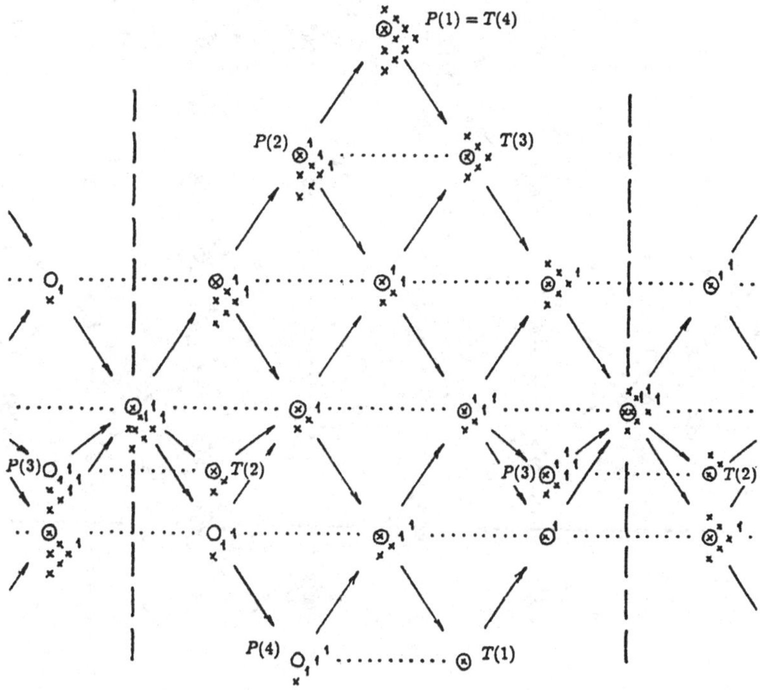

Table 2: The Auslander–Reiten quiver of $\mathcal{F}(\Delta)$ for A_4

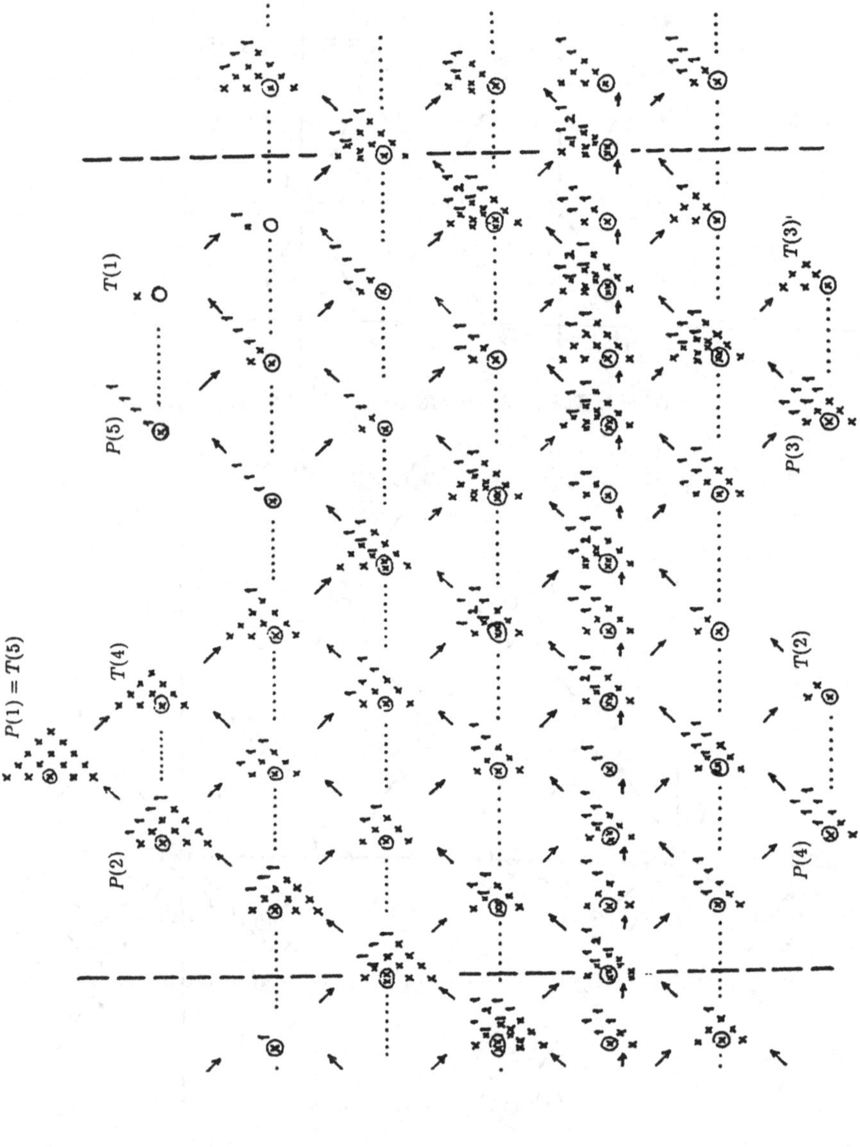

Table 3: The Auslander–Reiten quiver of $\mathcal{F}(\Delta)$ for A_5

References

[AR] Auslander, M.; Reiten, I.: Applications of contravariantly finite subcategories. Advances in Math. (To appear) Preprint 8/1989 Univ. Trondheim.

[BG] Bongartz, K.; Gabriel, P.: Covering spaces in representation theory. Invent. Math. 65 (1982), 331-378.

[CPS] Cline, E.; Parshall, B.J.; Scott, L.L.: Finite dimensional algebras and highest weight categories. J. reine angew. Math. 391 (1988), 85-99.

[DR1] Dlab, V.; Ringel, C.M.: The preprojective algebra of a modulated graph. In: Representation Theory II. Springer LNM 832 (1980), 216-231.

[DR2] Dlab, V.; Ringel, C.M.: Quasi-hereditary algebras. Illinois J. Math. 33 (1989), 280-291

[DR3] Dlab, V.; Ringel, C.M.: Auslander algebras as quasi-hereditary algebras. J. London Math. Soc. (2) 39 (1989), 457-466.

[DR4] Dlab, V.; Ringel, C.M.: Every semiprimary ring is the endomorphism ring of a projective module over a quasi-hereditary ring. Proc. Amer.Math.Soc. 107 (1989), 1-5.

[DR5] Dlab, V.; Ringel, C.M.: Filtrations of right ideals related to projectivity of left ideals. In: Sém. d'Algèbre Dubreil–Malliavin. Springer LNM 1404 (1989), 95–107.

[D] Donkin, St.: A filtration for rational modules. Math. Z. 177 (1981), 1-8.

[G] Gabriel, P.: Auslander–Reiten sequences and representation–finite algebras. Springer LNM 831 (1980), 1-71.

[GR] Gabriel, P.; Riedtmann, Chr.: Group representations without groups. Comment. Math. Helv. 54 (1979), 1-48.

[GP] Gelfand, I.M.; Ponomarev, V.A.: Model algebras and representations of graphs. Funkc. anal. i. prilož. 13 (1979), 1-12.

[HR1] Happel, D.; Ringel, C.M.: Tilted algebras. Trans. Amer. Math. Soc. 274 (1982), 399-443.

[HR2] Happel, D.; Ringel, C.M.: The derived category of a tubular algebra. In: Representation Theory I. Springer LNM 1177 (1986). 156-180.

[PS] Parshall, B.; Scott, L.L.: Derived categories, quasi-hereditary algebras, and algebraic groups. Proc. Ottawa–Moosonee Workshop. Carleton–Ottawa Math. LN 3 (1988), 1-105.

[Rm] Riedtmann, Chr.: Algebren, Darstellungsköcher, Überlagerungen, und zurück. Comment. Math. Helv. 55 (1980), 199-224.

[R1] Ringel, C.M.: *Tame Algebras and Integral Quadratic Forms.* Springer LNM 1099 (1984).

[R2] Ringel, C.M.: The category of modules with good filtrations over a quasi-hereditary algebra has almost split sequences. Math.Z. (To appear)

[R3] Ringel, C.M.: On contravariantly finite subcategories.*

[R4] Ringel, C.M.: The category of good modules over a quasi-hereditary algebra.*

[RV] Ringel, C.M.; Vossieck, D.: Hammocks. Proc. London Math. Soc. (3) 54 (1987), 216-246.

[Ro] Rojter, A.V.: Gelfand–Ponomarev algebra of a quiver. Abstract, 2nd ICRA (Ottawa 1979).

[S] Scott, L.L.: Simulating algebraic geometry with algebra I: The algebraic theory of derived categories. Proc. Symp. Pure Math. 47 (1987), 271-281.

[So] Soergel, W.: Construction of projectives and reciprocity in an abstract setting. (To appear)

V. Dlab
Department of Mathematics
Carleton University
Ottawa, Ontario
Canada K1S 5B6

C.M. Ringel
Fakultät für Mathematik
Universität
W–4800 Bielefeld 1
Germany

* These papers had been accepted for publication by the editors of the Proceedings of the Tsukuba Conference on Representations of Algebras and Related Topics (1990), Can. Math. Soc. Conf. Proc. However, the publisher of the Proceedings, the American Mathematical Society, has refused to print them, since they contain the following remark: "This paper is written in English in order to be accessible to readers throughout the world, but we would like to stress that this does not mean that we support any imperialism. Indeed, we were shocked when we heard that the Iraki military machinery was going to bomb Washington in reaction to the US invasion in Grenada and Panama, but maybe we were misinformed by the nowadays even openly admitted censorship." [C.M.Ringel]

Matrix problems, small reduction and representations of a class of mixed Lie groups

Yu. A. Drozd

In representation theory of Lie groups the cases of reductive and solvable groups are highly elaborated (cf. [K]). Much less seems to be known about mixed groups, i.e. those neither reductive nor solvable. Moreover, the simplest examples (cf. §1) show that in a sense the complete description of their representations is a hopeless problem. Nevertheless, in some cases it turns out to be possible to describe "almost all" of them in a rather appropriate way (cf. theorem 1.2), namely, they behave just like representations of a reductive group.

This lecture is an account of the author's joint work with A. Timoshin [DT] where we managed to apply to the investigation of representations of some mixed groups the method of so-called "matrix problems" which enabled us to obtain such an answer.

Thus, the lecture splits into two parts. The first (§§1-3) contains the formulation of the main theorem (theorem 3.1) with necessary preliminaries and its reduction to a matrix problem. The second part (§§4-6) is devoted to matrix problems which we treat in terms of representations of bocses (cf. [R]) and culminates in §6 with the proof of the main theorem. The most technical, but to my mind also the most important, is §5 where we present an algorithm elaborated in [BGOR]. I think that the importance of this algorithm (called "small reduction") is still far from being properly appreciated.

I am grateful to S. Ovsienko for friendly and fruitful discussions which were of great use to me, and to A. Timishin for his kind permission to use our joint results in this talk.

§1 Mackey Theorem. An Example.

In the calculation of representations of mixed groups a theorem of Mackey concerning representations of group extensions is widely used (cf. [K]). Recall it in the simplest case of a semi-direct product with normal abelian subgroup. From now on, for a locally compact group G we denote by $\hat{G}$ its dual space, i.e. the space of isomorphism classes of irreducible unitary representations (cf. [K]).

Theorem 1.1. *Suppose that a locally compact group is a semi-direct product, $G = H \ltimes N$, of a closed normal abelian subgroup N and a closed subgroup H and that all orbits of H on $\hat{N}$ are locally closed. Then there exists a surjection $p : G \to \hat{N}/H$ such that $p^{-1}(x^H) \cong \hat{H}_x$ for any orbit x^H, where $H_x = \{h \in H \mid x^h = x\}$.*

Here H acts on $\hat{N}$ naturally: $x^h(n) = x(hnh^{-1})$. The representation of G corresponding to the representation T of H_x, is $\mathrm{Ind}(G, G_x, xT)$, where $G_x = H_x N$ and $xT(hn) = x(n)T(h)$ for any $h \in H_x$, $n \in N$.

Thus in order to calculate representations of G we have to find orbits of H on $\hat{N}$, then to define the stabilizers H_x and finally to calculate the representations of each H_x. Moreover, we can obtain the Plancherel measure on $\hat{G}$ from those on $\hat{N}$ and $\hat{H}_x$ as is shown in [KL]. Sometimes it is possible to iterate this procedure, i.e. to decompose H_x into a similar semi-direct product and to apply Mackey's theorem again. But as a rule such iterations become more and more complicated as the following simple example shows.

Let K be a locally compact field and $G = G(m,n)$ be the group of invertible K-matrices of the form:

$$\begin{pmatrix} A_1 & A_{12} \\ 0 & A_2 \end{pmatrix}$$

with an $m \times m$ block A_1, an $n \times n$ block A_2 and an $m \times n$ block A_{12}. Clearly, $G = H \ltimes N$ with H consisting of the "diagonal" matrices (such that $A_{12} = 0$) and N of the "unipotent" ones (such that A_1 and A_2 are identities). Then $N = \mathrm{Mat}(m \times n)$, hence $\hat{N} = \mathrm{Mat}(n \times m)$ under the pairing $(x, n) = e(\mathrm{tr}(xn))$, e being a non-trivial character of the additive group of K (cf. [W]) and one easily obtains $x^h = A_2^{-1} x A_1$ for $h = \mathrm{diag}(A_1, A_2)$. Of course, its orbits are well-known: each of them contains a unique matrix x of the form:

$$x = \begin{pmatrix} 0 & I \\ 0 & 0 \end{pmatrix}$$

(I being an identity matrix). A simple calculation shows that H_x consists of all $h = \mathrm{diag}(A_1, A_2)$ with

$$A_1 = \begin{pmatrix} B_1 & B_{12} \\ 0 & B_2 \end{pmatrix}, \quad A_2 = \begin{pmatrix} B_2 & B_{23} \\ 0 & B_3 \end{pmatrix}$$

(the size of B_2 coincides with that of I in x).

Now again $H_x = H' \ltimes N'$ with H' "diagonal" and N' "unipotent". If k_i is the size of B_i, then

$$\begin{aligned} N' &= \mathrm{Mat}(k_1 \times k_2) \times \mathrm{Mat}(k_2 \times k_3), \quad \text{hence} \\ \hat{N}' &= \mathrm{Mat}(k_2 \times k_1) \times \mathrm{Mat}(k_3 \times k_2). \end{aligned}$$

Moreover, if $h = \mathrm{diag}(B_1, B_2, B_3) \in H'$, $x = (x_1, x_2) \in \hat{N}'$, then one can check that $x^h = (B_2^{-1} x B_1, B_3^{-1} x B_2)$. This means that elements of $\hat{N}'$ can be viewed as representations of the quiver

$$\bullet \longrightarrow \bullet \longrightarrow \bullet$$

and orbits correspond to isomorphism classes of these representations (cf. [G]). It is known then that any orbit contains a unique pair $x = (x_1, x_2)$ with

$$x_1 = \begin{pmatrix} 0 & I & 0 \\ 0 & 0 & I \\ 0 & 0 & 0 \\ 0 & 0 & 0 \end{pmatrix}, \quad x_2 = \begin{pmatrix} 0 & I & 0 & 0 \\ 0 & 0 & 0 & I \\ 0 & 0 & 0 & 0 \end{pmatrix}$$

the size of columns in x_1 being the same as that of rows in x_2. One can can calculate the stabilizer H'_x of such a pair: it consists of all triples (B_1, B_2, B_3) of the form:

$$B_1 = \begin{pmatrix} C_1 & C_{12} & \underline{C_{13}} \\ 0 & C_2 & C_{23} \\ 0 & 0 & C_3 \end{pmatrix}, \quad B_2 = \begin{pmatrix} C_2 & C_{23} & C_{24} & \underline{C_{25}} \\ 0 & C_3 & 0 & C_{35} \\ 0 & 0 & C_4 & C_{45} \\ 0 & 0 & 0 & C_5 \end{pmatrix},$$

$$B_3 = \begin{pmatrix} C_3 & C_{35} & \underline{C_{36}} \\ 0 & C_5 & C_{56} \\ 0 & 0 & C_6 \end{pmatrix}.$$

Let M be the normal subgroup of H'_x consisting of all triples with $C_i = I$ and $C_{ij} = 0$ except maybe for $(ij) = (13)$, (25) or (36) (the matrices underlined above). Write $F = H'_x/M$. Then again $F = H'' \ltimes N''$ with H'' "diagonal" and N'' "unipotent". But in this case the action of H'' on $\hat{N}''$ is described by the quiver:

$$\begin{array}{ccccccc} \bullet & \longrightarrow & \bullet & \longrightarrow & \bullet & & \\ & & \downarrow & & \downarrow & & \\ & & \bullet & \longrightarrow & \bullet & \longrightarrow & \bullet \end{array}$$

which is known to be wild (cf. [N]), i.e. the classification of its representations (or orbits of our action) contains the well-known unsolved problem of classification of pairs of linear operators on a finite-dimensional space. Thus we have no hope of obtaining a precise description of representations of groups $G(m, n)$.

Nevertheless, we can describe "almost all" their representations via the following simple observation. Consider the open dense subset $\tilde{N} \subseteq \hat{N}$ consisting of all matrices of maximal rank $r = \min(m, n)$. This subset forms an orbit of H. Namely, a matrix $x \in \tilde{N}$ is equivalent to one of the form

$$(0 \quad I) \quad \text{or} \quad \begin{pmatrix} 0 \\ I \end{pmatrix} \quad \text{or} \quad I$$

according as $n < m$ or $m < n$ or $m = n$, respectively. Hence its stabilizer H_x is isomorphic to the group of pairs (A_1, A_2) with

$$A_1 = \begin{pmatrix} B_1 & B_{12} \\ 0 & A_2 \end{pmatrix} \quad \text{or} \quad A_2 = \begin{pmatrix} A_1 & B_{12} \\ 0 & B_2 \end{pmatrix} \quad \text{or} \quad A_1 = A_2,$$

respectively. Thus H_x is isomorphic respectively to $G(m-n,n)$ or $G(m,n-m)$ or $GL(m,K)$. Now the iteration gives us the following

Theorem 1.2. *For $G = G(m,n)$ there exists an open dense subset $G^0 \subseteq \hat{G}$ homeomorphic to $\widehat{GL(d,K)}$, where d is the greatest common divisor of m and n.*

Moreover, using [KL] one can show that under this homeomorphism the restriction to G^0 of the Plancherel measure μ_G on G coincides with the Plancherel measure on $GL(d,K)$ and $\mu_G(G^0) = 1$.

An analogous result was proved in [L] for groups of block-triangular matrices with an arbitrary number of blocks. In this case $G^0 \cong \hat{D}$ for some direct product D of full linear groups over K.

We are going to generalise the latter result to a wider class of linear groups which will be defined in the next paragraph.

§2 Dynkin Algebras

The groups we will consider are linear groups over algebras. These are by definition the groups of the form $G(P,A) = \mathrm{Aut}_A(P)$, where A is a finite-dimensional algebra and P a finitely generated projective A-module. To obtain a good description of their representations we have to restrict the class of algebras to the so-called Dynkin algebras which we are going to define. We need some notions about categories in order to do so.

All categories considered are supposed to be linear over some field K and all functors will be K-linear. We write $\otimes$ and Hom instead of $\otimes_K$ and Hom_K. A module over a category A is by definition a functor $M : A \to \mathrm{Vect}$ (the category of vector spaces over K). Let A-mod denote the category of A-modules. An A-B-bimodule is a K-bilinear bifunctor $V : A^{\mathrm{op}} \times B \to \mathrm{Vect}$. Denote by add A the smallest category which contains A and is completely additive, i.e. additive and with all idempotents split. A category A will be called *skeletal* if its objects are pairwise non-isomorphic and there are no non-trivial idempotents in the endomorphism algebras $A(i,i)$. A *skeleton* of A is defined as a skeletal category equivalent to A. For instance, if A is a finite-dimensional algebra, a skeleton can be chosen as the full subcategory of A-mod consisting of a complete set of pairwise non-isomorphic indecomposable projective A-modules.

Let A be a skeletal category such that all spaces $A(i,j)$ are finite-dimensional and $A(i,i)/\mathrm{rad}\ A(i,i) = K$ for all $i \in \mathrm{Ob}A$ (in this case we call A *split* over K). Write

$$g(i,j) = \dim \mathrm{rad}\ A(i,j)/\mathrm{rad}^2\ A(i,j)$$

(in this case rad A can be defined as the two-sided ideal of A consisting of all non-invertible morphisms). Define the *graph* (Gabriel quiver) Q_A of A as the graph having Ob A for the set of points and $g(i,j)$ arrows leading from i to j for all i,j. Call A connected provided Q_A is connected.

If a category A has a skeleton A_0, it is called *split* if A_0 is split and *connected* if A_0 is connected. For instance, a finite-dimensional algebra is split if

$$A/\text{rad } A \cong \Pi\text{Mat}(n_m, K)$$

and is connected if it can be decomposed into a direct product of algebras (cf. [DK]).

Suppose A and $\tilde{A}$ are skeletal categories. Recall that a functor $F : \tilde{A} \to A$ is called a *covering* [BG] provided it is surjective on objects and for any $i \in$ Ob A, $j \in$ Ob A the following mappings induced by F are surjective:

$$\oplus_{Fk=i} \tilde{A}(k,j) \to A(i,Fj);$$
$$\oplus_{Fk=i} \tilde{A}(j,k) \to A(Fj,i).$$

A skeletal category A is called *simply connected* if it is connected and has no non-trivial connected coverings. A category A with a skeleton A_0 is called *simply connected* if A_0 is simply connected.

Now let A be skeletal, split and finite-dimensional over K (that means that Ob A is finite and all spaces $A(i,j)$ finite-dimensional). Denote by C_A its Cartan matrix, i.e. the $n \times n$ integral matrix, where $n = |\text{Ob } A|$, with entries $c_{ij} = \dim A(i,j)$, and by ${}^SC_A$ the symmetrisation of C_A, i.e. the symmetric $n \times n$ matrix with entries ${}^Sc_{ij} = (c_{ij} + c_{ji})/2$. Call A *positive definite* if the matrix ${}^SC_A$ is positive definite. If A is of finite global dimension (e.g. there are no oriented cycles in its graph Q_A), we can reformulate this notion. Namely, consider the n-dimensional real vector space $\mathbb{R}^n$ and define for a finite-dimensional A-module M its *vector-dimension* $\underline{\dim}\, M$ as the vector $(d_1, ..., d_n)$ where $d_i = \dim M(i)$. Then a bilinear (non-symmetric) form can be defined on $\mathbb{R}^n$ such that

$$(\underline{\dim}\, M, \underline{\dim}\, N)_A = \sum_m (-1)^m \dim \text{Ext}_A^m(M,N)$$

for all finite-dimensional A-modules M, N, and hence the quadratic form of A, $q_A(X) = (X,X)_A$ is defined on $\mathbb{R}^n$. One can easily check (cf. [Ri]) that in this case the dimensions of the representable functors $A(i,-)$ form a basis in $\mathbb{R}^n$ and that C_A is just the matrix of the form $(,)_A$ with respect to this basis. Thus A is positive definite if its quadratic form q_A is.

Of course, these definitions may be applied to any category A having a finite dimensional skeleton A_0, so we have for such categories the notions of Cartan matrix, positive definiteness, bilinear and quadratic forms $(,)_A$ and q_A, and vector-dimensions $\underline{\dim}\, M$. In the case of a finite-dimensional algebra all of them coincide with the usual ones as defined, e.g. in [Ri].

Definition. *A finite-dimensional K-algebra A is called* Dynkin *if it is split, positive definite, has no oriented cycles in its Gabriel quiver and all its connected direct factors are simply connected.*

One can verify that this definition is equivalent to that of Happel [H] (we will not use this equivalence).

We give some examples of Dynkin algebras A and linear groups $G(P, A)$ over them:

(i) Take for A the algebra of upper-triangular $n \times n$ matrices over K. Its quiver is

$$\bullet \longrightarrow \bullet \longrightarrow \dots \longrightarrow \bullet$$

(n points) and its quadratic form is just that of the quiver [G], thus known to be positive definite. Thus A is Dynkin. In this case the groups $G(P, A)$ which arise are just the groups of block-triangular matrices with n diagonal blocks, or, which is the same, parabolic subgroups of full linear groups.

(ii) Let S be a finite partially ordered set and $A = A(S)$ its incidence algebra, i.e. the subalgebra of $\mathrm{Mat}(n, K)$, where $n = |S|$, with a basis formed by matrix units e_{ij} with $i \leq j$ in S. One can check that if q_A is positive definite, this algebra is simply connected. Now as $G(P, A)$ we obtain a class of so-called net subgroups [B] (this was the starting point of the investigation in [DT]).

§3 Main Theorem. Reduction to a matrix problem

Theorem 3.1. *Let $G = G(P, A)$ be a linear group over a Dynkin K-algebra, where K is a locally compact field. Then there exists an open dense subset $G^0 \subseteq \hat{G}$ with $\mu_G(G^0) = 1$ such that $G^0 \cong \hat{D}$ and the restriction $\mu_G|_{G^0}$ coincides with μ_D (via this homeomorphism), where $D = \Pi_m GL(d_m, K)$ for some dimensions d_m (depending on A and P).*

Proof. Choose a complete set $P_1, \dots, P_n$ of pairwise non-isomorphic indecomposable projective A-modules. For each projective A-module P put $|P| = k_1 + \dots + k_n$ if $P = k_1 P_1 \oplus \dots \oplus k_n P_n$. Then an obvious induction reduces our theorem to the following statement.

Lemma 3.2. *Under the assumptions of theorem 3.1 suppose that $G \not\cong \Pi_m GL(d_m, K)$ for any d_m. Then there exists an open dense subset $\tilde{G} \subseteq \hat{G}$ with $\mu_G(\tilde{G}) = 1$ such that $\tilde{G} \cong \hat{G}'$ and $\mu_G|_{\tilde{G}} = \mu_{G'}$, where $G' = G(P', A')$ for some Dynkin algebra A' and some projective A'-module P' with $|P'| < |P|$.*

Proof of the lemma. As there are no cycles in Q_A, we can find an indecomposable projective, say P_1, such that $\mathrm{Hom}_A(P_1, P_i) = 0$ for all $i \neq 1$. Then $P = P^1 \oplus P^2$ with $P^1 = k_1 P_1$ and $P^2 = k_2 P_2 \oplus \dots \oplus k_n P_n$. Then

$\mathrm{Hom}_A(P^1, P^2) = 0$, so an element $g \in G$ can be written as a matrix:

$$g = \begin{pmatrix} g_1 & f \\ 0 & g_2 \end{pmatrix}$$

with $g_i \in G_i = G(P^i, A)$ and $f \in N = \mathrm{Hom}(P^2, P^1)$. Hence $G = H \ltimes N$ where $H = G_1 \times G_2$ and $\hat{N}$ is isomorphic to the dual vector space of N. It can happen, of course, that $N = 0$ and we obtain no real reduction. To avoid this case we need some additional results.

It is convenient to consider instead of A its skeleton, which has the form KQ_A/I, where KQ_A is the category of paths of the graph Q_A and I an ideal of KQ_A contained in J^2 (J is the ideal generated by all arrows). The points of this category, which we will also denote A, are just the indices $1, ..., n$ and $A(i,j) = \mathrm{Hom}_A(P_i, P_j)$. Let $c_{ij} = \dim A(i,j)$ (the entries of the Cartan matrix of A) and put $B = \mathrm{End}_A(P_2 \oplus ... \oplus P_n)$. Then B can be viewed as the full subcategory of A consisting of objects $2, ..., n$.

Call two objects i, j of A *joint* if either $A(i,j) \neq 0$, or $A(j,i) \neq 0$, and *disjoint* otherwise.

Lemma 3.3. *(i)* $c_{ij} \leq 1$ *for all* i, j.
(ii) There exists no chain $i_1, i_2, ..., i_m$ *with* m *even, each* i_k *joint with* i_{k+1}, i_m *joint with* i_1 *and all other pairs* i_k, i_l *disjoint.*
(iii) If i, j, k *are pairwise different and disjoint, then there exists at most one object joint to each of them.*
(iv) If $a \in A(i,1)$ *and* $b \in A(j,i)$ *are non-zero but* $ab = 0$, *then* $A(j,1) = 0$.
Proof. *(i)-(iii)* follow from the positive definiteness of A. *(iv)* will be proved in §6.

Lemma 3.4. *The algebra B is also Dynkin.*
Proof. Clearly, we have only to prove that each component of B is simply connected. For the sake of simplicity, suppose B is connected (it plays practically no rôle in the proof). Let $F : \tilde{B} \to B$ be a non-trivial connected covering. Consider the set S of all objects i of B such that there exists an arrow in Q_A leading from i to 1. It follows from lemma 3.3 (iv) that objects of S are pairwise disjoint. Define an equivalence relation $\vee$ on $F^{-1}(S)$ as the weakest one such that $i \vee j$ provided there exists an object k of $\tilde{B}$ and non-zero morphisms $b : k \to i$ and $c : k \to j$ with $a_{F(i)}F(b)$ and $a_{F(j)}F(c)$ both non-zero. Here, for any $s \in S$ we denote by a_s the only arrow leading from s to 1. Lemma 3.3 implies that $F(i) \neq F(j)$ for distinct equivalent $i, j \in F^{-1}(S)$. Let $S/\vee$ be the set of equivalence classes and s_T, for $s \in S$, $T \in S/\vee$, be the unique object in T with $F(s_T) = s$ if such an object exists. Now construct $\tilde{A}$ as the category with object set $\mathrm{Ob}\, B \cup S/\vee$ and arrow set the arrows of $\tilde{B}$ together with new arrows $a_{sT} : s_T \to T$ for each $s \in S$, $T \in S/\vee$ such that s_T exists. Extend F to $\tilde{A}$ by putting $F(T) = 1$ and $F(a_{sT}) = a_s$ whenever s_T exists. One can easily extend also all relations

from I to $\tilde{A}$ and hence obtain a non-trivial connected covering $F : \tilde{A} \to A$ which contradicts simple connectedness of A, q.e.d.

Now if $N = 0$, we can consider the group G_2 which is again a linear group over the Dynkin algebra B. So without loss of generality we may suppose that $N \neq 0$.

Put $A_i = \mathrm{End}_A(P^i)$. Then N is an A_1-A_2-bimodule and $\hat{N}$ is an A_2-A_1-bimodule. Now G_i is the group of invertible elements of A_i and if $h = (g_1, g_2) \in H$, $x \in \hat{N}$, we have $x^h = g_2^{-1} x g_1$ via this bimodule structure.

Define $\mathrm{End}(x)$ to be the subalgebra of $A_1 \times A_2$ consisting of all pairs (a_1, a_2) for which $a_2 x = x a_1$. Then H_x is just the group of invertible elements of $\mathrm{End}(x)$. So the last step will be:

Lemma 3.5. *If $N \neq 0$, there exists an open dense orbit x^H in $\hat{N}$ such that $\mathrm{End}(x) = A'$ is also a Dynkin algebra and $|A'| < |P|$.*

We then put $\tilde{G} = p^{-1}(x^H)$ (cf. theorem 1.1) and obtain $\tilde{G} \cong \hat{H}_x$. But $H_x = G(A', A')$ has just the necessary form.

In order to prove lemma 3.5 we have to elaborate some technical tools which allow us to investigate some kinds of "matrix problems", e.g. those arising from actions of linear groups on bimodules as above.

§4 Representations of bocses

Recall ([R], [D]) that a *bocs* is a pair $\underline{A} = (A, V)$ where A is a category and V an A-coalgebra, i.e. an A-bimodule supplied with bimodule homomorphisms $m : V \to V \otimes_A V$ (comultiplication) and $e : V \to A$ (counit) satisfying the usual rules (cf. [M]). We always suppose that the bocs is *normal*, i.e. for each object i there is an element $u_i \in V(i,i)$ such that $e(u_1) = 1_i$ and $m(u_i) = u_i \otimes u_i$. Let $\overline{V} = \mathrm{Ker}\ e$, the *kernel* of the bocs.

For any $a \in A(i,j)$ and $v \in \overline{V}(i,j)$ let $Da = au_i - u_j a$ and $Dv = m(v) - vu_i - u_j v$ (here and later we write xy instead of $x \otimes y$ for elements x, y of the bimodule V). One easily sees that $Da \in \overline{V}$ and $Dv \in \overline{V} \otimes_A \overline{V}$. The mapping D is called the *differential* of the bocs (it really induces a differential on the (graded) tensor category of the bimodule $\overline{V}$). It is clear that knowing the kernel and the differential we can reconstruct the bocs.

A *representation* of a bocs $\underline{A}$ is defined as a functor $M : A \to \mathrm{Vect}$ (the category of finite-dimensional vector spaces). Given another representation N, we define $\mathrm{Hom}_{\underline{A}}(M, N)$ as the set of all bimodule homomorphisms $V \to (M, N)$, where (M, N) is the A-bimodule such that $(M, N)(i, j) = \mathrm{Hom}(M(i), N(j))$ and left (right) multiplication by a is defined as left (right) multiplication by $N(a)$ (resp. by $M(a)$). If $f : M \to N$ and $g : N \to L$ are two such homomorphisms, their product gf is defined as composition:

$$V \xrightarrow{m} V \otimes_A V \xrightarrow{g \otimes f} (N, L) \otimes_A (M, N) \xrightarrow{\mathrm{mult}} (M, L)$$

the last arrow being induced by the usual multiplication of mappings. One can prove that in this way we obtain the category $R(\underline{A})$ of representations of the bocs $\underline{A}$.

A morphism from a bocs $\underline{A} = (A, V)$ to another bocs $\underline{B} = (B, W)$ is by definition a pair $F = (F_0, F_1)$ consisting of a functor $F_0 : A \to B$ and a homomorphism of A-bimodules $F_1 : V \to W$ compatible with the comultiplications and counits of V and W ([D]). Such a morphism induces a natural functor $F^* : R(\underline{B}) \to R(\underline{A})$.

Suppose given a bocs $\underline{A} = (A, V)$, a category B and a functor $F : A \to B$ such that any object of B is isomorphic to a direct summand of some Fi. Construct a new bocs $\underline{A}^F = (B, B \otimes_A V \otimes_A B)$ and a morphism $(F, F_1) : \underline{A} \to \underline{A}^F$ where $F_1(v) = 1_{Fi} \otimes v \otimes 1_{Fj}$ for $v \in V(i,j)$ (we denote this morphism also by F). The restriction imposed on F involves the surjectivity of the counit in $\underline{A}^F$ and then we obtain easily from general nonsense

Proposition 4.1. *In the above situation the induced functor $F^* : R(\underline{A}^F) \to R(\underline{A})$ is fully faithful and its image consists of those representations $M : A \to$ Vect which can be factored through F.*

A bocs $\underline{A} = (A, V)$ is called *free* provided A is a free category, i.e. the path category KQ of some graph Q and the kernel $\overline{V}$ is a free A-bimodule. Such a bocs is usually described by its bigraph $Q_{\underline{A}} = (Q_0, S_0, S_1)$ whose vertex set is the same as that of Q but as well as the set $S_0 = Q_1$ of arrows of Q (free generators of A) there is an additional set of arrows S_1 corresponding to free generators of the A-bimodule $\overline{V}$: if such a generator lies in $V(i,j)$, the corresponding arrow goes from i to j (usually the arrows of S_0 are called *solid* and those of S_1 *dashed*). To define $\underline{A}$ we need also to know its differential D and of course it has to be defined only for arrows, i.e. for free generators. The set $S = S_0 \cup S_1$ is called the set of free generators of $\underline{A}$.

Notice that in general we can change the set of free generators and a good choice of it sometimes plays an important rôle (e.g. this is the case in the definition of triangularity below). Nevertheless, the bigraph $Q_{\underline{A}}$ does not depend on this choice. We call $\underline{A}$ *connected* if its bigraph is connected. If $i, j \in \text{Ob } A$, let $S(i,j) = \{s \in S \mid s : i \to j\}$ and similarly for $S_0(i,j)$, $S_1(i,j)$.

A system $S = S_0 \cup S_1$ of free generators is said to be *triangular* provided there exists a function $h : S \to \mathbb{N}$ such that $h(Ds) < h(s)$ for any $s \in S$ where $h(Ds)$ denotes the maximum of $h(b)$ for all $b \in S$ which occur when we express Ds via generators from S (as $\underline{A}$ is free, such an expression is unique). A free bocs possessing a triangular system of free generators is called *triangular*. It is known (cf. [RK]) that if $\underline{A}$ is triangular, then any idempotent in $R(\underline{A})$ splits and a homomorphism $f \in \text{Hom}_{\underline{A}}(M, N)$ is invertible if and only if $f(u_i)$ is invertible for each $i \in \text{Ob } A$.

The notion of coverings can also be defined for bocses. Namely, a morphism

$$F = (F_0, F_1) : \underline{\tilde{A}} = (\tilde{A}, \tilde{V}) \to \underline{A} = (A, V)$$

is called a *covering* if F_0 is a covering of categories and for all objects $i \in \operatorname{Ob} A$, $j \in \operatorname{Ob} \tilde{A}$, F_1 induces bijections:

$$\bigoplus_{Fk=i} \tilde{V}(k,j) \to V(i, Fj) \quad \text{and} \quad \bigoplus_{Fk=i} \tilde{V}(j,k) \to V(Fj, i).$$

A connected bocs (free, triangular) will be called *simply connected* if it possesses no non-trivial connected covering.

The advantage of bocses is that they admit plenty of "reduction algorithms" based on proposition 4.1. A general scheme for producing such algorithms is the following one. Suppose that A' is a subcategory of A and $F' : A' \to B'$ is a functor. Consider the amalgamation (or push-out) B of B' and A with respect to F' and the inclusion of A' into A. Then universality of amalgamation and proposition 4.1 imply for the natural functor $F : A \to B$

Proposition 4.2. *The functor $F^* : R(\underline{A}^F) \to R(A)$ is an equivalence between $R(\underline{A}^F)$ and the full subcategory of $R(\underline{A})$ consisting of all representations $F : A \to$ Vect whose restrictions to A' can be factored through F'.*

In most cases (cf. [RK], [R], [D]) A' and F' are chosen in such a way that any functor $A' \to$ Vect can be factored through F', so F^* is an equivalence of categories. However for our present purposes an algorithm will be useful to deal with other cases.

Recall also some notions related to representations of bocses. If the category A contains finitely many (say, n) objects, then the vector-dimension $\underline{\dim} M = (d_1, ..., d_n)$ is defined as in §2 and we put $|M| = d_1 + ... + d_n$. Suppose now that $\underline{A}$ is free with a finite set of free generators. Denote by s_{ij} the cardinality of $S_0(i,j)$, and by t_{ij} that of $S_1(i,j)$. Then the bilinear form $(,)_{\underline{A}}$ on $\mathbb{R}^n$ is defined as follows:

$$(X, Y)_{\underline{A}} = \sum_i x_i y_i - \sum_{i,j} s_{ij} x_i x_j + \sum_{i,j} t_{ij} x_i y_j.$$

The corresponding quadratic form $q_{\underline{A}}(X) = (X, X)_{\underline{A}}$ is called the *Tits form* of the bocs $\underline{A}$.

Fixing a vector $\underline{d} \in \mathbb{R}^n$, denote by $R_{\underline{d}}(\underline{A})$ the set of all representations of $\underline{A}$ of vector-dimension $\underline{d}$. Fixing bases in all $M(i)$ we can consider these representations as sets of matrices, thus $R_{\underline{d}}(\underline{A})$ as an algebraic variety over K (really, an affine space, as our bocs is free and hence any set of matrices of prescribed dimensions defines a representation of $\underline{A}$).

§5 Small reduction

We are going to use proposition 4.2 in the following situation. Let $\underline{A}$ be a free triangular bocs given by its triangular system S of free generators and its differential D. Suppose there is $a \in S_0(i,j)$ with $i \neq j$ and $Da = 0$ (call such a a *minimal edge*). Denote by A' the subcategory of A consisting of two objects i, j and the single morphism a. Take for B' the trivial category $K\{i,j\}$ (i.e. the category consisting of the objects i and j and scalar multiples of the identity morphisms only) and for the functor $F' : A' \to B'$, that mapping i to $i \oplus j$, j to j and a to the morphism $i \oplus j \to j$ given by the matrix $(\,0 \quad 1_i\,)$ (we really consider rather the additive envelope add B' of the category B'). Now we can construct the amalgamation B and the functor $F : A \to B$ prolonging F' and hence construct a new bocs $\underline{A}^F$ and the functor $F^* : R(\underline{A}^F) \to R(\underline{A})$. Call $\underline{A}^F$ the *small reduction* of $\underline{A}$ via a in direction ij. The small reduction of $\underline{A}$ via a in direction ji is defined similarly: we only have to take for F' the functor mapping i to i, j to $i \oplus j$ and a to $\begin{pmatrix} 1 \\ 0 \end{pmatrix}$. Now proposition 4.2 implies

Proposition 5.1. *If $\underline{A}^F$ is the small reduction of a minimal edge a in direction ij (ji), then F^* establishes an equivalence between $R(\underline{A}^F)$ and the full subcategory of $R(\underline{A})$ consisting of all representations M with* rank $M(a) =$ dim $M(j)$ *(resp.* rank $M(a) =$ dim $M(i)$*).*
Moreover, in this case $\underline{A}^F$ is also a free triangular bocs and if both $M(i)$ and $M(j)$ are non-zero for some $M = F^*(N)$, then $|M| > |N|$.

(The last assertions are proved just as they are in [D] for the usual reduction of a minimal edge.)

Corollary 5.2. *Under the conditions of proposition 5.1 suppose that a vector-dimension $\underline{d}$ of representations of $\underline{A}$ is given such that $d_j \leq d_i$ (resp. $d_i \leq d_j$). Then there is an open and dense (in the Zariski topology) subset $U \subseteq R_{\underline{d}}(\underline{A})$ such that $U \subseteq \operatorname{Im} F^*$.*

We need a precise description of $\underline{A}^F$ given in [BGOR] (it can also be easily obtained using calculations similar to [D]) in terms of its bigraph $Q' = Q_{\underline{A}}F$. Suppose that the minimal edge a was reduced in direction ij (for the ji case the answer is analogous). Then Q' and the new differential D' are as follows. The set of vertices of Q' coincides with that of $Q = Q_{\underline{A}}$. The set of free generators (arrows) S' is obtained from S by deleting a, adding instead of it a new arrow $a' \in S'_1(j,i)$ (note that a' differs from a both in direction and in type: it is dashed while a was solid) and extra arrows b' corresponding to the arrows $b \in S$ with source or target i (for the sake of simplicity suppose that $S(i,i) = \emptyset$ as it is in the only case we need). Namely, if $b \in S_m(i,k)$ ($m = 0, 1$), then $b' \in S'_m(j,k)$ and if $b \in S_m(k,i)$, then $b' \in S'_m(k,j)$. In other words, corresponding to each arrow (other than a) incident on i, we adjoin another incident on j having the same type and direction.

To define D' we introduce some notation. If $w = ...ab_1...ab_2...$ is an (oriented) path in Q where all inclusions of a are shown, put $\tilde{w} = ...b'_1...b'_2...$ (all a are excluded and each right neighbour b of some a replaced by the corresponding b', if a does not occur in w, then $\tilde{w} = w$); if $w = pa$, i.e. a is the last arrow of w, then $\tilde{w} = 0$. For $w = pa$ put $w' = pa'$, otherwise $w' = w$; finally for $w = ap$ put $'w = a'p$, otherwise $'w = w$. Extend these three operations by linearity to all elements of A, $\overline{V}$ and $\overline{V} \otimes_A \overline{V}$. Now we are able to describe D':

- if $b \in S(k,l)$ and $l \neq i$, then $D'b = \widetilde{Db}$;
- if $l = i$, then $D'b = \widetilde{Db} + a'b'$ and $D'b' =' (\widetilde{Db})$;
- if $k = i$, then $D'b' = (\widetilde{Db})' + ba'$;
- $Da' = 0$.

A useful consequence of this description is that we can determine whether a bocs was really obtained from another one by a small reduction of some minimal edge.

Proposition 5.3. *Suppose a bocs $\underline{A}$ is given by its quiver and there is a dashed arrow $a' \in S_1(j,i)$ for $j \neq i$ and there exists an injective mapping $b \mapsto b'$ putting in correspondence to each $b \in S_m(i,k)$ or $S_m(k,i)$ an arrow $b' \in S_m(j,k)$ or $S_m(k,j)$, respectively, such that:*

- *if $b \in S(i,k)$, then $Db' = (Db)' + ba'$;*
- *if $b \in S(k,i)$, then $Db = w + a'b'$ and $Db' = {}'w$ for some element w (not necessarily a word);*

where the operations w' and $'w$ are defined as above. Then there exists a bocs $\underline{B}$ (with the same set of objects) and a minimal edge $a : i \to j$ in $\underline{B}$ such that $\underline{A}$ is the small reduction of $\underline{B}$ via a in direction ij.

(Of course, the analogous result is valid for a small reduction in direction ji). We shall call such an arrow a' *integrable* and the bocs $\underline{B}$ *integrated* from $\underline{A}$ via a' in direction ij (or ji as the case may be).

Recall also the algorithm of "regularisation" (cf. [RK], [D]). An arrow $a \in S_0(i,j)$ is called *irregular* provided $Da \in S_1(i,j)$ (this depends on the choice of free generators). Define A' and B' as for the algorithm of small reduction but put now $F'(i) = i$, $F'(j) = j$ and $F'(a) = 0$ (notice that now it is possible that $i = j$). Again we are able to construct the amalgamation B and the functor $F : A \to B$ prolonging F', but in this case any representation of $\underline{A}$ can easily be shown to be equivalent to some M with $M(a) = 0$, thus proposition 4.2 implies

Proposition 5.4. *In the above situation $F^* : R(\underline{A}^F) \to R(\underline{A})$ is an equivalence of categories.*

Small reductions and regularisation are compatible with the Tits forms of bocses.

Proposition 5.5. *If a bocs $\underline{B}$ is obtained from another one $\underline{A}$ by small reduction or regularisation, then their bilinear forms $(,)_{\underline{A}}$ and $(,)_{\underline{B}}$ are equivalent (hence so are also their Tits forms $q_{\underline{A}}$ and $q_{\underline{B}}$).*
Proof. For regularisation it is obvious that even $(,)_{\underline{A}} = (,)_{\underline{B}}$. For small reduction, say in direction ij, one can easily check using the above description of $\underline{B}$ that $(,)_{\underline{B}}$ is obtained from $(,)_{\underline{A}}$ if we change, in the standard basis $\{e_i\}$ of $\mathbb{R}^n$, the vector e_j to $e_j + e_i$.

Small reduction and regularisation are also compatible with coverings of bocses.

Proposition 5.6. *If a bocs $\underline{B}$ is a small reduction of another bocs $\underline{A}$, then $\underline{A}$ and $\underline{B}$ are either both simply connected or both not.*
Proof. Clearly, either both $\underline{A}$ and $\underline{B}$ are connected or neither is. Suppose they are connected and $F : \tilde{\underline{A}} \to \underline{A}$ is a non-trivial connected covering. Let $\underline{B}$ be the small reduction of a minimal edge $a : i \to j$, say in direction ij. Then one can choose for each object i' of $\tilde{\underline{A}}$ such that $Fi' = i$ an object j' with $Fj' = j$ and an arrow $a' : i' \to j'$ with $Fa' = a$ (hence $Da' = 0$). But then we can apply to $\tilde{\underline{A}}$ the small reduction of all a' (each in direction $i'j'$) and obtain a commutative diagram:

$$\begin{array}{ccc} \tilde{\underline{A}} & \to & \tilde{\underline{B}} \\ F\downarrow & & G\downarrow \\ \underline{A} & \to & \underline{B} \end{array}$$

in which G is again a covering (for more details, cf. [DOF]). The description of small reduction shows immediately that $\tilde{\underline{B}}$ is connected, thus $\underline{B}$ is not simply connected.

Conversely, let $G : \tilde{\underline{B}} \to \underline{B}$ be a non-trivial connected covering of $\underline{B}$ and $a' : j \to i$ the new dashed arrow in $\underline{B}$, hence integrable in the sense of proposition 5.3. Then again for any object i' of $\tilde{\underline{B}}$ such that $Gi' = i$ one can choose j' with $Gj' = j$ and an integrable arrow $a'' : j' \to i'$. Thus we can integrate all a'' and obtain a covering $F : \tilde{\underline{A}} \to \underline{A}$ which is also non-trivial and connected.

Analogous arguments give

Proposition 5.7. *If $\underline{B}$ was obtained from $\underline{A}$ by regularisation and $\underline{A}$ was simply connected, then each component of $\underline{B}$ is also simply connected.*

Of course, in this case $\underline{B}$ need not be connected. Moreover, easy examples show that $\underline{B}$ can be simply connected even if $\underline{A}$ was not.

§6 Proof of the main theorem

Recall that in order to accomplish the proof of Theorem 3.1, we have only to prove lemma 3.5 and lemma 3.3 (iv). To do this, we need a special type of bocs related to bimodules (cf. [D]).

Suppose given two categories A_1, A_2 and an A_2-A_1-bimodule U. It is convenient here to consider A_1 and A_2 additive but with finite-dimensional skeletons with object sets I_1 and I_2 respectively. Define the category $C(U)$ of "elements of the bimodule U" which has for objects the elements of all $U(i,j)$ with $i \in \mathrm{Ob}\, A_2$, $j \in \mathrm{Ob}\, A_1$. A homomorphism from $u \in U(i,j)$ to $u' \in U(i',j')$ is defined as a pair (a,b) with $a : i \to i'$ and $b : j \to j'$ such that $bu = u'a$.

Categories of type $C(U)$ coincide with those of bocs representations, as the following theorem, proved in [D], shows.

Theorem 6.1. *Suppose that, in the above situation, the categories A_1, A_2 are split and all spaces $U(i,j)$ are finite-dimensional. Then there exists a free triangular bocs $\underline{A}$ such that categories $C(U)$ and $R(\underline{A})$ are equivalent.*

We will need also the precise construction of the bocs $\underline{A}$ which will be given in terms of its bigraph $Q_{\underline{A}}$ and differential D. Namely, the vertex set of $Q = Q_{\underline{A}}$ will be $I_1 \cup, I_2$. All solid arrows go from some vertex $i \in I_2$ to some $j \in I_1$ and the solid arrows $a : i \to j$ are in 1-1 correspondence with a basis of the dual vector space $U(i,j)^*$. All dashed arrows are going from some i to j where i and j are both either in I_1 or in I_2 and the dashed arrows $b : i \to j$ are in 1-1 correspondence with a basis of $\mathrm{rad}(i,j)^*$, where rad denotes the radical of the appropriate category. The differential D is defined on a solid arrow $a : i \to j$ as $l^*(a) - r^*(a)$ and on a dashed arrow $b : i \to j$ as $m^*(b)$ where l^*, r^*, m^* are the mappings dual to l, r, m respectively, where

$$l : \bigoplus_k \mathrm{rad}(k,j) \otimes U(i,k) \to U(i,j)$$

$$r : \bigoplus_k U(k,j) \otimes \mathrm{rad}(i,k) \to U(i,j)$$

$$m : \bigoplus_k \mathrm{rad}(k,j) \otimes \mathrm{rad}(i,k) \to \mathrm{rad}(i,j)$$

are generated by multiplications.

It is convenient to consider elements of U as matrices with entries in $U(i,j)$. Namely, let $v \in U(x,y)$. Decompose $x = \oplus_{d_j} j$ with $j \in I_1$, $y = \oplus_{d_j} i$ with $i \in I_2$. Then v will be considered as a block matrix (v_{ij}), where v_{ij} is a matrix of size $d_i \times d_j$ with coefficients from $U(i,j)$. Then we obtain a representation of the bocs $\underline{A}$ corresponding to v by mapping each vertex $i \in I_1 \cup I_2$ to a d_i-dimensional vector space and any solid arrow

$a : i \to j$, i.e. an element of $U(i,j)^*$, to the linear map defined by the matrix $a(v_{ij})$ (applying a to a matrix component-wise). Thus the dimension of this representation is the vector $\underline{d} = (d_i)$.

Returning to the proof of the main theorem, recall that in lemma 3.5 we are considering the A_2-A_1-bimodule $\hat{N} = \mathrm{Hom}_A(P^2, P^1)^*$, where $A_m = \mathrm{End}_A(P^m)$, $P^1 = k_1P_1$ and $P^2 = k_2P_2 \oplus ... \oplus k_nP_n$; $P_1, ..., P_n$ are all non-isomorphic, indecomposable projective modules over a Dynkin algebra A and $\mathrm{Hom}_A(P_1, P_i) = 0$ if $i \neq 1$. Of course, the categories A_1 and A_2 are not additive, but we are able to replace them by additive ones, say, that of all modules of the form d_1P_1 and that of all modules of the form $d_2P_2 \oplus ... \oplus d_nP_n$ respectively. So we can define the bimodule U with $U(X,Y) = \mathrm{Hom}_A(Y,X)^*$ and obtain $N = U(P^1, P^2)$. Clearly, isomorphic elements of N in the category $C(U)$ are just those lying in one H-orbit, and the endomorphism ring $\mathrm{End}(x)$ of an element $x \in N$ defined before the lemma 3.5 is in fact its endomorphism ring in the category $C(U)$.

Applying theorem 6.1, we construct the corresponding bocs $\underline{A}$. Its vertices correspond to modules $P_1, ..., P_n$ (and will be denoted by $1, ..., n$). The only solid arrows are those from 1 to some $i \neq 1$ corresponding to a basis of $A(i,1) = \mathrm{Hom}_A(P_i, P_1) = U(P_1, P_i)^*$. Since $A(1,1) = K$ (there are no oriented cycles in Q_A), there are no dashed arrows $1 \to 1$. If $i, j \neq 1$, the dashed arrows $b : i \to j$ correspond to a basis of $\mathrm{rad}(i,j)^*$. Hence, there are no loops and at most one arrow $i \to j$ if $i \neq j$ (it exists if $A(i,j) \neq 0$). The differential of $\underline{A}$ arises from multiplication in A. Namely, using the above description, one can verify that for an arrow $a : i \to j$ it has the form: $Da = \sum b_k c_k$, where the sum is taken over all vertices k such that there are arrows $c_k : i \to k$, $b_k : k \to j$ for which the corresponding products in A are non-zero (if a is solid, hence $i = 1$, we need $A(j,1)A(k,j) \neq 0$ and if a is dashed, then $A(k,j)A(i,k) \neq 0$). As the algebra A can be reconstructed from the bocs $\underline{A}$, it is quite obvious that any covering of one of them provides a covering for the other, whence we have

Proposition 6.2. *The bocs $\underline{A}$ is simply connected.*

Consider the bilinear form $(,)_{\underline{A}}$ of the bocs $\underline{A}$. By definition, it is:

$$(X,Y)_{\underline{A}} = \sum_i x_i y_i - \sum_{i \neq 1} s_{1i} x_1 y_i + \sum_{i,j \neq 1} s_{ij} x_i y_j$$

where s_{ij} is the number of arrows from i to j (in our case the type of an arrow is prescribed by its ends). But $s_{1i} = \dim A(i,1) = c_{1i}$ and, for $i \neq 1$, $s_{ij} = \dim A(i,j) = c_{ij}$, where the c_{ij} are the entries of the Cartan matrix C_A of the algebra A. Then an easy calculation shows that $(,)_{\underline{A}}$ is equivalent to the bilinear form with Cartan matrix C_A. So we have

Proposition 6.3. *The bilinear forms $(,)_{\underline{A}}$ and $(,)_A$ are equivalent. Thus the Tits form $q_{\underline{A}}$ is positive definite.*

The next lemma is the key to the whole proof. Define a *road* in the bigraph $Q = Q_{\underline{A}}$ as a path in the graph obtained from Q by adding a new arrow a^{-1} for each a of Q (no matter of what type) going in the opposite direction. In other words, a road is a non-oriented chain of arrows in contrast with paths which are always supposed oriented. If the source of a road w coincides with its target, call w a *circle* (again in contrast to cycles which are supposed oriented). A road (in particular, a circle) will be called *clean* if all arrows with their sources and targets in the vertices of the road are themselves in the road. In particular, if a path p of Q occurs in Da for an arrow a (with non-zero coefficient, of course), then $a^{-1}p$ is a circle in Q and is called an *active circle* with *marked arrow a*. All other circles will be called *passive*.

Lemma 6.4. *If $\underline{A}$ is a simply connected bocs with positive definite Tits form $q = q_{\underline{A}}$, then every clean circle in its bigraph Q is active.*

Leaving the proof of this lemma (which is rather cumbersome) to the end of the section, we now show how it implies all the results needed. First of all, since the conclusion of the lemma is, of course, sufficient for $\underline{A}$ to be simply connected, we obtain

Corollary 6.5. *Any bocs $\underline{B}$ obtained from $\underline{A}$ by deleting some vertices (together with all arrows which they are ends of) is also simply connected.*

Proof of lemma 3.3 (iv). (Cf. the notation of this lemma.) If $ab = 0$, then, of course, $A(j,1)A(j,i) = 0$ as these spaces are 1-dimensional. Now if $A(j,1) \neq 0$, there are arrows $1 \to i$, $1 \to j$ and $j \to i$ in $\underline{A}$, which form a clean passive circle, so we have a contradiction to proposition 6.2.

Now, as $\hat{N}$ was identified with the set of representations of $\underline{A}$ of dimension $(k_1, ..., k_n)$, the proof of lemma 3.5 follows from the more general result:

Lemma 6.6. *Let $\underline{A}$ be a simply connected bocs with positive definite Tits form, and $\underline{d} = (d_1, ..., d_n)$ be a fixed dimension for representations of $\underline{A}$. Then in the set of all representations of $\underline{A}$ of dimension $\underline{d}$ there is an open dense subset consisting of isomorphic representations and, if M is one of them, then $\mathrm{End}_A(M)$ is a Dynkin algebra. Moreover, if there exists a solid arrow $a : j \to i$ with $d_i d_j \neq 0$, then $|\mathrm{End}_{\underline{A}}(M)| < |\underline{d}|$ if a is regular.*

Proof. Corollary 6.5 allows us to assume that all $d_i \neq 0$. Now use induction on $d = d_1 + ... + d_n$. If $\underline{A}$ has no solid arrows at all, then its representation is completely defined by its dimension, so there exists, up to isomorphism, only one representation M of dimension $\underline{d}$. Its endomorphism ring is, by definition,

$\mathrm{Hom}_{\underline{A}\text{-}\underline{A}}(V,(M,M))$. Since the category A is trivial, this endomorphism ring is isomorphic to:

$$\bigoplus_{i,j} \mathrm{Hom}(V(i,j), \mathrm{Hom}(M(i), M(j))) = \bigoplus_{i,j} M(j) \otimes V(i,j)^* \otimes M(i)^*.$$

Denote the last expression simply as $M \otimes V^* \otimes M^*$. One can check that the multiplication of $\underline{A}$-homomorphisms corresponds via this isomorphism to the composition

$$M \otimes V^* \otimes M^* \otimes M \otimes V^* \otimes M^* \xrightarrow{\mathrm{ev}} M \otimes V^* \otimes V^* \otimes M^* \xrightarrow{1 \otimes m^* \otimes 1} M \otimes V^* \otimes M^*,$$

where m is the comultiplication in V and ev is generated by the evaluation map $M^* \otimes M \to K$. But $m^* : V^* \otimes V^* \to V^*$ turns V^* into an algebra. Since $\underline{A}$ was simply connected, this algebra is easily shown to be simply connected too and one can check that the Cartan matrix of V^* is just that of the bilinear form $(,)_{\underline{A}}$. Hence V^* is a Dynkin algebra. To each vertex i of the bigraph $Q_{\underline{A}}$ corresponds a projective indecomposable V^*-module P_i' and if we put $P' = d_1 P_1' \oplus \ldots \oplus d_n P_n'$, then $M \otimes V^* \otimes M^*$ may be identified with $\mathrm{End}_{V^*}(P')$ which is therefore also Dynkin. Note that in this case $|\mathrm{End}_{\underline{A}}(M)| = |P'| = |\underline{d}|$.

If there are solid arrows in $\underline{A}$, take one of them, say $a : i \to j$ with $h(a)$ a minimal value of the function h used in the definition of triangular bocses. Then either $Da = 0$ or $Da = b$ for some $b \in S_1$. Note that in the first case $i = j$ is impossible as $\underline{A}$ is simply connected with positive Tits form. Thus we may apply to a either small reduction or regularisation. The latter does not change the dimension but reduces the number of arrows. As for the former, proposition 5.1 and corollary 5.2 allow us to choose the direction of reduction in such a way that all representations of $R_{\underline{d}}(\underline{A})$ lying in some open dense subset have the form $F^*(M)$ for some representation M of the reduced bocs $\underline{A}^F$ with dimension $\underline{d}'$ where $|\underline{d}'| < |\underline{d}|$. Propositions 5.5 and 5.6 show that the new bocs is also simply connected with positive definite Tits form, so an obvious induction completes the proof.

The proof of lemma 6.4 requires some topological methods, though these are rather simple and available from any elementary course, say [AB].

First of all define the fundamental group of the bocs $\underline{A}$ to be the group consisting of equivalence classes of circles containing some chosen vertex i under the equivalence relation

$$a^{-1}a \sim aa^{-1} \sim 1 \quad \text{for any arrow } a,$$
$$w \sim 1 \quad \text{for any active circle } w.$$

(Group multiplication is, of course, composition of circles.) If this group is non-trivial, then one can build a non-trivial covering of $\underline{A}$ using practically the same construction as in [BG]. But it is quite obvious that the group defined above coincides with the fundamental group of the cell complex Q^* obtained by attaching a 2-dimensional cell to each active circle of the bigraph Q. I am going to show that, if $\underline{A}$ has positive definite Tits form and contains a passive circle, then even $H_1(Q^*) \neq 0$.

On the contrary, suppose $H_1(Q^*) = 0$. Choose a minimal subcomplex S of Q^* such that: (i) $H_1(S) = 0$; (ii) S contains a clean passive circle w_0; (iii) with any two points, S contains all arrows connecting them, if such arrows exist in Q.

Recall that if X, Y are two subcomplexes of Q^*, there is a Mayer-Vietoris exact sequence for homologies:

$$\begin{aligned} 0 \longrightarrow & H_2(X \cap Y) \longrightarrow H_2(X) \oplus H_2(Y) \longrightarrow H_2(X \cup Y) \xrightarrow{d_2} \\ & H_1(X \cap Y) \longrightarrow H_1(X) \oplus H_1(Y) \longrightarrow H_1(X \cup Y) \xrightarrow{d_1} \\ & H_0(X \cap Y) \longrightarrow H_0(X) \oplus H_0(Y) \longrightarrow H_0(X \cup Y) \longrightarrow 0. \end{aligned} \qquad \text{(MV)}$$

We use it to show that $H_2(S) = 0$. Otherwise take a 2-dimensional cycle z on S and a cell X which really occurs in z. We can suppose that $z = X + \dots$. Put $Y = S - \mathring{X}$ (where $\mathring{X}$ is the interior of X). Then in (MV)

$$H_1(X \cap Y) = \mathbb{Z} = \operatorname{Im} d_2, \quad X \cup Y = S,$$

whence $H_1(Y) = 0$ in contradiction to the minimality of S.

We prove now that S contains only one passive circle. If there were two of them, w and w', then there would be an edge a belonging to w' but not to w. Let X be the closure of a 2-dimensional cell containing a. Put $Y = S - (\mathring{X} \cup \mathring{a})$. Then $X \cup Y = S$ and $X \cap Y = B - \mathring{a}$, where B is the boundary of X. Thus (MV) implies $H_1(Y) = 0$ and, since Y contains w, this contradicts the minimality of S. Thus w_0 is the only passive circle in S.

Now it is evident that each arrow a belonging to S lies either on w_0 or on some active circle such that the corresponding cell is in S (we shall say then that this circle is active in S). Otherwise apply (MV) for $X = a$, $Y = S - \mathring{a}$. Again $X \cup Y = S$, so $d_1 = 0$ while $X \cap Y$ consists of 2 points, thus $H_0(X \cap Y) = \mathbb{Z} \oplus \mathbb{Z}$ whence also $H_0(Y) = \mathbb{Z} \oplus \mathbb{Z}$. This means that Y consists of 2 connected components and we can diminish S replacing it by the component of Y containing w_0.

Call a circle of Q *even* if it contains an even number of dashed arrows and *odd* otherwise. I claim that Q contains no even clean circles. To see this, let w be such a circle. Consider the quadratic form q_w obtained from q by putting to 0 all variables except those corresponding to the vertices of w. It has the form:

$$q_w(x_1, \dots, x_m) = \sum_i x_i^2 + \sum_i (\pm x_i x_{i+1}), \qquad (x_{m+1} = x_1)$$

(+ corresponds to dashed arrows, − to solid ones). But as m is even, we can choose the values $x_i = \pm 1$ such that $q_w(x_1, ..., x_m) = 0$ which contradicts positive definiteness of q.

For the sake of simplicity, I suppose that Q contains no loops and not more than one arrow between any two points. In considering a road w, I shall always cross out of it parts $a^{-1}a$ where a is an edge, i.e. either an arrow or its inverse. If there are two circles $w' = pa$ and $w'' = a^{-1}q$, call the circle $w = pq$ their *simple composition* and similarly for the simple composition of several circles. Then it is evident that any circle in S is a simple composition of clean ones, each of the latter being either active or w_0. As $H_2(S) = 0$, it follows that any circle active in S is clean or contains w_0 as a simple component and the same is true for any circle which is a boundary in S (of some 2-dimensional chain).

The next claim is that S contains no tetrahedron, i.e. a figure consisting of four vertices denoted 1,2,3,4 and roads $p_{ij} : j \to i$, pairwise distinct, with $p_{ji} = p_{ij}^{-1}$. Note that the vertices 1,2,3,4 need not be distinct. Suppose, on the contrary, that such a tetrahedron exists. Suppose all its faces $p_{ij}p_{jk}p_{ki}$ to be clean (the general case can be reduced to this one). If all of them are active in S, we obtain a 2-dimensional cycle. Thus one of them, say for $(ijk) = (123)$, is w_0. The other three are active in S, so each of them has its marked arrow which may or may not lie on w_0. First suppose that no marked arrow lies on w_0. Take one of them, a. Suppose it lies on the road p_{14} in the direction from 4 to 1 (see Figures 1,2). It follows from the definition of differential that p_{34} and p_{13} are then paths and $p_{41} = pa^{-1}q$ for some paths p, q while $Da = qp_{13}p_{34}p + ...$ (some other terms not containing the same path). Consider now the marked arrow b of the face (234). If it lies on the path p_{34}, it should coincide with it and $Db = p_{32}p_{24} + ...$ But then the marked arrow c of the face (124) cannot lie on either p_{24} or p_{14} (Figure 1). Thus b lies on p_{24} going in the direction from 4 to 2. Let c lie on p_{14} (Figure 2), then it must coincide with a, and b with p_{24}, with $Db = p_{23}p_{34} + ...$ and $Da = qp_{12}bp + ...$ (of course here ... includes the term $qp_{13}p_{34}p$ cited above). Now we have:

$$D^2a = 0 = qp_{12}p_{23}p_{34}p + ...$$

So Da has to contain one more term x with $Dx = qp_{12}p_{23}p_{34}p$ and it can be neither of those mentioned above since the face (123) is passive. Hence we obtain a contradiction: $qp_{12}p_{23}p_{34}px^{-1}$ is an active but non-clean circle. The other cases are similar: Figure 3 shows an example with 2 marked arrows on w_0, while for the cases of 1 or 3 such arrows are impossible (see Figure 4 for 3).

We prove now that each arrow a can belong to at most 2 clean circles in S. Again we have to use the positive definiteness of q. Indeed, if a belonged to 3 clean circles and there were no extra arrows between their vertices, one could prove, just as in the case of even circles, that q is not positive definite.

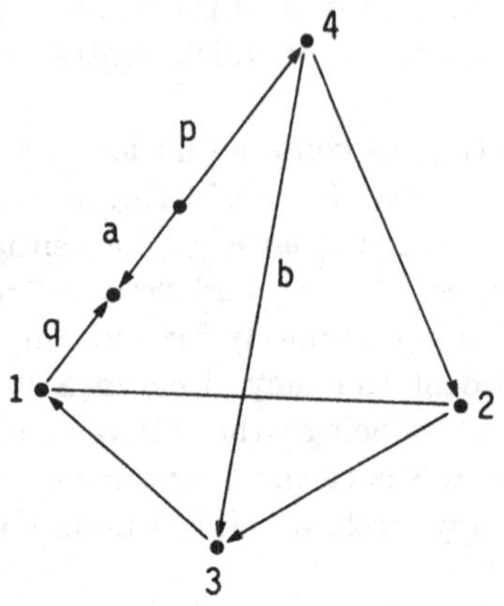

Fig.1

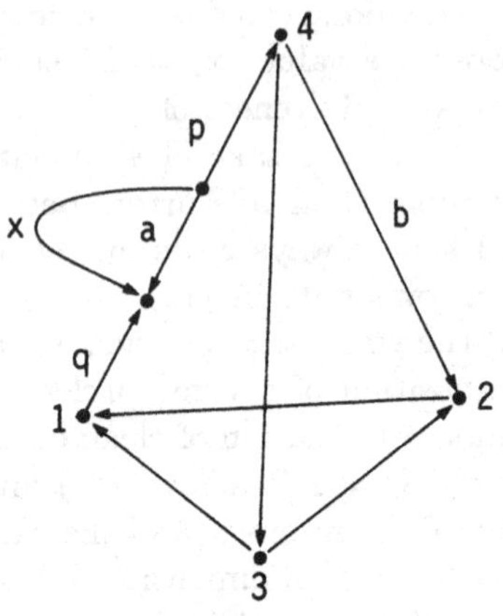

Fig.2

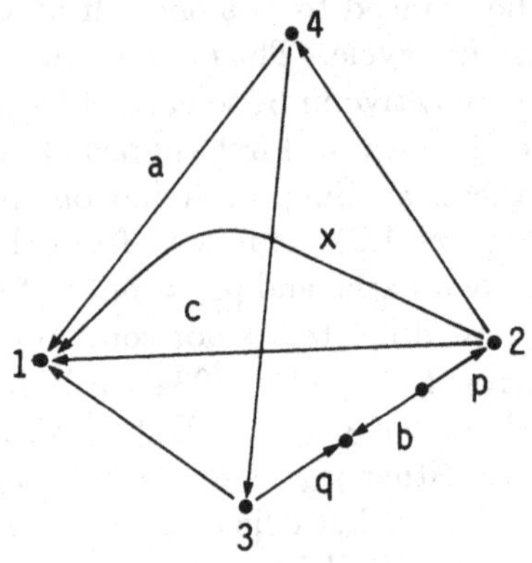

Fig.3

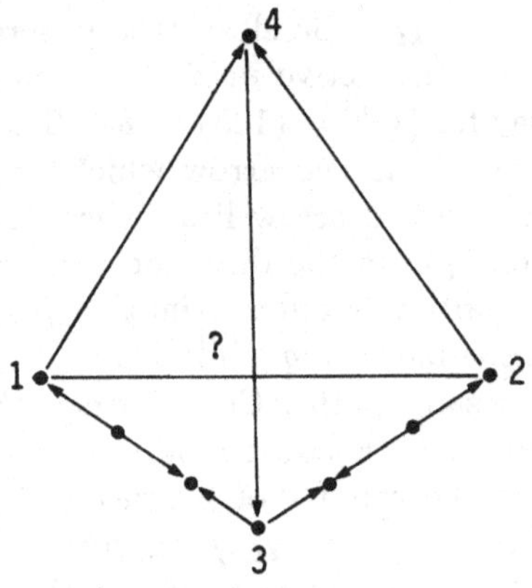

Fig.4

$$Db = qp_{34}p_{42}p + \dots$$

$$Da = p_{13}p_{34} + \dots$$

$$Dc = ap_{42} + \dots$$

$$D^2c = p_{13}p_{34}p_{42} + \dots$$

$\exists\, x \quad Dc = x + \dots$

$\qquad Dx = p_{13}p_{34}p_{42} + \dots$

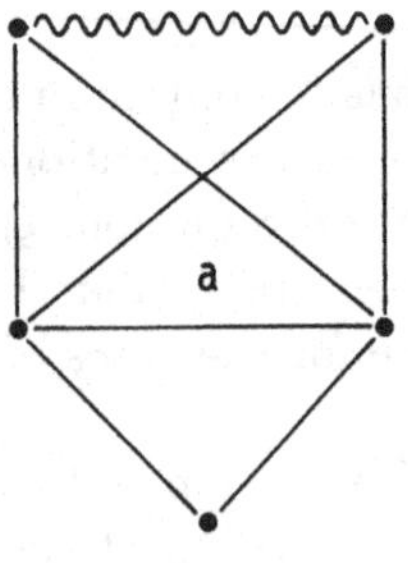

Fig.5

So there is at least one extra arrow, but then we obtain a tetrahedron (see Figure 5).

We now show that if a vertex i lies on two active circles, then the total number of active circles on which it lies is even. Suppose a belongs to 2 circles which are active in S and a is marked in one of them and non-marked in the other. Then the former is xa^{-1} with $Da = x + ...$, the latter is $paqb^{-1}$ with $Db = paq + ...$, whence $D^2b = pxq + ...$ so there is another term y in Db with $Dy = pxq + ...$ and we obtain again a non-clean active circle $pxqy^{-1}$. As any arrow of S lies at least in one active circle (otherwise w_0 would be the boundary of no 2-dimensional chain of S), we are able to speak about arrows marked in S. Note that if we replace all marked arrows by their inverses, all active circles become oriented and if two of them have a common arrow, their orientations are opposite. Let $C_1, ..., C_m$ all be active circles containing the vertex i. We can number them in such a way that C_t and C_{t+1} have a common arrow for $1 \leq t \leq m$ with $C_{m+1} = C_1$. Then C_t and C_{t+1} have opposite orientations and hence m must be even.

Finally, choose a minimal subcomplex S_0 of S containing w_0 and such that $H_1(S_0) = 0$ (with no other restriction). Then we may suppose that any arrow in S_0 lies in two circles, one of which is active and the second either active or w_0: otherwise we could delete the only active circle containing it without changing H_1. But nevertheless any arrow of Q joining two points of S_0 lies in S_0: as it lies in S, it would belong to 3 clean circles of S.

Of course, there is a vertex in S_0 not belonging to w_0. Let $w_1, ..., w_m$ be all clean circles containing this point. Then m is even. Arrange them in such a way that $w_i = q_i p_i q_{i-1}^{-1}$ (replacing 0 by m) where q_i is the maximal common part of w_i and w_{i+1} starting at our fixed vertex. Consider now the road $p_1 p_2 ... p_m = W$ (it is really a circle). It is clean as all p_i are clean and any arrow between vertices of p_i and p_j with $i \neq j$ would provide a tetrahedron. Now use the fact that any active circle is odd. (This may be seen as follows: Any element of the kernel $\overline{V}$ of the bocs $\underline{A}$ has the form, $\sum_t a_t b_t a'_t$, where $a_t, a'_t \in A$ and $b_t \in Q_1$. Thus $\overline{V}$ is spanned by all paths containing precisely 1 dotted arrow and $\overline{V} \otimes_A \overline{V}$ is spanned by the paths containing precisely 2 dotted arrows. Now let $\delta a = p + ...$ If $a \in S_0$, then $\delta a \in \overline{V} \otimes_A \overline{V}$ and pa^{-1} contains 3 dotted arrows). As m is even, it follows that W is even which contradicts positive definiteness and so completes our proof.

Remark. For the bocses arising in the proof of the main theorem, the lemma 6.4 could be obtained from the results of [Go]. I preferred to give an independent proof, first of all because the latter work is hardly accessible and the proofs in it are more complicated, and secondly because it does not imply lemma 6.4 for arbitrary bocses. This seems to be of interest too (e.g. together with [BGOR] it implies that a simply connected bocs with positive Tits form is Schurian, i.e. $\mathrm{End}(M) = K$ for any of its indecomposable representations.)

Appendix: An example

I shall give a simple example showing that the proof of our main theorem really allows us to calculate representations belonging to the open dense subset G^0. Take for A the algebra of all 4×4 matrices of the form:

$$\begin{pmatrix} * & * & * & * \\ 0 & * & 0 & * \\ 0 & 0 & * & * \\ 0 & 0 & 0 & * \end{pmatrix}$$

(any element of K in places marked $*$). Its quiver has the form:

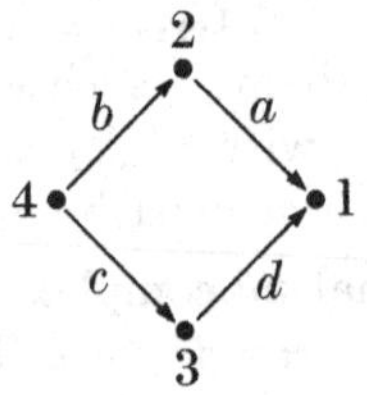

$$ab = dc$$

Let $P = 10P_1 \oplus 7P_2 \oplus 8P_3 \oplus 5P_4$. Then G is a net subgroup of $GL(30)$.

The corresponding bocs has the form:

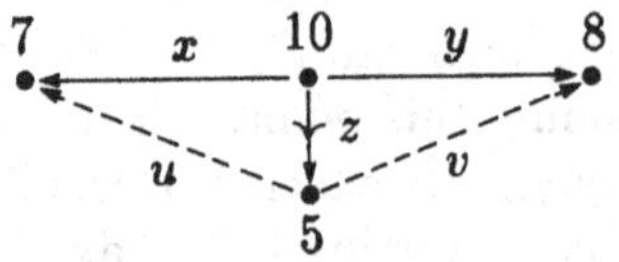

$$Dx = uz$$
$$Dy = \dot{v}z$$

(the dimension of representations corresponding to $\hat{N}$ is prescribed near the vertices). Apply the algorithm. The direction of small reduction at each step is marked on the reduced arrow:

(crossed arrows vanish under regularisation)

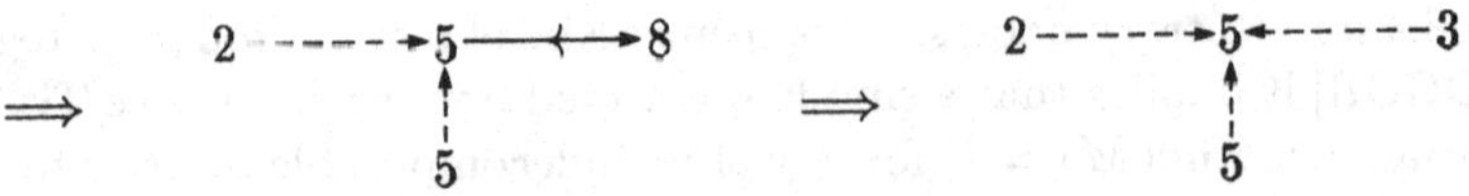

As there are no more solid arrows, we have just calculated $\tilde{G}$: it is $\hat{G}'$ with $G' = G(P', A')$ where A' is described by the last quiver (we need only to replace dashed arrows by solid ones) and P' is defined by the dimensions given at each vertex. So the new bocs is:

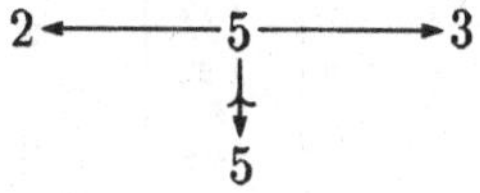

and is reduced as follows:

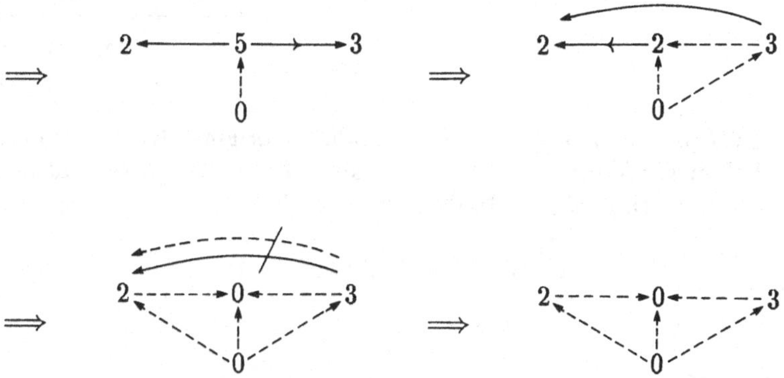

But now the stabiliser is just $D = GL(3) \times GL(2)$, so we have finished: $G^0 \cong \hat{D}$. Going back, we can obtain normal forms for elements $x \in \hat{N}$ and $x' \in \hat{N}'$, namely, they are:

$$x = \begin{pmatrix} I & 0 \\ 0 & 0 \\ \hline I & 0 \\ 0 & 0 \\ \hline 0 & I \end{pmatrix} \qquad x' = \begin{pmatrix} 0 & I \\ \hline I & 0 \\ \hline I & 0 \\ 0 & I \end{pmatrix}$$

(the division corresponding to that of matrices in $\hat{N}$ and $\hat{N}'$ is shown). So we can really describe representations from G^0 corresponding to those of D: we have to consider the subgroup D' of G consisting of all matrices g of the form:

$$\left(\begin{array}{cc|cc|cc|c|cc|c|cc}
g_1 & 0 & X_{11} & X_{12} & Z_{11} & & Z_{12} & Z_{13} & & Z_{14} & Z_{15} & \\
 & g_2 & X_{21} & X_{22} & & & & & & & & \\
\hline
 & & g_1 & 0 & Z_{21} & & Z_{22} & Z_{23} & & Z_{24} & Z_{25} & \\
 & & & g_2 & & & & & & & & \\
\hline
 & & & & g_1 & 0 & Y_{12} & 0 & & 0 & X_{11} & X_{12} \\
 & & & & & g_2 & Y_{22} & & & & X_{21} & X_{22} \\
 & & & & & & g_2 & 0 & & 0 & 0 & 0 \\
\hline
 & & & & & & & g_1 & 0 & Y_{11} & X_{11} & X_{12} \\
 & & & & & & & & g_2 & Y_{21} & X_{21} & X_{22} \\
\hline
 & & & & & & & & & g_1 & 0 & 0 \\
\hline
 & & & & & & & & & & g_1 & 0 \\
 & & & & & & & & & & & g_2
\end{array}\right)$$

with $g_1 \in GL(3)$ and $g_2 \in GL(2)$. Here again the original division of matrices of G and their subdivision in $G^{\cdot}$ are marked. Then the representation of G corresponding to that of D is $\mathrm{Ind}(G, D^{\cdot}, T)$ with

$$T(g) = x(g)T_1(g_1) \otimes T_2(g_2)$$

where

$$T_1 \in \widehat{GL(3)}, \quad T_2 \in \widehat{GL(2)} \quad \text{and}$$
$$x(g) = e(\mathrm{tr}(Z_{11} + Z_{13} + Z_{25}) + \mathrm{tr}(X_{11} + Y_{11}) + \mathrm{tr}(X_{22} + Y_{22}))$$

(e being a non-trivial character of the additive group of K).

References

[AB] Artin, E., Brown, H., *Introduction to Algebraic Topology*, Merril Publ. Co., Columbus, Ohio 1969.

[B] Borevich, Z. I., *Description of the subgroups of the general linear group that contain the group of diagonal matrices*, Zap. Nauchn. Semin. LOMI, 64 (1976) 12-29 (= J. Soviet Math. 17 (1981) 1718-1730).

[BG] Bongartz, K., Gabriel, P., *Covering spaces in representation theory*, Invent. math. 65 (1982), 331-378.

[BGOR] Bondarenko, V. M., Golovashchuk, N. S., Ovsienko, S. A., Roiter A. V., *On Schurian matrix problems*, Preprint, Kiev, Inst. Math. AN UkrSSR, 1980.

[D] Drozd, Yu. A., *Tame and wild matrix problems*, In: Representations and quadratic forms, Kiev, Inst. Math. AN UkrSSR (1979) 39-74 (= Amer. Math. Soc. Transl. (2), 128 (1986) 31-56).

[DK] Drozd, Yu. A., Kirichenko, V. V., *Finite-dimensional Algebras*, Vyshcha Shkola, Kiev, 1980.

[DOF] Drozd, Yu. A. Ovsienko, S. A., Furchin, B. Yu., *Categorical constructions in representation theory*, In: Algebraic structures and their applications, Kiev (1988) 17-43.

[DT] Drozd, Yu. A., Timoshin, A. S., *Representations of net subgroups and their generalisations*, Preprint, Kiev State Univ. 1990.

[G] Gabriel, P., *Unzerlegbare Darstellungen I*, Manus. Math. 6 (1972) 71-103.

[Go] Golovashchuk, N. S., *Universal coverings of a class of matrix problems*, Thesis, Kiev State Univ. 1989.

[H] Happel, D., *Dynkin algebras*, Lecture Notes in Math. 1220, Springer (1986), 1-14.

[K] Kirillov, A. A., *Elements of the Theory of Representations*, Nauka, Moscow (1972) (= Springer, 1976).

[KL] Kleppner, A., Lipsman, R. L., *The Plancherel formula for group extensions*, Ann. Sci. Ecole Norm. Sup. 5 (1972) 459-516.

[L] Lee Sung Ghen, *Representations of Lie groups of step matrices*, Thesis, Kiev State Univ., 1986.

[M] MacLane, S., *Homology*, Springer 1967.

[N] Nazarova, L. A., *Representations of quivers of infinite type*, Izv. Akad. Nauk USSR, 37 (1973) 752-791.

[Ri] Ringel, C. M., *Tame Algebras and Integral Quadratic Forms*, Lecture Notes in Math. 1099, Springer (1984).

[R] Roiter, A. V., *Matrix problems and representations of BOCS's*, Lecture Notes in Math. 832, Springer (1980), 288-324.

[RK] Roiter, A. V., Kleiner, M. M., *Representations of differential graded categories*, Lecture Notes in Math. 488, Springer (1975), 316-336.

[W] Weil, A., *Basic Number Theory*, Springer, 1967.

Kiev State University,
252 017 Kiev
USSR

CLASSIFICATION PROBLEMS FOR MODULAR GROUP REPRESENTATIONS

Karin Erdmann

Oxford

These notes are based on a lecture given at the Workshop preceding the International Conference of Representations of Algebras 1990 at Tsukuba, Japan. The aim is to give a survey over recent progress in the study of blocks of tame representation type. These are the 2–blocks whose defect groups are dihedral or semidihedral or quaternion. Over the last few years, a range of new results on a class of algebras including such blocks have been obtained. The algebras are essentially defined in terms of their stable Auslander–Reiten quivers, and it has been proved that any such algebra is Morita equivalent to one of the algebras in a small list which is explicitly given by quivers and relations. In particular, this describes tame blocks; and it allows to extend classical results on the arithmetic properties of such blocks. We also show how the methods may be applied to the study of blocks of finite type. More generally, we give an outline of the classification of all symmetric algebras which are stably equivalent to Nakayama algebras. This is a different proof of the well–known theorem of Gabriel and Riedtmann.

1. Introduction

Let G be a finite group and K a field of characteristic p; we assume that K is algebraically closed. One approach to modular group representation theory is to study KG, as an algebra, and its module category. The group algebra KG is a direct sum of blocks; where a block is an indecomposable direct summand of KG, as an algebra.

A block has a defect group. We define a defect group of B to be a maximal subgroup D of G such that every module of B is D–projective. (Recall that the KG–module M is D–projective if the multiplication map $M \otimes_{KD} KG \to M$ splits.) It is well–known that the defect groups are always p–groups, and that they form one conjugacy class of subgroups of G. The category mod B is therefore to some extent determined by mod KD; for example, B and KD have the same representation type. It is well–known that the block is of finite type if and only if D is cyclic. In case $p = 2$ and D is a dihedral or semidihedral or quaternion group, the block B is of tame type; and otherwise the representation type of B is wild.

On the other hand, two blocks with the same defect group usually do not have the same algebra structure. Properties of a block can be influenced by various blocks of p–local subgroups. An example for this is the existence of 3–tubes for tame blocks, see Theorem 2.2 below; or, in general, the dimension of the center of a block.

The study of modular group representations started with a functional approach, here R. Brauer's work is still extremely influential. This approach deals, for example, with matrix representations, characters or functions on the group; and here multiplicities are very important. One main problem here is to determine the character values of all characters belonging to the block (see [A]). Accordingly, defect groups were first defined here, by properties of ordinary characters belonging to a block. This lead to an approach "Fix D and study properties of all blocks B with defect group D". Incidentally, a complete answer to this was achieved for blocks with cyclic defect groups [B_1, D]. Afterwards, blocks with dihedral or semidihedral or quaternion defect groups were studied [B_2, O]; the choice of defect group was done independent of considerations of representation type. Amongst other things, Brauer and Olsson showed that the number of simple modules is at most three; however the results are not as complete as the corresponding theorems for cyclic defect groups. For example, it was not possible to determine decomposition numbers. For other types of defect groups, the situation appears to be more difficult.

After new method in the representation theory of algebras were developed, one was trying to apply them also to group representations. Possibly inspired by Ringel who had classified tame local algebras [Rn_1], Donovan obtained a list of all tame symmetric algebras with two simple modules; and by using arithmetic results of Brauer [B_2] on blocks with dihedral defect groups, he determined which of these algebras could be Morita equivalent to such a block [Do]. In the following, Auslander–Reiten theory, covering techniques and other methods were used with great success for the classification of various types of algebras of finite and tame type. Encouraged by this, one might study the problem:

Fix D, study possible basic algebras for blocks with defect group D.

Ideally, determine a list.

Here we want to give an outline on results concerning this problem, for finite and tame type.

2.On the stable Auslander–Reiten quiver of blocks

Suppose that B is a block with basic algebra Λ; then B and Λ have the same Auslander–Reiten quiver. The algebras B and Λ are symmetric and therefore self–injective, hence the Auslander–Reiten quiver $\Gamma(\Lambda)$ of Λ is the union of $\Gamma_s(\Lambda)$ and the indecomposable projective modules. The Auslander–Reiten translation is isomorphic to Ω^2. Moreover, let θ be a connected component of $\Gamma_s(\Lambda)$; then for K algebraically closed, the tree class of θ is either one of the infinite trees A_∞, A_∞^∞, D_∞; or else it is one of a few finite Dynkin diagrams and Euclidean diagrams. This has been proved in [W] and also in [O], by exploiting [HPR].

One would like to know which of these trees occur and where. As far as finite type is concerned, one has:

2.1 THEOREM *Suppose B is a block of finite type then*

$\Gamma_s(B) \cong \mathbb{Z}A_n/<\tau^q>$.

We note that no other Dynkin diagrams occur as the tree class of an AR–component of a block. For blocks of tame type, the structure of $\Gamma_s(B)$ has also been determined.

2.2 THEOREM *Suppose B is a block of tame type, with defect group D. Then the components of* $\Gamma_s(B)$ *are as follows:*

D	dihedral	semidihedral	quaternion
tubes	rank 1, 3	rank ≤ 3	rank ≤ 2
	$\leq$ two 3–tubes	$\leq$ one 3–tube	
others	$\mathbb{Z}A_\infty^\infty/\Pi$	$\mathbb{Z}A_\infty^\infty$, $\mathbb{Z}D_\infty$	–

In the case when D is dihedral, the tree classes of non–periodic components is $\tilde{A}_{12}$, $\tilde{A}_5$ or A_∞^∞.

The proof is not easy, except in the quaternion case. First, one reduces to group algebras of small groups; here results of [K_1, K_2] are extremely useful. The main work then is to study KD and group algebras of a few groups such as A_4 or S_4. For the dihedral case, one can use results on the stable AR–quiver of special biserial algebras [BS, BR, DS, others]. In the semidihedral case, the problem of determining the tree class of non–periodic components was unsolved for some time. This could only be finished after the indecomposable representations of the local semidihedral algebra had been classified [CB]. Using his list of indecomposable representations, one can show, without calculating AR–sequences, that there is an additive function on non–periodic components which is uniformly bounded. This excludes A_∞^∞; and the proof is complete, by applying the general structure theorems on $\Gamma_s(B)$ [W, Li]. In the quaternion case, the proof is easy due to homological results from group theory. The variation in the number of 3–tubes arises from blocks of 2–local subgroups. One has to take into account tame blocks of groups $N_G(V)$ where V is a Klein 4–group. Such blocks are Morita equivalent to either KV or KA_4 or KS_4. These are well–known special biserial algebras, with no or two or one 3–tube. For details, we refer to [E_2].

We note that for blocks it is natural to consider the stable AR–quiver (and not the whole AR–quiver), since one uses Green correspondence and control over projective modules is lost.

3. An approach to classification problems

Suppose B is a block of finite type. It is well–known that B is stably equivalent to a (connected) Nakayama algebra (a block of a group algebra $KN_G(D)$ where D is a group of order p). Recall that blocks are symmetric algebras.

In [GR], Gabriel and Riedtmann classified, more generally, all symmetric algebras which are stably equivalent to some Nakayama algebra. Now, stable equivalence preserves the stable Auslander–Reiten quiver; and it is well–known that the stable AR–quiver of a connected Nakayama algebra is isomorphic to $\mathbb{Z}A_n/<\tau^q>$.Hence the work in [GR] consists essentially of

the classification of all symmetric algebras Λ *with* $\Gamma_s(\Lambda) \cong \mathbb{Z}A_n/<\tau^q>$.

Later, C. Riedtmann classified more generally all self–injective algebras whose stable AR–quiver has tree class A_n, without any further condition [Ri].

Inspired by this approach, we try to follow a similar strategy with the aim of classifying blocks of tame type. Let Λ be a symmetric algebra, over some algebraically closed field of arbitrary characteristic. Consider the following three problems:

Classify all symmetric algebras Λ *with* $\Gamma_s(\Lambda)$ *as in Theorem* <u>2.2</u> *whose Cartan matrix is non–singular.*

We say that such an algebra Λ is of dihedral or semidihedral or quaternion type, according to the structure of $\Gamma_s(\Lambda)$. The results are as follows:

An algebra Λ *is of dihedral type if and only if* Λ *belongs to a small list of algebras which is explicitly given by quiver and relations.*

Suppose Λ *is of semidihedral or quaternion type and in addition assume that* Λ *is tame. Then* Λ *belongs to a small list of algebras which is explicitly given by quiver and relations.*

A complete list of these algebras may be found in $[E_2]$; we will comment on the results later. The main idea in this work is to exploit ends of components in order to study small modules closely related to projectives. To do this, general criteria were found for certain modules which ensure that they have only one predecessor in $\Gamma_s(\Lambda)$. These results actually hold for more general algebras. In particular, they can also be applied for finite type; this gives a different proof for [GR], in which the approach is similar to that for the dihedral type. We will outline the classification for finite type and also for dihedral type, emphasizing the common features.

For finite type, the fact needed, apart from the structure of $\Gamma_s(\Lambda)$, is the following.

Assume Λ is stably equivalent to a connected Nakayama algebra. Then

3.1 *Suppose M, X are indecomposable Λ-modules, and assume that X lies at the end of $\Gamma_s(\Lambda)$. Then*

(a) $\underline{Hom}_\Lambda(M, X) \neq 0$ if and only if a diagonal through M ends at X.

(b) $\underline{Hom}_\Lambda(X, M) \neq 0$ if and only if a diagonal through M starts at X.

Moreover, if there is a diagonal from M to some module M' then $\underline{Hom}_\Lambda(M, M') \neq 0$.

It is straightforward to prove that a Nakayama algebra has these properties, and they are preserved under stable equivalence.

The condition that the Cartan matrix is non–singular holds for arbitrary blocks and is a natural hypotheses in group representation theory. Without this condition, there are too many algebras. For example, the number of simple modules is not bounded, see 4.3.

4. Ends of components and projective modules

We will first describe the general results on ends of components. Let Λ be an algebra such that $\Gamma_s(\Lambda)$ is known, and let P be an indecomposable projective Λ–module. We would like to know how P looks like, but P does not belong to the stable AR–quiver! Instead, we consider modules which are closely related to P. For appropriate submodules X of P, we study conditions which ensure that P/X lies at the end of some component of $\Gamma_s(\Lambda)$.

4.1 THEOREM [E_2, BR] *P/X lies at the end if one of the following holds:*

(1) Λ is self–injective and rad P/ soc P $\cong$ X/soc P $\oplus$ U (both $\neq$ 0).

(2) Λ is self–injective and there is an AR–sequence

$0 \to X/\mathrm{soc}\,P \to \mathrm{rad}\,P/\mathrm{soc}\,P \to U \to 0$ (and X/soc P, U not simple).

(3) Λ arbitrary and $X = \alpha\Lambda$ where α is an arrow.

The situation of (2) occurs at ends of components with tree class D_∞ (or $\tilde{D}_n$ or D_n). We note that a consequence of (3) is the fact that the AR–quiver of an arbitrary algebra has "ends", a Theorem which was first proved in [M].

4.2 *In the following, we assume that Λ is symmetric, basic and connected, and either*

(F) $\Gamma_s(\Lambda) \cong \mathbb{Z}A_n/\langle\tau^q\rangle$, and Λ satisfies 3.1, or

(D) Λ is of dihedral type; that is, $\Gamma_s(\Lambda)$ has the structure as in 2.2 for the dihedral case, and the Cartan matrix of Λ is non–singular.

The aim is to determine the quiver and relations for Λ.

4.3 THEOREM *In 4.2(F), Λ has q simple modules. In 4.2(T), Λ has at most 3 simple modules.*

Proof: Consider first the dihedral case, and suppose Λ has more than one simple module. The quiver is connected, hence it contains some arrow $e \xrightarrow{\alpha} f$ where $e \neq f$. By [BR], the module $\alpha\Lambda$ lies at the end of some component. Now, by the hypothesis here, the stable AR–quiver of Λ consists only of tubes and components in which each vertex has two predecessors, and therefore the only ends occur at tubes. It follows that $\alpha\Lambda$ lies at the end of some tube. Suppose this is a tube of rank 1, then $\Omega^2(\alpha\Lambda) \cong \alpha\Lambda$, and there is an exact sequence $0 \to \alpha\Lambda \to P_1 \to P_0 \to \alpha\Lambda \to 0$ where P_0 and P_1 are projective. By exactness, $\underline{\dim}\, P_0 = \underline{\dim}\, P_1$, and since the Cartan matrix of Λ is assumed to be non–singular, it follows that $P_0 \cong P_1$. Now, $P_0 \cong f\Lambda$ and $P_1 \cong e\Lambda$ and $e = f$, a contradiction. Hence $\alpha\Lambda$ lies at the end of a 3–tube and therefore there are at most six such arrows; which limits the size of a connected quiver. By refinements of this type of arguments one obtains the bound 3, and this is best possible as examples show. For details we refer to [E_4]; and the proofs for the other two tame types are given in [E_1] and [E_2].

Now consider the finite type. The result is well–known; but we include a short proof which is taken from [Ri]. Let $\mathscr{C} := \{\ \Omega S$: S is a simple Λ–module$\}$; it suffices to show that

(*) *Each diagonal in $\Gamma_s(\Lambda)$ contains one module in $\mathscr{C}$.*

Since every module lies in two diagonals and since there are 2q diagonals in $\Gamma_s(\Lambda)$, it follows form (*) that $|\mathscr{C}| = q$.

Let W be any indecomposable non–projective Λ–module. Then there is some X in $\mathscr{C}$ such that Hom (W, X) $\neq$ 0 (For example, take X = ΩS where S occurs in top W). In particular, if W lies at the end of some diagonal Δ then by 3.1, any such X must lie in Δ; and this shows that each diagonal contains some member of $\mathscr{C}$. Suppose X, Y are in $\mathscr{C}$ and that X and Y lie in the same diagonal. Then by 3.1, Hom (X, Y) $\neq$ 0 (say), and then Hom (Ω^{-1}X, Ω^{-1}Y) $\neq$ 0; and since Ω^{-1}X, Ω^{-1}Y are simple, it follows that X $\cong$ Y.

Remark: The theorem does not hold without the hypothesis that the Cartan matrix is non–singular. For example, consider a special biserial algebra with quiver

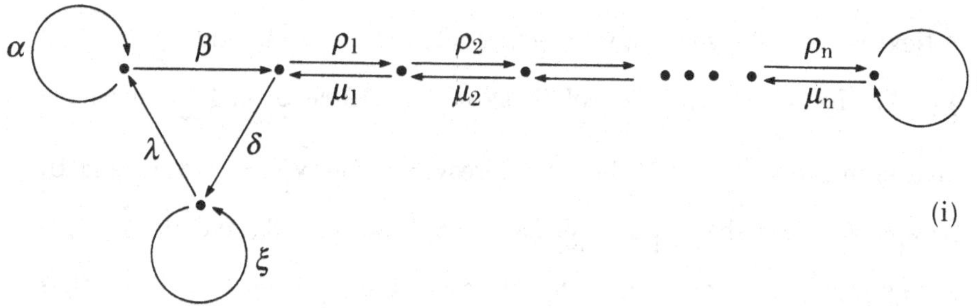

(i)

and where the relations are given as follows. Let π = $(\beta\ \delta\ \lambda)(\alpha)(\xi)(\mu_1\ \rho_1)(\mu_2\ \rho_2) \dots (\mu_n\ \rho_n)(\epsilon)$ and let also $\pi^* = (\alpha\ \beta\ \rho_1\ \rho_2 \cdots \rho_n\ \epsilon\ \mu_n\ \mu_{n-1} \cdots \mu_1\ \delta\ \xi\ \lambda)$. Then all products of two successive arrows in π are zero. For any arrow γ, define ω_γ to be the product over all arrows in π^* starting with γ. Then for some k $\geq$ 1,

$$\omega_\alpha{}^k = \omega_\beta{}^k, \omega_\lambda{}^k = \omega_\xi{}^k, \omega_\delta{}^k = \omega_{\rho_1}{}^k, \omega_{\mu_i}{}^k = \omega_{\rho_{i+1}}{}^k, \omega_{\mu_n}{}^k = \omega_\epsilon{}^k.$$

Then the algebra Λ is symmetric, and $\Gamma_s(\Lambda)$ has the structure as in 4.2(D); but the Cartan matrix is singular.

We denote by P_0, P_1, ... or $e_0\Lambda$, $e_1\Lambda$, ... the indecomposable projective Λ–modules and by S_0, S_1, ... the corresponding simple modules.

4.4 *Let $H_i = e_iJ/S_i$. Then H_i is a direct sum $H_i = U_i \oplus V_i$ with U_i indecomposable, and V_i indecomposable or zero; and $U_i \ncong V_i$. Moreover, $V_i = 0$ if and only if e_iJ lies at the end of $\Gamma_s(\Lambda)$.*

Proof: Consider the AR–sequence of $e_i\Lambda/S_i$; the middle is $H_i \oplus e_i\Lambda$, and the summands of H_i are the predecessors of $e_i\Lambda/S_i$ in Γ_s. By the hypothesis 4.2(F), a vertex in Γ_s has either two distinct predecessors; or it lies at the end and has only one predecessor.

Next we consider some other modules related to these U_i and V_i.

4.5 We fix a decomposition of H_i as in 4.4. Define $\underset{\sim}{U}_i$ and $\underset{\sim}{V}_i$ to be the inverse image if U_i and V_i in $e_i\Lambda$. Moreover, define $\widetilde{V}_i := e_i\Lambda/U_i$ and $\widetilde{U}_i := e_i\Lambda/V_i$. Note that $\underset{\sim}{U}_i$ and $\underset{\sim}{V}_i$ have simple socles $\cong S_i$, and $\widetilde{U}_i$, $\widetilde{V}_i$ have simple tops also isomorphic to S_i. Moreover top $\underset{\sim}{U}_i \cong$ top U_i and soc $\widetilde{U}_i \cong$ soc U_i, similarly for the V_i.

PROPOSITION *Assume 4.2 (F) holds. Let $\mathcal{E} := \{$ X: X lies at the end of $\Gamma_s(\Lambda)\}$. Then $\mathcal{E}$ is invariant under Ω. Moreover*

(i) $\mathcal{E} = \{ \widetilde{U}_i, \widetilde{V}_i \} = \{ \underset{\sim}{U}_i, \underset{\sim}{V}_i \}$.

(ii) The modules U_i and V_i have simple socles and tops.

(iii) We have that top $U_i \ncong$ top V_i and soc $U_i \ncong$ soc V_i.

Proof: For self–injective algebras, Ω induces a graph isomorphism of Γ_s; in particular it preserves ends. By 4.1, the modules $\widetilde{U}_i$ and $\widetilde{V}_i$ have one

predecessor in $\Gamma_s(\Lambda)$, that is, they belong to $\mathcal{E}$. On the other hand, we claim that $\#\{\widetilde{U}_i, \widetilde{V}_i\} = 2q$ which is also the number of elements of $\mathcal{E}$.
If $i \neq j$, then $\widetilde{U}_i$ and $\widetilde{U}_j$ or $\widetilde{V}_j$ have different tops and are therefore non–isomorphic. Moreover, since $U_i \ncong V_i$ we deduce rad $\widetilde{U}_i \cong U_i \ncong V_i \cong$ rad $\widetilde{V}_i$ and hence $\widetilde{U}_i \ncong \widetilde{V}_i$. Similarly one obtains the other part of (i); alternatively, one may apply Ω.

Each module in $\{ \underset{\sim}{U}_i, \underset{\sim}{V}_i\}$ has a simple socle, and hence soc $\widetilde{U}_j$ is also simple; and this is soc U_j, hence (ii) holds. Part (iii) is clear since Λ is of finite type.

Now consider an algebra of dihedral type. The analogue result is the following:

PROPOSITION *Assume 4.2 (D) holds. Let $\mathcal{E} := \{ X$: X lies at the end of some 3–tube$\}$. Then $\mathcal{E}$ is invariant under Ω. Moreover*

(i) If top $V_i \ncong S_i$ then $\widetilde{U}_i$ and $\underset{\sim}{V}_i$ belong to $\mathcal{E}$; and if top $U_i \ncong S_i$ then $\widetilde{V}_i$ and $\underset{\sim}{U}_i$ belong to $\mathcal{E}$.

(ii) The modules U_i, V_i have simple socles and tops.

(iii) If Λ is not local then top $U_i \ncong$ top V_i and soc $U_i \ncong$ soc V_i, for any i.

Proof: (1) First, we claim that for a module X with $\Omega^2(X) \cong X$ we must have that top X $\cong$ top $\Omega(X)$ and soc X $\cong$ soc $\Omega(X)$: If $\Omega^2(X) \cong X$ then there is an exact sequence $0 \to X \to Q \to P \to X \to 0$ where P and Q are projective, with top P $\cong$ top X and top Q $\cong$ top $\Omega(X)$; and also soc X $\cong$ soc Q and soc $\Omega(X)$ $\cong$ soc P. By the exactness, P and Q have the same composition factors; and since the Cartan matrix is assumed to be non–singular, we deduce that P $\cong$ Q.

(i) By 4.1, we know that $\widetilde{U}_i$ lies at the end of some component. The only ends for Λ as in (D) occur at tubes; in particular $\widetilde{U}_i$ is periodic. Suppose $\widetilde{U}_i$ does not belong to $\mathcal{E}$, then $\widetilde{U}_i$ lies at the end of a 1–tube. Apply (1) with X $= \widetilde{U}_i$, then top $\Omega(X) =$ top $\underset{\sim}{V}_i =$ top V_i. Moreover, if $\widetilde{U}_i$ belongs to $\mathcal{E}$ then so does $\underset{\sim}{V}_i$ since $\underset{\sim}{V}_i \cong \Omega(\widetilde{U}_i)$.

The proofs of (ii) and (iii) are given in [E_2, VI.4.3 and VI.3.1]. They also use the results described in 4.1 and arguments as in (1) above.

Next, we use the following results from [E_2] which hold more generally.

4.6 *Assume A is a symmetric basic algebra satisfying the following conditions:*

(I) For each vertex f of Q, the number of arrows starting at f equals the number of arrows ending at f and this number is 1 or 2.

(II) If two arrows start at f then rad $f\Lambda$/soc $f\Lambda$ is a direct sum of two modules.

In [E_2, VI.1] we proved the following results for such an algebra: First, we have

LEMMA *Let f be a vertex of Q. Then for some choice of arrows α, β ending at f and γ, δ starting at f we have that $\Omega(\alpha A)$/soc $fA \cong \gamma A$/soc fA and $\Omega(\beta A)$/soc $fA \cong \delta A$/soc fA.*

Here we identify $\Omega(\alpha A)$ wirh the annihilator $\{ x \in fA: \alpha x = 0 \}$. One chooses arrows which give rise to direct sum decompositions of rad fA/soc fA, according to (II). Then one shows, by comparing lengths, that also rad fA $= \Omega(\alpha A) + \Omega(\beta A)$; and clearly $\Omega(\alpha A) \cap \Omega(\beta A) =$ soc fA. The statement follows now by the Krull–Schmidt theorem.

This allows to define a permutation which describes the Ω–action on modules of the for $\Omega(\alpha A)$ as follows.

4.6.1 For a vertex f at which only one arrow starts we introduce a "loop of length zero" ω_f by choosing for ω_f a generator of soc fA. Denote by Q' the set of these ω_f. Now we define a permutation π on the set $Q_1 \cup Q_1'$ by setting

$\pi(\alpha) = \gamma$ where $\Omega(\alpha A)/\text{soc } fA \cong \gamma A/\text{soc } fA$.

That is, if $\alpha = \omega_f$ then $\Omega(\alpha A)$ is generated by an arrow; and if $\Omega(\alpha A)$ is simple then $\gamma = \omega_f$. This is a permutation, by [E_2, VI.1.2].

This gives rise another permutation on the set of arrows, namely set

$\pi^*(\alpha) = \delta$ if $\pi(\alpha) = \gamma$ (or ω_f).

PROPOSITION *The module αA is uniserial; and the succession of composition factors of αA is determined by the cycle of* π^* *containing α. Moreover, there is an integer m which depends only on σ such that for each e* $\in Q_0$, *dim* $\alpha\Lambda e = m.\#\{\gamma \text{ in } \sigma\text{: } e\gamma = \gamma\}$. *Hence all indecomposable projective A–modules are biserial.*

4.6.2 *The Brauer graph G(A) for A:* We define this to be the graph whose vertices are the cycles of π^*, including loops of length zero; and the edges are the primitive idempotents of A, or equivalently the simple A–modules. The edge e is joined to the cycle σ if for some γ in σ we have $e\gamma = \gamma$. The hypothesis (II) ensures that each edge is joined to two vertices in G(A).

4.7 In 4.5 we have proved that algebras with 4.2 (F) satisfy (I) and (II), hence we may apply 4.6; in particular we deduce that the projective modules of Λ are biserial. We will now completely determine quiver and relations for these algebras. The results in 4.5 show that there are many zero relations. The permutation π consists of one cycle of length 2q, and our aim is to prove that two successive elements in it have product zero.

The modules in 4.5 have simple socles and tops, therefore they are of the form $\omega\Lambda$ for $\omega \in e\Lambda f$, for primitive idempotents. Recall that for such a module, we identify $\Omega(\omega\Lambda)$ with $\{ x \in f\Lambda\colon \omega x = 0\}$; this gives a zero relation in case $\Omega(\omega\Lambda)$ is generated by an arrow. Moreover, we have enough modules available for finding all arrows.

There must be some vertex of $\mathcal{Q}$, e_0 say, at which only one arrow starts: Otherwise, the separated quiver of Λ/J^2 would contain A_n for some n, and Λ/J^2 and Λ would be of infinite type. Then, with the notation of 4.5, we have that $\underset{\sim}{U}_0 \cong \text{rad } e_0\Lambda$. Choose generators of the Ω–translates of $\underset{\sim}{U}_0$ as follows:

$$\Omega^{2k}(\underset{\sim}{U}_0) = \alpha_k\Lambda \text{ and } \Omega^{2k+1}(\underset{\sim}{U}_0) = \beta_k\Lambda.$$

LEMMA *Each of the elements α_k, β_k is either an arrow, or a generator of soc $e_i\Lambda$; and the latter happens in the case when $\alpha_k\Lambda$ (or $\beta_k\Lambda$) is isomorphic to $\underset{\sim}{V}_\nu$ with $V_\nu = 0$. The set $\{\alpha_i, \beta_i\}$ generates rad Λ. If e is a vertex and α_i ends at e then β_i starts at e, and if β_i ends at e then α_{i+1} starts at e. Moreover*

$$(*) \quad \alpha_i\beta_i = 0 \text{ and } \beta_i\alpha_{i+1} = 0, \text{ for } 0 \le i \le q-1.$$

In the case when some α_k or β_k lies in soc Λ we will call the element an arrow of length zero.

To prove the lemma, let $\gamma \in \{ \alpha_i, \beta_j\}$; then $\gamma\Lambda \cong \underset{\sim}{U}_\nu$ or $\underset{\sim}{V}_\nu$, and since

$e_\nu J/\mathrm{soc}\, e_\nu\Lambda \cong U_\nu \oplus V_\nu$, either γ is an arrow or in case $\gamma\Lambda \cong \underset{\sim}{V}_\nu$ and $V_\nu = 0$ we have that $\gamma\Lambda \subseteq \mathrm{soc}\,\Lambda$. Moreover, $e_\nu J = \underset{\sim}{U}_\nu + \underset{\sim}{V}_\nu$ for each ν, and therefore $\{\alpha_i, \beta_j\}$ generate J. The length of J/J^2 is equal to the number of U_ν, V_ν which are non–zero, by 4.5, and this is the same as the number of elements of α_i, β_j which do not lie in soc Λ.

Note that also $\beta_{q-1}\alpha_0 = 0$: We started with $\underset{\sim}{U}_0 \cong \Omega(S_0)$, and then β_{q-1} is a generator of soc $e_0\Lambda$. In the cycle $(\alpha_0\ \beta_0\ \alpha_1\ \beta_1\ \cdots\ )$, each vertex of Q occurs twice as starting point of some α_i or β_j. Moreover, the arrows can be divided into an "α–camp" and a "β–camp" [GR]:

LEMMA *Suppose f is a vertex of Q. Then we have*

(a) There is a unique i such that $\alpha_i = \alpha_i f$, and there is a unique j such that $\beta_j = \beta_j f$.

(b) There is a unique k such that $\alpha_k = f\alpha_k$, and there is a unique l with $\beta_l = f\beta_l$.

In particular, Λ is special biserial.

Proof: Part (b) follows from (a), with $k = j+1$ and $l = i$. Consider the position of $fJ = \mathrm{rad}\, f\Lambda$ in $\Gamma_s(\Lambda)$; and suppose the diagonals ending at fJ start at X and Z. By <u>3.1</u> we have that X and Z are characterized amongst the modules in $\mathscr{E}$ by the property that $\underline{\mathrm{Hom}}_\Lambda(X, fJ) \neq 0$ and $\underline{\mathrm{Hom}}_\Lambda(Z, fJ) \neq 0$. Now, X and Z lie in the Ω–orbit of $\underset{\sim}{U}_0$ but in different τ–orbits. Therefore, say, $X \cong \Omega^{2j}(\underset{\sim}{U}_0) = \alpha_j\Lambda$ and $Z \cong \Omega^{2k+1}(\underset{\sim}{U}_0) = \beta_k\Lambda$. On the other hand, if $fJ/\mathrm{soc}\, f\Lambda \cong U \oplus V$ then $\underline{\mathrm{Hom}}_\Lambda(\underset{\sim}{U}, fJ) \neq 0$, since there is an inclusion map, and similarly $\underline{\mathrm{Hom}}_\Lambda(\underset{\sim}{V}, fJ) \neq 0$, and $\underset{\sim}{U}$ and $\underset{\sim}{V}$ belong to $\mathscr{E}$. This implies that the modules $\underset{\sim}{U}$, $\underset{\sim}{V}$ coincide with $\alpha_j\Lambda$, $\beta_k\Lambda$.

By 4.6 the projectives are biserial, and the succession of composition factors is given by the α–cycles and the β–cycles. Each cycle has a multiplicity, which only depends on the cycle. It is easy to see that at most one of the cycles of π^* can have a multiplicity > 1 if the algebra is of finite type.

Similarly, it is not difficult to show that in the case when Λ is of finite type that then the Brauer graph $G(\Lambda)$ which we defined in 4.6 must be a tree (except possibly for loops). This implies that the quiver of Λ can be drawn flat, or equivalently, that the quiver together with the arrows of length zero is the Brauer quiver introduced in [GR]. We note that in case Λ is Morita equivalent to a block then the Brauer graph is the same as the Brauer tree; and the exceptional vertex corresponds to the cycle of π^* with multiplicity > 1 if it exists.

The quiver of Λ is formed by q vertices and the α_i, β_j which are of lenght > 0. Then the zero relations in 4.7(*) together with the obvious socle relations give a complete set of relations for Λ. One obtains the list of algebras, as described in [GR].

4.8 Now consider an algebra Λ of dihedral type, that is, an algebra which satisfies 4.2(D). We wish to determine the quiver and relations for Λ. By 4.3, the quiver has at most three vertices. Moreover, we have seen in 4.5 that Λ satisfies the hypotheses (I) and (II) of 4.6. Hence we deduce from 4.5 that the indecomposable projective Λ–modules are biserial. There is also the permutation π as defined in 4.5, associated to Λ. We claim that π consists of either $\leq$ two 3–cycles, or of one 6–cycle and otherwise only fixed loops; unless, possibly, Λ is local: Let α: e $\rightarrow$ f be an arrow and e $\neq$ f. Then with the notation as in 4.5, $\alpha\Lambda = \underset{\sim}{U}$, say. Moreover, top $\underset{\sim}{U} \cong S_f \not\cong S_e$. Hence by 4.5 we have that V and U belong to $\mathcal{E}$. That is, α occurs in a cycle of π of length 3 or 6.

Now one determines all connected quivers with $\leq$ three vertices which satisfy 4.6(I), excluding the ones which only lead to algebras of finite type, such as $\bullet \rightleftarrows \bullet$.

There are only two such quivers with two vertices, namely

(2$\mathcal{A}$) 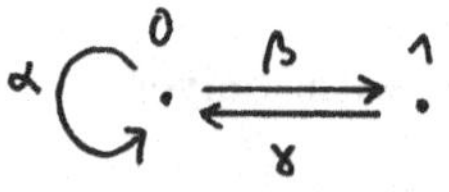

(2$\mathcal{B}$) 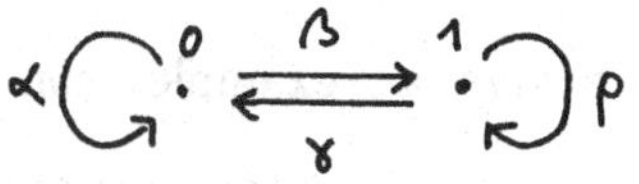

For three vertices, one finds that there are seven possible quivers, see [E_2, VI.5].

Next, one writes down all possible permutations π which consist of either $\leq$ two 3–cycles, or one 6–cycle and otherwise only fixed loops. To give some examples, let $\mathcal{Q}$ be of the form (2$\mathcal{A}$).

Then e_1 must occur in a non–trivial cycle, and hence the only possibility for π is $(\beta\ e_1\ \gamma)(\alpha)$. Similarly, there is only one possible π for (2$\mathcal{B}$); and for the quiver

(3$\mathcal{Z}$) 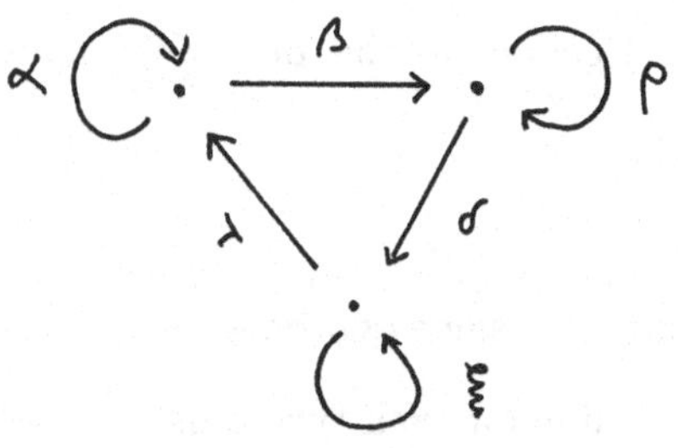

one obtains two possible permutations, namely $\pi_1 = (\beta\ \delta\ \lambda)(\alpha)(\rho)(\xi)$ and $\pi_2 = (\alpha\ \beta\ \rho\ \delta\ \xi\ \lambda)$.For each π, we write down the corresponding π^*. In some cases, we see at once that any algebra with (I), (II) where this occurs has a singular Cartan matrix. For example, take π_1 for the quiver (3$\mathcal{Z}$); then the corresponding π_1^* consists of one 6–cycle, and by the Proposition in 4.6 we see that for each i, the composition factors of $e_i\Lambda$ are the same.

For each of the quivers and each remaining π, one determines now relations. Note that the definition of π shows that for some choice, at each vertex there are some zero relations; but one has to choose arrows cosistently. The quiver is very small, and this can be done, except that problems can arise at loops but only when the field has characteristic 2.

To give an example, consider algebras Λ with quiver $(3R)$ and permutation π_2. Here the relations are as follows.

Any two successive arrows occuring in π_2 *have product zero. There are integers k, s, t, u with* $k \geq 1$ *and* $s, t, u \geq 2$ *such that* $\alpha^s = (\beta\delta\lambda)^k$, $\rho^t = (\delta\lambda\beta)^k$ *and* $\xi^u = (\lambda\beta\delta)^k$.

The modules $\omega\Lambda$ for ω in π_2 form an Ω–orbit of length 6, and hence Ω permutes the 3–tubes.

Originally, the definition "dihedral type" also included the condition that the 3–tubes should be permuted by Ω. This has now been omitted; and there are some more algebras. To classify these as well, creates no difficulty. One of the new algebras is the one with quiver $(3\mathcal{I})$ and the second possibility for π.

5. Tame type, results and some questions

As a result, one obtains for each type a list of algebras, explicitly given by quiver and relations. Each of these lists consists of a few families; they may be found in $[E_2]$.

Comparing these lists wc make the following observations.

5.1 Consider first algebras of dihedral type. For any algebra Λ in the list,

$\Lambda/\text{soc}\ \Lambda$ is special biserial. In particular, this implies that Λ is tame. If Λ itself is not special biserial then char K = 2, and there is a loop α such that α^2 is a non–zero element in soc Λ. If this happens then there are two isomorphism types of algebras whose socle factors are isomorphic.

Now consider algebras Λ of semidihedral or quaternion type; then in most cases $\Lambda/\text{soc}_2\Lambda$ is special biserial. However, there are exceptions, for example the quiver

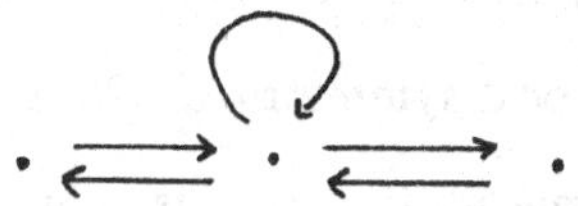

occurs. We have organized the lists, such that the algebras belonging to one family have the same quiver and there are integer parameters which determine the multiplicities. There are sometimes also scalar factors occuring in socle relations. For a fixed family and fixed multiplicities, in most cases the algebra $\Lambda/\text{soc}\ \Lambda$ is uniquely determined. However, it happens that $\Lambda/\text{soc}\ \Lambda$ is uniquely determined but for Λ there are infinitely many algebras. An example for this are the algebras of quaternion type where the quiver is of the form (2A), and the relations are

$$\gamma\beta\gamma = (\gamma\alpha\beta)^{k-1}\gamma\alpha,\quad \beta\gamma\beta = (\alpha\beta\gamma)^{k-1}\alpha\beta,\quad \alpha^2\beta = 0,$$
$$\alpha^2 = (\beta\gamma\alpha)^{k-1}\beta\gamma + c(\beta\gamma\alpha)^k \quad \text{where } k \geq 2 \text{ and } c \in K.$$

If char K $\neq$ 2 then one may assume c = 0. On the other hand, for char K = 2 there are infinitely many non–isomorphic algebras.

5.2 We will now compare the algebras of the three different types. We fix one of the quivers occuring. Then there is in many cases a one–one correspondence between these families per type, such that

(i) Algebras in corresponding families are isomorphic modulo the second socle, if the ground field is the same, and

(ii) Algebras in corresponding families have the same Cartan matrix, and the centre has the same dimension. In fact, the centres are isomorphic as rings whenever the quiver and the multiplicities are the same.

For example, consider algebras with one simple module. In [Rn_1], Ringel determines more generally a list of all tame local algebras; and in [E_2, III] there is a list of all tame local symmetric algebras. In each case, Λ is of the form $K\langle X, Y\rangle/I$ for some ideal I. If Λ is commutative then $\Lambda/\mathrm{soc}\ \Lambda$ is special biserial. Supoose Λ is non–commutative, then $\mathrm{soc}_2\Lambda$ contains X^2 and Y^2, and hence $\Lambda/\mathrm{soc}_2\Lambda$ is special biserial. Moreover, soc Λ = $<(XY)^k>$, and we must have that $(XY)^k = (YX)^k$ since Λ is symmetric. The Cartan matrix of Λ is here [dim Λ] = [4k], and the centre of Λ has dimension k+3.

For the case k+3 = 5 there is a connection with work of [CK]. They study arbitrary local symmetric algebras whose centre has dimension 5, and they prove that any such algebra which is non–commutative must have dimension 8.

We observe also that any tame local symmetric algebra is either of dihedral type, or of semidihedral type, or of quaternion type. Consider more generally self–injective tame algebras. We ask whether there are larger classes of algebras to which the algebras of semidihedral and quaternion type belong, comparable with the class of special biserial algebras which contains algebras of dihedral type (and also blocks of tame type) such that many tame symmetric algebras belong to these classes. For example, to generalize the semidihedral type, the non–periodic components of $\Gamma_s(\Lambda)$ should have tree classes D_∞, A_∞^∞ or possibly $\tilde{D}_n$, and may be also D_n, including algebras of finite type.

<u>5.3</u> We consider now converse. As we noted, for an algebra Λ of dihedral type, we have that $\Lambda/\mathrm{soc}\ \Lambda$ is special biserial; and this allows to prove the converse. By applying general results on special biserial algebras, for example [BS, DS, BR], one verifies easily that the stable Auslander–Reiten quiver of any listed algebra has the right structure.

As far as the converse for the other types is concerned, here we would like to know whether the algebras in the lists are tame and have the right Auslander–Reiten quiver. This is only answered in some cases:

(1) The local semidihedral algebra (using [CB]).

(2) Any algebra Λ such that $\Lambda/\mathrm{soc}\ \Lambda$ is Morita equivalent to $B/\mathrm{soc}\ B$ for some tame block B.

(3) Algebras in some other families, for the smallest multiplicities (by work of A. Skowronski, using covering techniques, see [E_3]).

To make progress, one would like a good description of the indecomposable modules. In particular, the problem of determining a list of indecomposable modules for the local quaternion algebra is still unsolved.

<u>5.4</u> A more general recent result allows also to determine the growth type of algebras of dihedral type and also to decide whether $\Gamma_s(\Lambda)$ has Euclidean components. We have

THEOREM [ES] *Suppose* Λ *is special biserial and self–injective. Then the following are equivalent:*

(1) $\Gamma_s(\Lambda)$ *has a component of the form* $\mathbb{Z}\tilde{A}_{p,q}$.

(2) $\Gamma_s(\Lambda)$ *is infinite but has no component of the form* $\mathbb{Z}A_\infty^\infty$.

(3) There are positive integers m, p, q such that $\Gamma_s(\Lambda)$ *is the union of m components of the form* $\mathbb{Z}\tilde{A}_{p,q}$, *m components of the form* $\mathbb{Z}A_\infty/<\tau^p>$, *m components of the form* $\mathbb{Z}A_\infty/<\tau^q>$ *and infinitely many components of the form* $\mathbb{Z}A_\infty/<\tau>$.

(4) All but finitely many components of $\Gamma(\Lambda)$ *are of the form* $\mathbb{Z}A_\infty/<\tau>$.

(5) Λ *is representation–infinite domestic.*

For a special biserial algebra Λ of infinite type, there is only one alternative, namely

THEOREM [ES] *Suppose* Λ *is special biserial and self–injective, then the following are equivalent:*

(1) $\Gamma_s(\Lambda)$ *has a component of the form* $\mathbb{Z}A_\infty^\infty$.

(2) $\Gamma_s(\Lambda)$ *has infinitely many components of the form* $\mathbb{Z}A_\infty^\infty$.

(3) $\Gamma_s(\Lambda)$ *is a disjoint union of a finite number of components of the form* $\mathbb{Z}A_\infty/<\tau^n>$ *with* $n > 1$, *infinitely many components of the form* $\mathbb{Z}A_\infty/<\tau>$ *and infinitely many components of the form* $\mathbb{Z}A_\infty^\infty$.

(4) Λ *is not of polynomial growth.*

This may be applied to decide whether a given special biserial algebra has a Euclidean component, or equivalently, is domestic. Firstly, one determines what the ranks of tubes are. This is easily done since the ends of tubes of rank > 1 are formed by the modules $e\Lambda/\alpha\Lambda$ for α an arrow, by [BR]. In the case there are more than two integers t > 1 such that Λ has a tube of rank t then Λ is not domestic and does not have Euclidean components. Suppose there are at most two integers $t > 1$ such that Λ has tubes of rank t. Say these integers are p, q. Then one calculates in some non–periodic component a part of the form

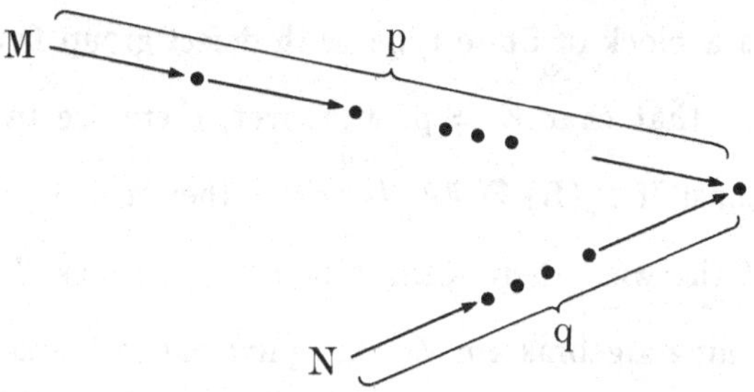

(ii)

This is easy using the algorithm for calculating irreducible maps [BR]. If $M \cong N$ then Λ is domestic and has only finitely many non–periodic component. Otherwise, there are infinitely many non–periodic components, and Λ does not have Euclidean components.

In particular, let Λ be of dihedral type. The only possible Euclidean components are $\tilde{A}_{1,3}$ or $\tilde{A}_{3,3}$; and it is easy to determine the ones which are domestic. The results may be found in [ES].

This appears to have some analogue for other tame algebras: In our work, some algebras which occurred during the work for semidihedral type were excluded since they had a component $\mathbb{Z}\tilde{D}_n$ (or even $\mathbb{Z}D_n/\langle \tau^k \rangle$). It is not surprising that they appear since we only use a small part of any AR–component. Alternatively we could have weakened the definition of "semidihedral type" and include other algebras which would probably be domestic.

6. Blocks

Suppose now B is a block of finite or tame type; then B is Morita equivalent to some algebra in the list. We would like to know which of the algebras in the list occur as basic algebras of blocks.

6.1.1 Suppose B is a block of finite type, with defect group D of order p^n; then we must have that char K = p. Moreover, there are two necessary arithmetic conditions: If $\Gamma_s(B) \cong \mathbb{Z}A_n/<\tau^q>$ then n + 1 = |D| and q must divide p–1. Otherwise, there seem to be no restrictions. The question, which of the remaining algebras are Morita equivalent to blocks is the same as the question, which Brauer trees occur. In $[F_2]$, W. Feit proved that only very few occur, by using the classification of finite simple groups. Examples may be found in [HL].

6.1.2 Now consider a block B of tame type; then we require that char K = 2. The defect group of B can be recognized from $\Gamma_s(\Lambda)$, for B tame, by 2.2. For example, if B is an algebra of dihedral type then the defect groups of B must be dihedral. Moreover, in the case when Λ is of dihedral type, we only have to consider those algebras where Ω permutes the 3–tubes, this is the case for blocks with dihedral defect groups.

To continue the search for blocks, one may use the following general principles. The determinant of the Cartan matrix C of B is a power of p where p = char K, and moreover, the largest elementary divisor of det C is the order of a defect group of the block. Important is that the dimension of the centre of the algebra B is the same as the number of ordinary irreducible characters of the block, denoted by k(B).

For each algebra in our list, it is easy to determine the centre. On the other hand, by block theoretic methods, one can calculate a formula for k(B); and combining these one obtains necessary conditions. For the converse, one has to exhibit examples. There are a few open cases.

As an illustration, consider the case when Λ is an algebra in one of the three lists, with 3 simple modules. The Cartan matrix C of Λ and the dimension of $Z(\Lambda)$ are as follows:

C		det C	dim $Z(\Lambda)$
$\begin{bmatrix} 4k & 2k & 2k \\ 2k & k+s & k \\ 2k & k & k+t \end{bmatrix}$	(I)	4kst	$k+s+t+1$
$\begin{bmatrix} a+b & a & b \\ a & a+b & c \\ b & c & a+c \end{bmatrix}$	(II)	4abc	$a+b+c+1$
$\begin{bmatrix} k+1 & k-1 & k \\ k-1 & k+1 & k \\ k & k & k+s \end{bmatrix}$	(III)	4ks	$k+s+2$

(Here we have excluded the algebras of dihedral type where Ω permutes the 3–tubes.) We see that the formula for dim $Z(\Lambda)$ does not depend on the type of the algebra. Actually, in [E_2], we gave the centres explicitly. From this description one can see that the centres are isomorphic as algebras whenever the parameters are the same. A typical use of this is as follows:

PROPOSITION *Suppose Λ is of semidihedral type with Cartan matrix of the form I above. Assume also that Λ is Morita equivalent to some tame block B. Then $p = 2$, and one of the following holds:*

(i) B is Morita equivalent to the principal block of $U_3(q)$ for $q \equiv 1 \bmod 4$.

(ii) B is Morita equivalent ro the principal block of $L_3(q)$ *for* $q \equiv 3 \bmod 4$.

(iii) B belongs to a family of blocks which contains the principal block of M_{11}.

(iv) B belongs to one possible family for which no examples are known.

Proof: If Λ is Morita equivalent to a tame block then p = 2 and a defect group must be semidihedral, by 2.2. Also, det C = 4kst is a power of 2. By methods of local block theory one can show that the number k(B) of irreducible characters is $2^{n-2} + 4$ where 2^n is the order of a defect group. On the other hand, k(B) = dim Z(Λ) = k + s + t + 1 and hence {k, s, t} = {1, 2, 2^{n-2}}. This gives essentially three solutions for k, s, t, and there are four families of algebras where this occurs. To obtain the conclusions, one compares with the known examples.

6.2 The results on algebras allow to recover arithmetic properties of group representations. As a main new result, we obtain all possible decomposition matrices for all tame blocks, and this improves [B_2, O]. In particular, given the algebra structure of the block there is a unique decomposition matrix.

Suppose Λ is the basic algebra of some tame block; then Λ is explicitly given, and then we also know the Cartan matrix of B and dim Z(B) = k(B); and also ℓ(B), the number of simple modules. A block has a decomposition matrix D which records the composition factors of the ordinary irreducible representations reduced modulo p; and there is a reciprocity theorem, that is, $D^TD = C$. The matrix D has size k(B) × ℓ(B), its entries are non–negative integers, and it does not have a row of zeros. Knowing k(B), ℓ(B) and the shape of C, it is for the tame blocks easy to solve the matrix equations for D, this may be found in [E_2]. One can also determine some generalized decomposition numbers.

Apart from this, it is also possible, by using some general principles to recover $k_0(B)$, the number of characters of B of height zero, and of course elementary divisors for the Cartan matrix.

In special cases information about Green correspondents of simple modules is available; but not in general.

6.3 There is an old conjecture, namely that the entries of the Cartan matrix of a block should be bounded by a function of the defect group (Conjecture K in [A]). This was strengthened by P. Donovan to:

Fix a p-group D. Then up to Morita equivalence, there are only a finite number of block algebras with a defect group isomorphic to D.

We have now an affirmative answer for dihedral and semidihedral type, but the question is open for quaternion type. A typical obstruction is the family of algebras given in 5.1. Nevertheless, we have a complete list of Cartan matrices, and there are only finitely many possibilities.

As far as the other types are concerned, there is still the question whether from any given family for fixed multiplicities more than one algebra can occur as a block. This question also arises for blocks with dihedral defect groups, with at most two simple modules. It can be proved that these blocks are special biserial, that is, the scalar can be taken as 0. We do not know how to settle this for the algebras of the other types.

6.4 We hope these results will be useful for other problems, may be to give evidence for conjectures. There is a conjecture due to Flanigan, which says that any group algebra of some finite group can be deformed into a semisimple algebra. This has been proved by M. Schaps [S] for finite type. There is also a proof for dihedral defect groups, for blocks with three simple modules.

More recently, there has been much interest in derived categories of blocks, in connection with perfect isometries for ordinary characters. It was proved in [R] that any two blocks of finite type with the same multiplicities are derived equivalent. Moreover, M. Linckelmann has proved recently that any two dihedral blocks with three simple modules, with the same multiplicity are derived equivalent [L]. We remark that one can also determine the derived equivalence classes for algebras of dihedral type.

7. Wild type

7.1 Consider now blocks of wild representation type, viewed as finite–dimensional algebras. For particular types of groups, Loewy series of projective modules have been determined, for example [ML, Ko]. Actually, already some time ago, Jennings has determined the Loewy series for arbitrary p–group algebras, see [H, Be]. There are also results on the quiver of some blocks of wild type, for example [Ma].

In the context of local block theory, there is much interest in algebras from a different point of view. Here one studies blocks over complete discrete valuation rings whose residue field has characteristic p. For example, in [P], L. Puig has introduced the source algebras, these are Morita equivalent to the given block and still carry an action of a p–local subgroup; but they are not basic in general. This is a different approach as ours.

7.2 We are interested in the Auslander–Reiten theory for blocks of wild type; one question is whether the stable AR–quiver may be used to study indecomposable projective modules. One could try to apply the result of [BR], which we described in 4.1. This raises the question whether the stable AR–quiver of a block of wild type has any distinguished "ends", or other distinguished τ–orbits. For example, one might hope for finiteness conditions.

It has been proved that for a block B of wild type, $\Gamma_s(B)$ has always infinitely many components of the form $\mathbb{Z}A_\infty$ [E_5]. Also, there are no Euclidean components [Be, O, ES]. It is well–known that $\Gamma_s(B)$ has always tubes. The rank is bounded, but this bound can be arbitrary large [BC].

On the other hand, it is not known whether tree class A_∞^∞ or D_∞ occurs. This question makes sense for algebras more generally, namely one may ask whether tree class A_∞^∞ or D_∞ only occurs for tame type.

One could also hope that the number of tubes for some particular rank could be finite (and non–zero). Recently we discovered for some classes of blocks, that for each rank which occurs there are infinitely many tubes of this rank, and probably this is generally true, for blocks of wild type.

7.3 We have seen that the defect group of a block determines to some extent its module category. Conversely, one may ask how good are defect groups in separating Morita equivalence classes. For blocks of finite or tame type, it is true that blocks which are Morita equivalent have isomorphic defect groups. However, consider blocks in general; one may ask the following. Suppose B_1 and B_2 are blocks with defect groups D_1 and D_2. If B_1 and B_2 are Morita equivalent, does this imply that D_1 and D_2 are isomorphic? Consider a special case, namely when $B_i \cong KD_i$ for i = 1, 2. Then B_1 and B_2 are Morita equivalent if and only if they are isomorphic; and we ask whether the fact that KD_1 and KD_2 are isomorphic algebras implies that D_1 and D_2 are isomorphic groups. This question is a variation of the isomorphism problem and is not solved.

7.4 Finally, we would like to mention some recent evidence for P. Donovan's question (see 6.3). It has been proved in [Sc] that for arbitrary symmetric groups, there are only finitely many Morita equivalence classes of blocks with a fixed defect group; and there is an explicit (small) bound.

References

[A] J.L. Alperin, Local Representation Theory, Proc. Sympos. Pure Math. Math. 37(1980), 369–375

[AR] M. Auslander, I. Reiten, Representation theory of Artin algebras III: Almost split sequences, Comm. Alg. 3(1975), 239–294

[Au] M.Auslander, Applications of morphisms determined by objects, Proc. Conf. on Representation Theory, Philadelphia (1976), M. Dekker 1978, 245–327

[B] D.J. Benson, Modular representation theory: New trends and methods, Springer Lecture Notes in Mathematics 1081(1984)

[BC] D.Benson, J.F. Carlson, Cohomological inflation, extraspecial subgroups and modules with large period, preprint (1990)

[Be] C. Bessenrodt, The Auslander–Reiten quiver of a modular group algebra revisited, Math. Z. 206(1991),25–34

[B_1] R. Brauer, On groups whose order contains a prime number to the first power, I and II, Amer. J. Math. 64(1942), 401–420 and 421–440

[B_2] R. Brauer, On 2–blocks with dihedral defect groups, Symposia Math. 13(1974), 366–394

[BR] M.C.R. Butler, C.M. Ringel, Auslander–Reiten sequences with few middle terms and applications to string algebras, Comm. Alg. 15(1987), 145–179

[BS] M.C.R. Butler, M. Shahzamanian, The construction of almost split sequences III: modules over two classes of tame local algebras, Math. Ann. 247(1980), 111–122

[CB] W.W. Crawley–Boevey, Functorial filtrations III: Semidihedral algebras, Journal L.M.S.(2)40 (1990), 31–39

[CK] M. Chlebowitz, B. Külshammer, Symmetric local algebras with 5–dimensional center, to appear in Trans. A.M.S.

[D] E.C. Dade, Blocks with cyclic defect groups, Ann. of Math. 84(1966), 2–48

[Do] P. Donovan, Dihedral defect groups, J. Algebra 56(1979), 184–206

[DS] P. Dowbor, A. Skowronski, Galois coverings of representation–infinite algebras, Comment. Math. Helv. 62(1987), 311–337

[E_1] K. Erdmann, On the number of simple modules in certain tame blocks and algebras, Arch. Math. 51(1988), 34–38

[E_2] K. Erdmann, Blocks of tame representation type and related algebras, Springer Lecture Notes in Mathematics 1428(1990)

[E_3] K. Erdmann, Algebras and quaternion defect groups I and II, Math. Ann. 281(1988), 545–560 and 561–582

[E_4] K. Erdmann, On the local structure of tame blocks, Colloque sur les representations des groupes finis (Luminy, 1988), Asterisque 181–182(1990), 173–190.

[E_5] K. Erdmann, On Auslander–Reiten components for wild blocks, Progress in Mathematics (Birkhauser), 95(1991) 371–387

[ES] K. Erdmann, A. Skowronski, On Auslander–Reiten components of blocks and self–injective biserial algebras, to appear in Trans. A.M.S.

[F_1] W. Feit, The Representation Theory of Finite Groups, North Holland 1982

[F_2] W. Feit, Possible Brauer trees, Ill. J. Math. 28(1984), 43–56

[G] P. Gabriel, Auslander–Reiten sequences and representation finite algebras, Representation Theory I, Springer Lecture Notes in Mathematics 831(1980), 1–71

[GR] P. Gabriel, C. Riedtmann, Group representations without groups, Comm. Math. Helv. 54(1979), 240–287

[H] B. Huppert, N. Blackburn, Finite groups II, Grundlehren Bd. 242, Springer 1982

[HL] G. Hiss, K. Lux, Brauer trees of sporadic simple groups, Oxford University Press 1990

[HPR] D. Happel, U. Preiser, C.M. Ringel, Vinberg's characterization of Dynkin diagrams using subadditive functions with applications to DTr–periodic modules, in Representation Theory II, Springer Lecture Notes in Mathematics 832(1981), 280–294

[K_1] S. Kawata, The Green correspondence and Auslander–Reiten sequences, J. Algebra 123(1989), 1–5

[K_2] S. Kawata, Module correspondence in Auslander–Reiten quivers for finite groups, Osaka J. Math. 26(1989), 671–678

[Ko] S. Koshitani, On the Loewy series of the group algebra of a finite p–solvable group with p–length > 1, Comm. Algebra 13(1985), 2175–2198

[LM] P. Landrock, G.O. Michler, The block structure of the smallest Janko group, Math. Ann. 232(1978), 205–238

[L] M. Linckelmann, A derived equivalence for blocks with dihedral defect groups, preprint 1990

[Li] P. Linnell, The Auslander–Reiten quiver of a finite group, Arch. Math. 45(1985), 289–295

[M] R. Martinez–Villa, Almost projective modules and almost split sequences with indecomposable middle terms, Comm. Algebra 8(1980), 1123–1150

[Ma] S. Martin, On the ordinary quiver of the principal block of certain symmetric groups, Quarterly J. Math., Oxford 40(1989), 209–223

[O] T. Okuyama, On the Auslander–Reiten quiver of a finite group, J. Algebra 110(1987), 425–430

[Ol] J.B. Olsson, On 2–blocks with quaternion and quasidihedral defect groups, J. Algebra 36(1975), 212–241

[P] L. Puig, Pointed groups and construction of characters, Math. Z. 176(1981), 265–292

[R] J. Rickard, Derived categories and stable equivalence, J. Pure Appl. Algebra 61(1989), 303–317

[Ri] C. Riedtmann, Representation–finite selfinjective algebras of class A_n, in Representation Theory II, Springer Lecture Notes in Mathematics 832(1981)

[Rn_1] C.M. Ringel, The representation type of local algebras, Springer Lecture Notes in Mathematics 488(1975), 282–305

[Rn_2] C.M. Ringel, The indecomposable representations of the dihedral 2–groups, Math. Ann. 214(1975), 19–34

[S] M. Schaps, A modular version of Maschke's theorem for group algebras of finite representation type and for blocks of cyclic defect group, preprint.

[Sc] J. Scopes, Cartan matrices and Morita equivalence for blocks of the symmetric groups, to appear in J. Algebra

[W] P. Webb, The Auslander–Reiten quiver of a finite group, Math. Z. 179(1982), 97–121

K. Erdmann
Mathematical Institute,
24–29 St. Giles, Oxford OX1 3LB, England

Lie Algebras

Arising in Representation Theory

Claus Michael Ringel

One of the reasons for the introduction of the Hall algebras for finitary algebras in [R1, R2, R3] was the following: Let A be a finite dimensional algebra which is hereditary, say of Dynkin type Δ. Let $\mathbf{g}$ be the simple complex Lie algebra of type Δ, with triangular decomposition $\mathbf{g} = \mathbf{n}_- \oplus \mathbf{h} \oplus \mathbf{n}_+$. The degenerate Hall algebra $\mathcal{H}(A)_1$ of A is the free abelian group on the set of isomorphism classes of A-modules of finite length. The Grothendieck group $K(A\text{-mod})$ of all A-modules of finite length modulo split exact sequences may be identified with the free abelian group on the set of isomorphism classes of indecomposable A-modules of finite length, thus with a subgroup of $\mathcal{H}(A)_1$. Now, with respect to the degenerate Hall multiplication, the subgroup $K(A\text{-mod})$ becomes a Lie subalgebra of $\mathcal{H}(A)_1$, so that $K(A\text{-mod})$ is isomorphic to the Chevalley $\mathbb{Z}$-form of $\mathbf{n}_+$, and $\mathcal{H}(A)_1$ to the corresponding Kostant $\mathbb{Z}$-form of the universal enveloping algebra $U(\mathbf{n}_+)$.

With $U(\mathbf{n}_+)$ also $\mathcal{H}(A)_1$ is a bialgebra. The papers mentioned before have concentrated on the definition of a multiplication using the evaluation of certain polynomials at 1. This approach was first presented at the Antwerp Conference in 1987, and the discussion there helped to direct the further investigations. In particular, M. van den Bergh proposed to consider instead of the Hall polynomials the Euler characteristic of corresponding varieties. This idea was developed in detail by Schofield [Sc] and Riedtmann [Rm], and also Lusztig's presentation [L] of the Hall algebras proceeds in this way. Schofield has considered the complete coalgebra structure. The aim of this short note is to point out the nature of the comultiplication of $\mathcal{H}(A)_1$. Of course, these considerations also may be used in the Euler characteristic approach.

1. The Comultiplication

An additive category (not necessarily with finite sums) will be called a *Krull–Schmidt category* provided any object can be written as a finite direct sum of objects with local endomorphism rings. Let $\mathcal{A}$ be a Krull–Schmidt category. For any object M in $\mathcal{A}$, we denote by $[M]$ its isomorphism class, and we assume that the isomorphism classes of objects in $\mathcal{A}$ form a set. Given M in $\mathcal{A}$, we denote by $d(M)$ the number of indecomposable direct summands in any direct decomposition of M, this is an invariant of M, according to the Krull–Schmidt theorem.

We denote by $C(\mathcal{A})$ the free abelian group with basis $(u_{[M]})_{[M]}$ indexed by the set of isomorphism classes $[M]$ of objects M in $\mathcal{A}$, with the following comultiplication:

$$\Delta(u_{[M]}) = \sum_{D(M)} u_{[M_1]} \otimes u_{[M_2]}$$

where $D(M)$ is the set of pairs $([M_1],[M_2])$ such that $[M_1 \oplus M_2] = [M]$, and with the counit

$$\epsilon(u_{[0]}) = 1, \quad \epsilon(u_{[M]}) = 0, \text{ for } [M] \neq [0].$$

Let $C(\mathcal{A})_{(n)}$ be the free abelian group with basis $(u_{[M]})_{[M]}$ indexed by the set of isomorphism classes $[M]$ of objects M with $d(M) = n$. Thus $C(\mathcal{A}) = \bigoplus_{n\geq 0} C(\mathcal{A})_{(n)}$.

Proposition 1. $C(\mathcal{A})$ is a strictly graded cocommutative $\mathbb{Z}$–coalgebra.

Recall that a graded $\mathbb{Z}$–coalgebra $C = \bigoplus_{n\geq 0} C_{(n)}$ is called *strictly graded* provided $C_{(0)} = \mathbb{Z}$, and $C_{(1)}$ is the set of primitive elements of C (an element $x \in C$ being called *primitive* in case $\Delta x = x \otimes 1 + 1 \otimes x$.)

Proof. Let $\Delta_2(u_{[M]}) = \sum u_{[M_1]} \otimes u_{[M_2]} \otimes u_{[M_3]}$, where the sum is taken over all triples $([M_1],[M_2],[M_3])$ such that $[M_1 \oplus M_2 \oplus M_3] = [M]$. Since

$$(1 \otimes \Delta)\Delta(u_{[M]}) = \Delta_2(u_{[M]}) = (\Delta \otimes 1)\Delta(u_{[M]}),$$

Δ is coassociative. If we apply $1 \otimes \epsilon$ to $\Delta(u_{[M]}) = \sum_{D(M)} u_{[M_1]} \otimes u_{[M_2]}$, then only $u_{[M]} \otimes 1$ remains, similarly, if we apply $\epsilon \otimes 1$ to $\Delta(u_{[M]})$, then only $1 \otimes u_{[M]}$ remains. Thus ϵ is a counit. This shows that $C = C(\mathcal{A})$ is a $\mathbb{Z}$–coalgebra, and, of course, it is cocommutative. Also,

$$\Delta(C_{(n)}) \subseteq \bigoplus_{0\leq i\leq n} C_{(i)} \oplus C_{(n-i)},$$

and, for $n \geq 1$, $\epsilon(C_{(n)}) = 0$, thus we deal with a coalgebra grading. It remains to be seen that any primitive element belongs to $C_{(1)}$. Consider an element x, say $x = \sum_{i=1}^{n} x_i u_{[M_i]}$ with non-zero integers x_i, and pairwise different isomorphism classes $[M_i]$. We can assume that $d(M_1) \leq d(M_2) \leq \cdots \leq d(M_n)$. We remark that $C \otimes C$ is the free abelian group with basis $u_{[M']} \otimes u_{[M'']}$, where both $[M']$ and $[M'']$ run through the isomorphism classes of objects in $\mathcal{A}$. If x is primitive, then $[M_1] \neq [0]$, since otherwise $\Delta(x_1 u_{[0]}) = x_1 u_{[0]} \otimes u_{[0]}$ shows that x_1 is the coefficient of $\Delta(x)$ at $u_{[0]} \otimes u_{[0]}$, whereas the coefficient of $x \otimes u_{[0]} + u_{[0]} \otimes x$ at $u_{[0]} \otimes u_{[0]}$ is $2x_1$. Since the elements in $C_{(1)}$ are primitive, we may assume that $d(M_1) \geq 2$. Take a direct decomposition $[M_1' \oplus M_1''] = [M_1]$, with an indecomposable object M_1', then x_1 is the coefficient at $u_{[M_1']} \otimes u_{[M_1'']}$ for $\Delta(x)$. On the other hand, the coefficient of $x \otimes u_{[0]} + u_{[0]} \otimes x$ at $u_{[M_1']} \otimes u_{[M_1'']}$ is zero. This shows that any primitive element is in $C_{(1)}$. Thus $C = \bigoplus C_{(n)}$ is strictly graded.

Remark. In case $\mathcal{A}$ is a length category, there is a different grading on $C(\mathcal{A})$, which is of interest in representation theory, namely let $C(\mathcal{A})_n$ be the free abelian group with basis $(u_{[M]})_{[M]}$ indexed by the set of isomorphism classes $[M]$ of objects M of length n. Then $\Delta(C_n) \subseteq \bigoplus_{0 \leq i \leq n} C_i \oplus C_{n-i}$, and, for $n \geq 1$, $\epsilon(C_n) = 0$, whereas $C_0 = C_{(0)} = \mathbb{Z}$. Thus, we deal with a coalgebra grading, however, $C(\mathcal{A}) = \bigoplus C(\mathcal{A})_n$ is not strictly graded.

Example. Let $\mathcal{A} = \mathcal{A}(k)$ be the category of finite dimensional vector spaces over the field k, or, more generally, a Krull Schmidt category which contains (up to isomorphism) just one indecomposable object X, and all its finite direct sums nX, with $n \in \mathbb{N}_0$. Let $u_n = u_{[nX]}$. Then $\Delta(u_n) = \sum_{i=0}^{n} u_i \otimes u_{n-i}$, and, for $n \geq 1$, $\epsilon(u_n) = 0$ (this is Example 2.2 in [A]).

We should remark that the coalgebra $C(\mathcal{A})$ only depends on the free commutative semigroup $S(\mathcal{A})$ of all isomorphism classes of objects in $\mathcal{A}$ with multiplication $\oplus$. Given any free commutative semigroup S, the free abelian group $C(S)$ with basis S is a coalgebra with respect to the following operations:

$$\Delta(d) = \sum_{d'd''=d} d' \otimes d'', \quad \epsilon(1) = 1, \quad \epsilon(d) = 0, \text{ for } d \neq 1,$$

and we have $C(\mathcal{A}) = C(S(\mathcal{A}))$.

2. Bialgebras

Let $C(\mathcal{A})$ be endowed with an associative and unitary multiplication $\circ$. Let $c_{ZX}^{Y} \in \mathbf{Z}$ be the structure constants, for $X, Y, Z \in \mathcal{A}$; thus

$$u_{[Z]} \circ u_{[X]} = \sum_{[M]} c_{ZX}^{M} u_{[M]}.$$

Proposition 2. *$(C(\mathcal{A}), \circ)$ is a bialgebra if and only if $u_{[0]}$ is the unit element, and the following condition is satisfied for all objects $X, Z, M_1, M_2 \in \mathcal{A}$*

$$c_{ZX}^{M_1 \oplus M_2} = \sum c_{Z_1 X_1}^{M_1} c_{Z_2 X_2}^{M_2},$$

where the sum on the right ranges over all pairs $([Z_1],[Z_2]) \in D(Z)$, and all pairs $([X_1],[X_2]) \in D(X)$.

Proof: We have

$$\begin{aligned}\Delta(u_{[Z]} \circ u_{[X]}) &= \Delta\Big(\sum_{[M]} c_{ZX}^{M} u_{[M]}\Big) = \sum_{[M]} c_{ZX}^{M} \Delta(u_{[M]}) \\ &= \sum_{[M]} c_{ZX}^{M} \sum_{D(M)} u_{[M_1]} \otimes u_{[M_2]},\end{aligned}$$

thus the coefficient at $u_{[M_1]} \otimes u_{[M_2]}$ in $\Delta(u_{[Z]} \circ u_{[X]})$ is just $c_{ZX}^{M_1 \oplus M_2}$.

On the other hand, since

$$\begin{aligned}\Delta(u_{[Z]}) \circ \Delta(u_{[X]}) &= \Big(\sum_{D(Z)} u_{[Z_1]} \otimes u_{[Z_2]}\Big)\Big(\sum_{D(X)} u_{[X_1]} \otimes u_{[X_2]}\Big) \\ &= \sum_{D(Z),D(X)} u_{[Z_1]} u_{[X_1]} \otimes u_{[Z_2]} u_{[X_2]} \\ &= \sum_{D(Z),D(X)} \sum_{[M_1]} \sum_{[M_2]} c_{Z_1 X_1}^{M_1} c_{Z_2 X_2}^{M_2} u_{[M_1]} \otimes u_{[M_2]},\end{aligned}$$

the coefficient at $u_{[M_1]} \otimes u_{[M_2]}$ in $\Delta(u_{[Z]}) \circ \Delta(u_{[X]})$ is $\sum c_{Z_1 X_1}^{M_1} c_{Z_2 X_2}^{M_2}$. Of course, Δ is a ring homomorphism if and only if the coefficients of $\Delta(u_{[Z]} \circ u_{[X]})$ and $\Delta(u_{[Z]}) \circ \Delta(u_{[X]})$ at any $u_{[M_1]} \otimes u_{[M_2]}$ coincide.

We denote by $K(\mathcal{A}) = C(\mathcal{A})_{(1)}$ the subgroup of $C(\mathcal{A})$ generated by the elements $u_{[M]}$, where M is an indecomposable object of $\mathcal{A}$. Since $K(\mathcal{A})$ is the free abelian group on the set of isomorphism classes of indecomposable objects in $\mathcal{A}$, we may consider it as the Grothendieck group of $\mathcal{A}$ with respect to split exact sequences.

Corollary. *Assume that* $(C(\mathcal{A}), \circ)$ *is a bialgebra. Then* $K(\mathcal{A})$ *is a Lie subalgebra (with respect to the Lie bracket* $[z, x] = z \circ x - x \circ z$.)

Proof: We have to show that for indecomposable objects $X, Z \in \mathcal{A}$, the commutator $[u_{[Z]}, u_{[X]}]$ is a linear combination of elements $u_{[M]}$, with M indecomposable. Clearly, it is a linear combination of elements $u_{[M]}$, with M non–zero, thus consider an object $M = M_1 \oplus M_2$, with both M_1, M_2 non–zero, and let us calculate c^M_{ZX} and c^M_{XZ}. Since X and Z are indecomposable, the sets $D(Z), D(X)$ both have just two elements, namely $([Z],[0]),([0],[Z])$, and $([X],[0]),([0],[X])$, respectively. Since the objects M_1, M_2 are non–zero, we have $c^{M_1}_{00} = 0 = c^{M_2}_{00}$, thus

$$c^M_{ZX} = c^{M_1}_{Z0} c^{M_2}_{0X} + c^{M_1}_{0X} c^{M_2}_{Z0} = c^M_{XZ}.$$

Consequently, the coefficient of $[u_{[Z]}, u_{[X]}]$ at $u_{[M]}$ is 0.

3. Hall algebras

Let A be a k–algebra, and X, Y, Z A–modules of finite length. We denote by $\mathcal{M}^Y_{ZX}$ the set of submodules U of Y which are isomorphic to X such that M/U is isomorphic to Z. Also, let $\mathcal{A} = A$–mod be the category of A–modules of finite length. In order to introduce a multiplication on $C(A\text{–mod})$, we will work with $\mathcal{M}^Y_{ZX}$.

Let us assume that k is a finite field, say with q elements. Let X, Y, Z be A–modules of finite length, and define

$$c^Y_{ZX} = |\mathcal{M}^Y_{ZX}|.$$

Proposition 3. *Let* X, Z, M_1, M_2 *be* A*–modules. Then* $q-1$ *divides*

$$c^{M_1 \oplus M_2}_{ZX} - \sum c^{M_1}_{Z_1 X_1} c^{M_2}_{Z_2 X_2}.$$

where the sum on the right ranges over all pairs $([Z_1],[Z_2]) \in D(Z)$, *and all pairs* $([X_1],[X_2]) \in D(X)$.

For the proof of Proposition 3, we need the following Lemma (see [R1] and [Sc]): Given a direct sum $M = M_1 \oplus M_2$, we may describe its endomorphisms by 2×2–matrices. For an endomorphisms f of M_1, let $*f$ be the action of the matrix $\begin{bmatrix} f & 0 \\ 0 & 1 \end{bmatrix}$ on $M_1 \oplus M_2$. Note that any element $\alpha \in k$ yields an endomorphism of any A–module, using multiplication. We denote by k^* the set of non–zero elements of k.

Lemma. *Let U be a submodule of $M = M_1 \oplus M_2$. The following assertions are equivalent:*

(i) $U = (U \cap M_1) \oplus (U \cap M_2)$,

(ii) For all $\alpha \in k$ *we have* $U * \alpha \subseteq U$,

(iii) There exists $\alpha \in k^*$ *with* $U * (1 + \alpha) \subseteq U$.

Proof: We only have to show that (iii) implies (i): Let $u \in U$. Write $u = (m_1, m_2)$ with $m_i \in M_i$. By assumption, $u' = u * (1 + \alpha)$ belongs to U, thus also $u' - u = m_1\alpha$. Since α is invertible, we see that m_1, and then also m_2 belong to U.

Proof of Proposition 3. Consider $\mathcal{M}_{ZX}^{M}$ for $M = M_1 \oplus M_2$, and we fix this decomposition. We have defined above an operation $*$ of k^* on $\mathcal{M}_{ZX}^{M_1 \oplus M_2}$, and we want to use it now. Let $\mathcal{M}'$ be the subset of $\mathcal{M}_{ZX}^{M}$, consisting of all submodules U of M which satisfy the equivalent conditions of Lemma, thus $\mathcal{M}'$ consists of fix points of the action $*$, and the elements of $\mathcal{M}_{ZX}^{M} \backslash \mathcal{M}'$ have trivial stabilizers. It follows that the k^*–orbits in $\mathcal{M}_{ZX}^{M} \backslash \mathcal{M}'$ are of length $q - 1$. Consequently

$$|\mathcal{M}_{ZX}^{M}| \equiv |\mathcal{M}'| \pmod{q-1}.$$

On the other hand, we may identify $\mathcal{M}'$ with the disjoint union of the products $\mathcal{M}_{Z_1X_1}^{M_1} \times \mathcal{M}_{Z_2X_2}^{M_2}$, the union being indexed by the pairs in $D(Z) \times D(X)$. For, given a submodule $U = (U \cap M_1) \oplus (U \cap M_2)$, isomorphic to X, and with $(M_1 \oplus M_2)/U$ isomorphic to Z, let $U_1 = U \cap M_1$, and $U_2 = U \cap M_2$. Then U_1 belongs to some $\mathcal{M}_{Z_1X_1}^{M_1}$, and U_2 to some $\mathcal{M}_{Z_2X_2}^{M_2}$, where $[Z_1 \oplus Z_2] = [Z]$, and $[X_1 \oplus X_2] = [X]$. Thus $|\mathcal{M}'|$ is just the sum of the cardinalities of the various $\mathcal{M}_{Z_1X_1}^{M_1} \times \mathcal{M}_{Z_2X_2}^{M_2}$. This completes the proof.

Given an extension field E of k, and a k–algebra A, we may consider the E–algebra $A^E = A \otimes E$. A field extension E of k will be said to be *conservative* for the k–algebra A provided for any indecomposable A–module M of finite length, the algebra $(\operatorname{End} M / \operatorname{rad} \operatorname{End} M)^E$ is a field. Given a representation–finite k–algebra A, there are infinitely many finite field extensions of k which are conservative. Given a k–algebra A with infinitely many finite field extensions which are conservative, we say that A has *Hall polynomials* provided for all A–modules X, Y, Z of finite length, there exists a polynomial $\varphi_{ZX}^{Y} \in \mathbf{Z}[T]$, such that for any conservative field extension E of k, we have

$$\varphi_{ZX}^{Y}(|E|) = |\mathcal{M}_{Z^E, X^E}^{Y^E}|.$$

For representation–directed algebras, the existence of Hall polynomials has been shown in [R1]. Of course, there also is the classical example: any local uniserial algebra has Hall polynomials [H,M]. One may conjecture that any representation–finite algebra has Hall polynomials.

Assume that the algebra A has Hall–polynomials. Then the *degenerate Hall–algebra* $\mathcal{H}(A)_1$ is defined as the coalgebra $C(A\text{–mod})$ with the multiplication

$$u_{[N_1]}u_{[N_2]} = \sum_{[M]} \varphi^{M}_{N_1N_2}(1)u_{[M]},$$

where N_1, N_2 are arbitrary A–modules of finite length.

Theorem. *$\mathcal{H}(A)_1$ is a bialgebra, and the subgroup $K(A\text{–mod})$ is a Lie–subalgebra.*

As a $\mathbb{Q}$–bialgebra, $\mathcal{H}(A)_1 \otimes \mathbb{Q}$ is isomorphic to the universal enveloping algebra of $K(A\text{–mod}) \otimes \mathbb{Q}$.

Proof: Let E be a conservative field extension of k, and assume that $|E| = q^n$. Then, according to Proposition 3, $q^n - 1$ divides $\varphi^{M_1\oplus M_2}_{ZX}(q^n) - \sum_{D(Z),D(X)} \varphi^{M_1}_{Z_1X_1}(q^n)\varphi^{M_2}_{Z_2X_2}(q^n)$, and therefore, $T-1$ divides $\varphi^{M_1\oplus M_2}_{ZX}(T) - \sum_{D(Z),D(X)} \varphi^{M_1}_{Z_1X_1}(T)\varphi^{M_2}_{Z_2X_2}(T)$, (see [R1]), thus

$$\varphi^{M_1\oplus M_2}_{ZX}(1) = \sum_{D(Z),D(X)} \varphi^{M_1}_{Z_1X_1}(1)\varphi^{M_2}_{Z_2X_2}(1).$$

It follows that we may apply Proposition 2 and its Corollary, thus $\mathcal{H}(A)_1$ is a bialgebra and $K(A\text{–mod})$ is a Lie subalgebra.

The remaining assertion follows from general Hopf algebra theory: Any graded coalgebra $C = \bigoplus_{n\geq 0} C_n$ with $C_0 = \mathbb{Z}$ is irreducible (there is just one group–like element). But a bialgebra over a field which is irreducible as a coalgebra is always a Hopf algebra ([Sw], Theorem 9.2.2). And an irreducible cocommutative Hopf algebra H over a field of characteristic zero with $P(H)$ the set of primitive elements is just the universal enveloping algebra $U(P(H))$ of the Lie algebra $P(H)$ ([Sw], Theorem 13.0.1). We apply this to the $\mathbb{Q}$–bialgebra $H = \mathcal{H}(A)_1 \otimes \mathbb{Q}$, its set of primitive elements being $P(H) = C(A\text{–mod})_{(1)} \otimes \mathbb{Q} = K(A\text{–mod}) \otimes \mathbb{Q}$.

4. Subbialgebras

Let C be a bialgebra, and let C' be a subalgebra of C generated by a subset S such that $\Delta(S) \subseteq C' \otimes C'$, then C' is a subbialgebra. Indeed, assume that we deal with elements $x, y \in C'$ so that $\Delta(x), \Delta(y) \in C' \otimes C'$. Then also $\Delta(x+y), \Delta(xy)$ belong to $C' \otimes C'$, since Δ is a ring homomorphism. Of course, with C also C' is irreducible. Also, if C is cocommutative, then also C' is cocommutative.

Assume we have endowed $C(\mathcal{A})$ with a multiplication so that it is a bialgebra. Let $\mathcal{B}$ be a full subcategory of $\mathcal{A}$ which is closed under direct summands, let $C(\mathcal{A};\mathcal{B})$ be the subalgebra of $C(\mathcal{A})$ generated by the elements u_X, with $X \in \mathcal{B}$. Then $C(\mathcal{A};\mathcal{B})$ is a subbialgebra.

We may apply these considerations to the category $\mathcal{A} = A$–mod of finite length A-modules, where A is some algebra, and to the full subcategory $\mathcal{B}$ either of all simple, or of all semisimple modules. In the Hall algebra case, we will obtain the corresponding composition algebra, and the corresponding Loewy algebra, respectively (see [R4],[R5]).

References

[A] Abe, E.: Hopf algebras. Cambridge University Press. 1980.

[H] Hall, P.: The algebra of partitions. Proc. 4th Canadian Math. Congress. Banff (1959), 147–159.

[L] Lusztig, G.: Canonical bases arising from quantized enveloping algebras. J. Amer.Math.Soc. 3 (1990), 447-498.

[M] Macdonald, I.G.: Symmetric functions and Hall polynomials. Clarendon Press (1979).

[Rm] Riedtmann, Chr.: Lie algebras generated by indecomposables. Prép. l'Inst. Fourier. Grénoble 170 (1991)

[R1] Ringel, C.M.: Hall algebras. In: Topics in Algebra. Banach Centre Publ. 26. Part I. Warszawa (1990), 433-447.

[R2] Ringel, C.M.: Hall polynomials for the representation–finite hereditary algebras. Adv. Math. 84 (1990), 137-178.

[R3] Ringel, C.M.: Hall algebras and quantum groups. Invent. Math. 101 (1990), 583-592

[R4] Ringel, C.M.: From representations of quivers via Hall and Loewy algebras to quantum groups. In: Proc. Novosibirsk Conf. Algebra 1989. (To appear)

[R5] Ringel, C.M.: The composition algebra of a cyclic quiver. Towards an explicit description of the quantum group of type $\tilde{\mathbb{A}}_n$. Proc. London Math. Soc. (To appear)

[Sc] Schofield, A.: Quivers and Kac–Moody algebras. (To appear).

[Sw] Sweedler, M. E.: Hopf algebras. Benjamin, New York. 1969.

C.M. Ringel
Fakultät für Mathematik
Universität
W–4800 Bielefeld 1
Germany

Printed in the United States
By Bookmasters